Lecture Notes in Mathematics

Edited by J.-M. Morel, F. Takens and B. Teissier

Editorial Policy
for the publication of monographs

1. Lecture Notes aim to report new developments in all areas of mathematics – quickly, informally and at a high level. Monograph manuscripts should be reasonably self-contained and rounded off. Thus they may, and often will, present not only results of the author but also related work by other people. They may be based on specialized lecture courses. Furthermore, the manuscripts should provide sufficient motivation, examples and applications. This clearly distinguishes Lecture Notes from journal articles or technical reports which normally are very concise. Articles intended for a journal but too long to be accepted by most journals, usually do not have this "lecture notes" character. For similar reasons it is unusual for doctoral theses to be accepted for the Lecture Notes series.

2. Manuscripts should be submitted (preferably in duplicate) either to one of the series editors or to Springer-Verlag, Heidelberg. In general, manuscripts will be sent out to 2 external referees for evaluation. If a decision cannot yet be reached on the basis of the first 2 reports, further referees may be contacted: the author will be informed of this. A final decision to publish can be made only on the basis of the complete manuscript, however a refereeing process leading to a preliminary decision can be based on a pre-final or incomplete manuscript. The strict minimum amount of material that will be considered should include a detailed outline describing the planned contents of each chapter, a bibliography and several sample chapters.
Authors should be aware that incomplete or insufficiently close to final manuscripts almost always result in longer refereeing times and nevertheless unclear referees' recommendations, making further refereeing of a final draft necessary.
Authors should also be aware that parallel submission of their manuscript to another publisher while under consideration for LNM will in general lead to immediate rejection.

3. Manuscripts should in general be submitted in English.
Final manuscripts should contain at least 100 pages of mathematical text and should include
– a table of contents;
– an informative introduction, with adequate motivation and perhaps some
  historical remarks: it should be accessible to a reader not intimately familiar
  with the topic treated;
– a subject index: as a rule this is genuinely helpful for the reader.

Continued on inside back-cover

Lecture Notes in Mathematics 1638

Editors:
J.-M. Morel, Cachan
F. Takens, Groningen
B. Teissier, Paris

Springer
Berlin
Heidelberg
New York
Barcelona
Hong Kong
London
Milan
Paris
Singapore
Tokyo

Pol Vanhaecke

Integrable Systems
in the realm
of Algebraic Geometry

Second Edition

Springer

Author

Pol Vanhaecke
Département de Mathématiques
UFR Sciences SP2MI
Université de Poitiers
Téléport 2
Boulevard Marie et Pierre Curie
BP 30179
86962 Futuroscope Chasseneuil Cedex, France

E-mail: Pol.Vanhaecke@mathlabo.univ-poitiers.fr

Cataloging-in-Publication Data applied for

Die Deutsche Bibliothek - CIP-Einheitsaufnahme

Vanhaecke, Pol:
Integrable systems in the realm of algebraic geometry / Pol Vanhaecke. - 2.
ed.. - Berlin ; Heidelberg ; New York ; Barcelona ; Hong Kong ; London ;
Milan ; Paris ; Singapore ; Tokyo : Springer, 2001
 (Lecture notes in mathematics ; 1638)
 ISBN 3-540-42337-0

Mathematics Subject Classification (2000): 14K20, 14H70, 17B63, 37J35

ISSN 0075-8434
ISBN 3-540-42337-0 Springer-Verlag Berlin Heidelberg New York
ISBN 3-540-61886-4 (1st edition) Springer-Verlag Berlin Heidelberg New York

Springer-Verlag Berlin Heidelberg New York
a member of BertelsmannSpringer Science+Business Media GmbH

http://www.springer.de

© Springer-Verlag Berlin Heidelberg 1996, 2001
Printed in Germany

Typesetting: Camera-ready T$_E$X output by the author
SPIN: 10844943   41/3142/LK - 543210 Printed on acid-free paper

Preface to the second edition

The present edition of this book, five years after the first edition, has been spiced with several recent results which fit naturally in the point of view that had been adapted in the original text and with some new examples and constructions that will help the reader to appreciate better our approach to integrable systems.

On this occasion I wish to thank my collaborators from the last five years, to wit Christina Birkenhake, Peter Bueken, Rui Fernandes, Masoto Kimura, Vadim Kuznetsov, Marco Pedroni, Michael Penkava, Luis Piovan and Claude Roger for a fruitful interaction and for their warm friendship. Most of the results that have been added are taken from, or are inspired by, joint work with some of them; I acknowledge their permission to add these, sometimes unpublished, results.

The colleagues at my newest working environment, the University of Poitiers (France), created for me a pleasant and stimulating working environment. I wish to acknowledge the support of all of them. Special thanks go to Marc van Leeuwen, Claude Quitté and Patrice Tauvel for sharing their insights with me, which usually led to a real improvement of parts of the text.

Last but not least, Yvette Kosmann-Schwarzbach, who was not acknowledged in the first version of this book — most probably because my gratitude to her was too big and too obvious! — is thanked here in all possible superlatives, for her constant support and for her sincere friendship. Merci Yvette!

Acknowledgments

The help of many people was indispensable for establishing and presenting the results which are contained in this work. Not enough credit can be given to those who created at home, at the Max-Planck-Institut in Bonn, at the University of Lille and finally at the University of California at Davis a pleasant and stimulating atmosphere. Even some people I don't know by name should be thanked here.

Special thanks are due to Mark Adler and Pierre van Moerbeke, whose fundamental work on a.c.i. systems was the starting point for the research contained in this book. Stimulating discussions with them have led to an improvement of many of the results and to a better understanding of the subject. Also Michèle Audin deserves a special place here for sharing her insights with me through long discussions and letters. Extremely helpful for a thorough understanding were several algebraic-geometric explanations by Laurent Gruson.

I wish to thank my collaborators José Bertin and Marco Pedroni for a fruitful interaction. I have also benefited from discussions with my colleagues at Lille, in particular Jean d'Almeida, Robert Gergondey, Johannes Huebschmann, Raphaël Freitas, Armando Treibich, Gijs Tuynman and Alberto Verjowski and at UC Davis, in particular Josef Mattes, Motohico Mulase, Michael Penkava, Albert Schwarz and Craig Tracy.

I also acknowledge my other friends scattered around the globe, to wit, Christina Birkenhake, Robert Brouzet, Peter Bueken, Jan Denef, Paul Dhooghe, Jean Fastré, Ljubomir Gavrilov, Luc Haine, Horst Knörrer, Franco Magri, Askold Perelomov, Luis Piovan, Elisa Prato and Taka Shiota for their interest in my work and helpful related discussions.

For useful comments on the manuscript I am indebted to Michèle Audin, an anonymous referee and several students in my graduate course in UC Davis (Spring 1996).

Last but not least, special thanks to my wife Lieve for her constant assistance through this adventure.

Table of Contents

Chapter I

Introduction

Integrable systems first appeared as mechanical systems for which the equations of motion are solvable by quadratures, i.e., by a sequence of operations which included only algebraic operations, integration and application of the inverse function theorem. Apart from some non-trivial examples which were constructed before, the first main result (due to Liouville, but essentially an application of a result due to Hamilton) was that if a mechanical system with n degrees of freedom of the form

$$\frac{dq_i}{dt} = \frac{\partial H}{\partial p_i}, \qquad \frac{dp_i}{dt} = -\frac{\partial H}{\partial q_i}, \qquad (i = 1, \ldots, n)$$

(H any function in the coordinates q_i, p_i) has n independent functions in involution, one of which is H, then it can be solved by quadratures; two functions f and g are said to be in involution if their Poisson bracket

$$\{f, g\} = \sum_{i=1}^{n} \left(\frac{\partial f}{\partial q_i} \frac{\partial g}{\partial p_i} - \frac{\partial f}{\partial p_i} \frac{\partial g}{\partial q_i} \right)$$

vanishes, $\{f, g\} = 0$ and f is called a first integral of the system if f and H are in involution.

Mechanical systems which satisfy the conditions of Liouville's Theorem are called Liouville integrable or integrable in the sense of Liouville. A quite short — but important — list of (non-trivial) examples of Liouville integrable systems were found during the last century: a few integrable tops (the Euler top, the Lagrange top, Kowalevski's top, the Goryachev-Chaplygin top), free motion of a particle on an ellipsoid (Jacobi), motion of a rigid body in an ideal fluid (Clebsch, Kirchhoff and Steklov case), motion in the field of a central potential (Newton) and a few others. Both finding these systems (i.e., showing that enough first integrals in involution exist, which was done by constructing them) and solving them explicitly

1

by quadratures required a lot of ingenuity and often quite long calculations. In the more complicated cases the solution was written down in terms of two-dimensional theta functions by a non-trivial use of the rich analytical properties of these functions. In turn it motivated the research in theta functions and Abelian varieties, which originated in the beginning of that century in the works of Riemann and Abel. A rich interaction between integrable systems and complex analysis[1] was about to develop from it. But it did not happen.

There were two reasons for this. The first one is that Poincaré showed at the beginning of this century that a general mechanical system of the above Hamiltonian form is not Liouville integrable. In particular he also showed that the famous three body problem is not integrable. This declined the interest in integrable systems for physicists, astronomers and mathematicians (as far as a distinction between these three groups could be made). The second reason is that complex analysis and algebraic geometry started to develop in completely different directions and the theory of theta functions and Abelian varieties, which is to be situated on their intersection where analytic and algebraic objects come together, also faded away from the picture. For about 60 years there was neither progress nor interest in integrable systems.[2]

The renewed interest was motivated by the discovery of a new approach to solve nonlinear evolution equations, due to Gardner, Greene, Kruskal and Miura in 1967. This method has been known as the inverse scattering method and was originally designed for studying evolution equations such as the Korteweg-de Vries equation. It became apparent that these "integrable" evolution equations possessed a Hamiltonian formulation, having an infinite number of independent first integrals in involution and that they could therefore be interpreted as integrable systems with an infinite number of degrees of freedom; the inverse scattering method provided the first integrals and led to explicit solutions. Later the method was succesfully applied in the context of classical (finite-dimensional) integrable systems, first by Flaschka (see [Fla1] and [Fla2]), Manakov (see [Man]) and Moser (see [Mos1] and [Mos2]) and later by many others. In a short period of time a connection was discovered with many branches of mathematics, especially Lie theory, representation theory, algebraic and differential geometry. The revival of the interest in integrable systems was as much present in physics as in mathematics: many physical interesting systems were found to be well-enough described by integrable systems and (infinite-dimensional) integrable systems play nowadays a dominant role in modern field theories.

Traditionally integrable systems are considered as differential geometric objects. The phase space is a smooth (or analytic) manifold, equipped with a symplectic structure and the functions in involution are smooth (or analytic) functions. Many important constructions relate to the existence of certain coordinates or transformations and are of a transcendental nature; we think here of the construction of Darboux coordinates, canonical transformations, separating variables, generating functions, etc. In most examples however the different elements which constitute the integrable system are algebraic: the phase space has the structure of (the real part of) an affine algebraic variety, the Poisson bracket of two regular functions is a regular function and the functions in involution are also regular functions. This suggests that algebraic geometric tools may be helpful in studying and solving integrable systems. For example the n-dimensional manifolds which are obtained by fixing the values of the n

---

[1] At that time the theory of theta functions was considered as a chapter in complex analysis.

[2] Basically everything that was known about integrable systems before the revival is contained in Whittaker's classical book [Whi].

first integrals are n-dimensional affine algebraic varieties whose precise nature influences the possible types of flows: if such a fixed level manifold admits a compactification to which the n integrable vector fields extend in a holomorphic way and remain independent then the level manifold must be a complex torus and the flows of these vector fields are linear on it.

Algebraic-geometric tools were already used in the study of integrable systems at the end of the 19-th century, but it was Adler and van Moerbeke (see [AM2]-[AM9] and [MM]) who clarified the meaning of the integrability of most classical systems by introducing the concept of an algebraic completely integrable system (a.c.i. system) and developed new tools to analyse and solve such systems. The basic conditions which they impose for an integrable system to be a.c.i. is that the general fiber of the momentum map is an affine part of an Abelian variety and that the flows of the integrable vector fields are linear on them. The main tool they developed for studying a.c.i. systems is the asymptotic analysis of integrable systems and goes back to Kowalevski. The results which were obtained from it include explicit embeddings of Abelian varieties (and related varieties such as Kummer varieties) in projective space, a detailed analysis of divisors and their singularities on Abelian varieties, a classification of integrable flows on $SO(4)$, the construction of Lax equations, a theorem which allows one to conclude from the asymptotic analysis that a given integrable system is a.c.i., some intricate relations to Lie theory, ...

The present work originated and was largely influenced by the work of Adler and van Moerbeke. It was presented as a "Habilitation à diriger des recherches" at the Université de Lille (France) and contains some of the author's work ([BV2], [Van3] and [Van5]; part of [Van6] and the main result of [Van4] are sketched). In order to present our work in a coherent way we did the effort to rewrite the relevant theory (and the above papers) completely in the language of algebraic geometry. Although trivial at many points, extra work was often needed to do this. In the rest of this introduction we explain in more detail the different notions which are introduced and the main results which are established.

1. Affine Poisson varieties

We start Chapter II by introducing the concept of an affine Poisson variety. An affine Poisson variety $(M, \{\cdot, \cdot\})$ is an affine variety M (defined over the field of complex numbers) with a Poisson algebra structure $\{\cdot, \cdot\}$ on its algebra $\mathcal{O}(M)$ of regular functions, i.e., $\mathcal{O}(M)$ has a Lie algebra structure $\{\cdot, \cdot\}$

$$\{\cdot, \cdot\} : \mathcal{O}(M) \times \mathcal{O}(M) \to \mathcal{O}(M)$$
$$(f, g) \qquad \mapsto \{f, g\}$$

which is a derivation in each of its arguments. The last condition implies that there is associated to each function $f \in \mathcal{O}(M)$ a vector field X_f (called its Hamiltonian vector field) defined by $X_f = \{\cdot, f\}$. At every point the Poisson bracket has a rank which is constant on a Zariski open subset of M; this constant, which is always even in view of the skew symmetry of the Poisson bracket, is called the rank of the affine Poisson variety and is denoted by $\text{Rk}\{\cdot, \cdot\}$. A regular function whose associated Hamiltonian vector field is zero is called a Casimir. The Casimirs form a subalgebra $\text{Cas}(M)$ of $\mathcal{O}(M)$ and lead to an important decomposition of M, the Casimir decomposition, which is given by the fibers of the morphism

$$\pi : M \to \text{Spec}\,\text{Cas}(M),$$

3

induced by the inclusion $\text{Cas}(M) \subset \mathcal{O}(M)$; when $\text{Cas}(M)$ is finitely generated, $\text{Spec Cas}(M)$ is an affine algebraic variety and π is a morphism. Picking a fiber may be interpreted as fixing some of the values of the Casimirs; the fibers over closed points are the ones which correspond to fixing all values of all Casimirs, while the fibers over other points in $\text{Spec Cas}(M)$ correspond to fixing only the values of the constants of the Casimirs which belong to some subalgebra. A general point in the spectrum being by definition a closed point which does not belong to a certain divisor, a general fiber corresponds to picking "generic" values for all Casimirs. Said differently the fibers of π are just the level sets of the Casimirs.

We will show that the Poisson bracket on M restricts[3] to the fiber over any point $c \in \text{Spec Cas}(M)$ and that all vector fields are tangent to these fibers; the rank of the restriction of the Poisson structure to each fiber is less than or equal to the rank of the Poisson structure, with equality for a general fiber (Proposition II.2.38). Moreover we have that

$$\dim \text{Cas}(M) \leq \dim(M) - \text{Rk}(M),$$

where $\dim \text{Cas}(M)$ is the Krull dimension of $\text{Cas}(M)$ (Proposition II.2.40). When we have equality in this equation we will say that the Poisson bracket is maximal; maximality is preserved by restriction to the general fiber which implies that the general fiber has a rank equal to its dimension. Examples will be given which show that not *all* fibers (over closed points) need to have the same dimension or rank and that the algebra of Casimirs of the restricted Poisson structure needs not be maximal.

Restricting the Poisson structure to a level set of the Casimirs is one obvious way to produce new affine Poisson varieties from old ones. To give a complete description of some other constructions it is useful to introduce first the concept of a morphism between affine Poisson varieties $(M_1, \{\cdot, \cdot\}_1)$ and $(M_2, \{\cdot, \cdot\}_2)$: it is a morphism $\phi : M_1 \to M_2$ which preserves the Poisson structure, i.e., the following diagram is commutative.

$$
\begin{array}{ccc}
\mathcal{O}(M_2) \times \mathcal{O}(M_2) & \xrightarrow{\{\cdot,\cdot\}_2} & \mathcal{O}(M_2) \\
\Big\downarrow{\phi^* \times \phi^*} & & \Big\downarrow{\phi^*} \\
\mathcal{O}(M_1) \times \mathcal{O}(M_1) & \xrightarrow[\{\cdot,\cdot\}_1]{} & \mathcal{O}(M_1)
\end{array}
$$

A Poisson morphism $\phi : M_1 \to M_2$ does not necessarily map the algebra of Casimirs of M_2 in the one of M_1. Conditions for this to happen will be given. Also the image of M_1 by a morphism needs not be an affine Poisson variety since the image needs not even be an affine variety. When the image *is* an affine subvariety of M_2 then it inherits a Poisson bracket from M_2 and the map to this image is Poisson (Proposition II.2.16).

An important property of a morphism $\phi : (M_1, \{\cdot, \cdot\}_1) \to (M_2, \{\cdot, \cdot\}_2)$ of affine Poisson varieties is that the rank at a point of M_1 is always higher than the rank at the image of this point,

$$\forall m \in M_1 : \text{Rk}_x\{\cdot, \cdot\}_1 \geq \text{Rk}_{\phi(x)}\{\cdot, \cdot\}_2.$$

---

[3] The fibers of $\pi : M \to \text{Spec Cas}(M)$ are algebraic but need not be irreducible; the Poisson structure gives (by restriction) each irreducible component the structure of an affine Poisson variety.

In particular one has equality for an isomorphism. This leads to an invariant polynomial for affine Poisson varieties as follows. The definition of the rank at a point gives a second natural decomposition of M into algebraic varieties which we call the rank decomposition. Each element of this decomposition (more precisely its closure) is given by

$$M_i = \{p \in M \mid \mathrm{Rk}_p\{\cdot,\cdot\} \leq 2i\},$$

where i ranges between 0 and $\frac{1}{2}\mathrm{Rk}\{\cdot,\cdot\}$. They are determinantal varieties and equations for them are quite easily obtained since they are given by

$$M_i = \{p \in M \mid \text{all determinants of order } 2i+1 \text{ of } g \text{ vanish at } p\},$$

where g is a matrix which collects the Poisson brackets of any system of generators of $\mathcal{O}(M)$. In particular this description shows that their irreducible components are affine varieties. Then an invariant polynomial with integer coefficients is associated to $(M,\{\cdot,\cdot\})$ by

$$\rho(M,\{\cdot,\cdot\}) = \sum \rho_{ij} R^i S^j, \qquad \rho_{ij} = \#\text{irred. comp. of dimension } j \text{ of } M_i.$$

We also give a refinement of the invariant: we associate to each point in $\mathrm{Spec}\,\mathrm{Cas}(M)$ the invariant of the Poisson structure of the fiber over it.

Some new affine Poisson varieties are obtained from old ones as follows. As we already pointed out, each irreducible component of a level set of the Casimirs is an affine Poisson variety; moreover the inclusion map is a Poisson morphism. If two affine Poisson varieties are given, then their product also has a natural Poisson structure; the projection map on each component is also a Poisson morphism (Proposition II.2.21). We prove that the invariant polynomial of the product is the product of the invariant polynomials of the two factors (Proposition II.2.57). Further the zero locus of any regular function f can be removed from an affine Poisson variety $(M,\{\cdot,\cdot\}_M)$. More precisely, there exist an affine Poisson variety $(N,\{\cdot,\cdot\}_N)$ and a morphism $(N,\{\cdot,\cdot\}_N) \to (M,\{\cdot,\cdot\}_M)$ which is surjective on the complement of the zero locus of f (Proposition II.2.35). Finally, if we have a (reductive) group acting (in such a way as to preserve the Poisson structure) then we can construct a quotient, which is again an affine Poisson variety, the quotient map is a morphism and the fixed point variety, which is in general not a Poisson subvariety, inherits a Poisson structure. These constructions have their counterparts for integrable Hamiltonian systems as will be discussed below.

2. Integrable Hamiltonian systems

In the classical definition of integrability in the sense of Liouville the existence of a sufficient number of first integrals in involution is demanded, sufficient meaning equal to the degrees of freedom of the system, i.e., half the dimension of the phase space when the latter is assumed symplectic. In order to get a good definition of a morphism of integrable Hamiltonian systems it is clearly better to consider the algebra of functions generated by the first integrals; if the first integrals are in involution then this whole algebra is involutive. Note also that the datum of the given function only (the Hamiltonian) does not suffice to determine the whole algebra, which confirms that the integrable Hamiltonian system should consist of an algebra and not of a single function. Having a sufficient number of functions in involution corresponds to this algebra having maximal dimension (in a sense which will be specified below). A last thing which we want from our algebra is that it is complete in the sense that every function which is in involution with all elements of the algebra is actually contained in it. This technical condition is demanded because we do not want two algebras of maximal dimension which lead to the same algebra after completion to be considered different. The final definition is the following.

Definition 2.1 If $(M \{\cdot,\cdot\})$ is an affine Poisson variety whose algebra of Casimirs is maximal and $\mathcal{A}$ is a complete involutive subalgebra of $\mathcal{O}(M)$, which is finitely generated, then $\mathcal{A}$ is called integrable if

$$\dim \mathcal{A} = \dim M - \frac{1}{2} \operatorname{Rk}\{\cdot,\cdot\}.$$

The triple $(M, \{\cdot,\cdot\}, \mathcal{A})$ is then called an integrable Hamiltonian system and each non-zero vector field in

$$\operatorname{Ham}(\mathcal{A}) = \{X_f \mid f \in \mathcal{A}\}$$

is called an integrable vector field. We call M the phase space of the system, $\operatorname{Spec} \mathcal{A}$ its base space and the natural map $M \to \operatorname{Spec} \mathcal{A}$ its momentum map.

The dimension of $\mathcal{A}$ being $\dim M - \frac{1}{2} \operatorname{Rk}\{\cdot,\cdot\}$ is the maximum possible, as we show in Proposition II.3.4. As in the case of the algebra of Casimirs we consider the level sets which correspond to this algebra, i.e., the fibers of the momentum map

$$\pi_{\mathcal{A}} : M \to \operatorname{Spec} \mathcal{A}.$$

The irreducible components of these fibers are again affine varieties and their study will be central throughout this text.

Many integrable algebras have the property of being integrable (involutive) with respect to two different Poisson structures $\{\cdot,\cdot\}_1$ and $\{\cdot,\cdot\}_2$, moreover these Poisson structures have the special property of being compatible, i.e., any linear combination of them is also a Poisson bracket. We call the two integrable Hamiltonian systems $(M, \{\cdot,\cdot\}_1, \mathcal{A})$ and $(M, \{\cdot,\cdot\}_2, \mathcal{A})$ obtained in this way compatible integrable Hamiltonian systems and prove some basic properties of these (see Paragraph II.3.4). This notion should not be confused with the notion of an integrable bi-Hamiltonian system, which corresponds to the special case in which some non-zero integrable vector field Y is Hamiltonian with respect to both Poisson structures,

$$Y = \{\cdot, f\}_1 = \{\cdot, g\}_2$$

where $f, g \in \mathcal{A}$.

With our definition of integrable Hamiltonian systems it is easy to define morphisms between two such systems.

Definition 2.2 Let $(M_1, \{\cdot,\cdot\}_1, \mathcal{A}_1)$ and $(M_2, \{\cdot,\cdot\}_2, \mathcal{A}_2)$ be two integrable Hamiltonian systems. A morphism $\phi : (M_1, \{\cdot,\cdot\}_1, \mathcal{A}_1) \to (M_2, \{\cdot,\cdot\}_2, \mathcal{A}_2)$ is a Poisson morphism $\phi : (M_1, \{\cdot,\cdot\}_1) \to (M_2, \{\cdot,\cdot\}_2)$ which leads to the following commutative diagram:

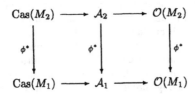

In parallel with what we did for affine Poisson varieties we describe some constructions which allow us to construct new integrable Hamiltonian systems from given ones. Some of these are quite obvious but one has to be careful when proving that the constructed algebra is complete. The least trivial construction which we discuss is that of taking the quotient of an integrable Hamiltonian system by a finite group which is acting in a Poisson way (Proposition II.3.25). It is only demanded here that the integrable algebra $\mathcal{A}$ is stable for the action of the group and not that the individual elements of $\mathcal{A}$ are stabilized. In the special case that each element of $\mathcal{A}$ *is* stabilized, all fibers of the momentum map are mapped to themselves and the quotient will be a quotient on each of the individual fibers. At the other extreme, if no element of the group stabilizes some element of $\mathcal{A}$ then a whole collection of isomorphic fibers will be identified in one. In the first case the fibers will be typically singular, in the second case they will only be singular if the original fibers were singular. Several examples of the quotient construction will be discussed in Chapters VI and VII.

In a separate section (Section II.4) we discuss the extension of our definitions to a more general class of spaces and we compare (integrable Hamiltonian systems on) affine Poisson varieties with (integrable Hamiltonian systems on) Poisson manifolds.

3. *A large family of integrable Hamiltonian systems*

In Chapter III we construct for any integer $d \geq 1$ a large family of integrable Hamiltonian systems on $\mathbf{C}^{2d}$, namely we show how every (non-zero) polynomial $\varphi(x, y)$ in two variables determines a Poisson structure of maximal rank $2d$ on $\mathbf{C}^{2d}$ and we explain how every second polynomial $F(x, y)$ (which we suppose to depend on y) determines an algebra which is integrable for each of these Poisson structures. Our construction generalizes a construction which is due to Mumford; his systems correspond to the choices $\varphi(x, y) = 1$ and $F(x, y) = y^2 - f(x)$ where $f(x)$ is a monic polynomial of degree $2d + 1$.

On $\mathbf{C}^{2d}$ with coordinates $(u_{d-1}, \ldots, u_0, v_{d-1}, \ldots, v_0)$ we show in Paragraph III.2.2 that there corresponds in a natural way to any non-zero polynomial $\varphi(x, y) \in \mathbf{C}[x, y]$ a Poisson bracket $\{\cdot, \cdot\}_d^\varphi$, which is given by

$$\{u(\lambda), u_j\}_d^\varphi = \{v(\lambda), v_j\}_d^\varphi = 0,$$

$$\{u(\lambda), v_j\}_d^\varphi = \{u_j, v(\lambda)\}_d^\varphi = \varphi(\lambda, v(\lambda)) \left[\frac{u(\lambda)}{\lambda^{j+1}}\right]_+ \bmod u(\lambda), \qquad 1 \leq j \leq d, \tag{3.1}$$

where $u(\lambda) = \lambda^d + u_{d-1}\lambda^{d-1} + \cdots + u_0$ and $v(\lambda) = v_{d-1}\lambda^{d-1} + \cdots + v_0$; also $[R(\lambda)]_+$ denotes the polynomial part of a rational function $R(\lambda)$ and $f(\lambda) \bmod g(\lambda)$ is the rest obtained when dividing $f(\lambda)$ by $g(\lambda)$. For fixed d, the map $\varphi \mapsto \{\cdot, \cdot\}_d^\varphi$ is clearly a linear map, showing that all our brackets are compatible; moreover this map is injective since the Poisson structures obtained are of maximal rank. If $\varphi(x, y)$ is a constant, say $\varphi(x, y) = 1$, then the bracket $\{\cdot, \cdot\}_d = \{\cdot, \cdot\}_d^1$ is given by the following matrix P of Poisson brackets:

$$P = \begin{pmatrix} 0 & U \\ -U & 0 \end{pmatrix} \quad \text{where} \quad U = \begin{pmatrix} 0 & 0 & \cdots & 0 & 1 \\ 0 & 0 & \cdots & 1 & u_{d-1} \\ \vdots & \vdots & & \vdots & \vdots \\ 0 & 1 & \cdots & u_3 & u_2 \\ 1 & u_{d-1} & \cdots & u_2 & u_1 \end{pmatrix}.$$

Thus, (3.1) provides us with a large class of affine Poisson varieties $\left(\mathbf{C}^{2d}, \{\cdot, \cdot\}_d^\varphi\right)$.

Chapter I. Introduction

It is remarkable that there exists a lot of sets of independent functions $\{H_0, \ldots, H_{d-1}\}$ which are in involution with respect to each of our Poisson structures. To describe these, let $F(x, y)$ be any polynomial in $\mathbf{C}[x, y] \setminus \mathbf{C}[x]$ and expand $F(\lambda, v(\lambda)) \bmod u(\lambda)$ as a polynomial in λ (of degree less than d):

$$F(\lambda, v(\lambda)) \bmod u(\lambda) = H_{d-1}\lambda^{d-1} + H_{d-2}\lambda^{d-2} + \cdots + H_0.$$

Notice that $H_0, \ldots, H_{d-1}$ are polynomials in u_i and v_j. The main result of Chapter III is that these polynomials are in involution for all brackets $\{\cdot, \cdot\}_d^\varphi$ on $\mathbf{C}^{2d}$, that is

$$\{H_i, H_j\}_d^\varphi = 0 \quad \text{for all } 0 \leq i, j \leq d-1 \text{ and } \varphi(x, y) \in \mathbf{C}[x, y].$$

More precisely, if we define for given F and d the algebra $\mathcal{A}_{F,d} = \mathbf{C}[H_0, \ldots, H_{d-1}]$ then for any non-zero φ we have an integrable Hamiltonian system $\left(\mathbf{C}^{2d}, \{\cdot, \cdot\}_d^\varphi, \mathcal{A}_{F,d}\right)$.

In the special case where $F(x, y)$ is of the form $F(x, y) = y^2 - f(x)$ the curve $F(x, y) = 0$ defines a hyperelliptic curve and a lot of simplifications occur. For example, the vector fields $X_{H_i}^\varphi$ of the integrable Hamiltonian system can in this case (even for arbitrary φ and d) be written as Lax equations

$$X_{H_i}^\varphi A(\lambda) = \left[A(\lambda), [B_i(\lambda)]_+\right],$$

where

$$A(\lambda) = \begin{pmatrix} v(\lambda) & u(\lambda) \\ -\left[\frac{F(\lambda, v(\lambda))}{u(\lambda)}\right]_+ & -v(\lambda) \end{pmatrix} \quad \text{and} \quad B_i(\lambda) = \frac{\varphi(\lambda, v(\lambda))}{u(\lambda)} \left[\frac{u(\lambda)}{\lambda^{i+1}}\right]_+ A(\lambda);$$

see Paragraph III.2.4.

The meaning of the polynomials $F(x, y)$ and $\varphi(x, y)$ becomes apparent in Section III.3, when we study the fibers of the momentum map. Namely we will show in Paragraph III.3.2 that the fiber over 0,

$$\mathcal{F}_{F,d} = \{(u(\lambda), v(\lambda)) \in \mathbf{C}^{2d} \mid H_{F,d}(u(\lambda), v(\lambda)) = 0\}$$

is smooth if and only if the plane algebraic curve $\Gamma_F \subset \mathbf{C}^2$, defined by $F(x, y) = 0$ is smooth and that in this case $\mathcal{F}_{F,d}$ is isomorphic to an affine part of the d-fold symmetric product of the curve Γ_F; a similar description of the structure of the other fibers of the momentum map (lying over other closed points in $\operatorname{Spec} \mathcal{A}$) follows at once.

We also look at the real parts of these fibers, i.e., the fixed points on $\mathcal{F}_{F,d}$ of the complex conjugation map. We show in Paragraph III.3.3 that the real part $\mathcal{F}_{F,d} \cap \mathbf{R}^{2d}$ of $\mathcal{F}_{F,d}$ can be described as the set of all d-tuples in Γ_F, consisting only of real points and points which appear in complex conjugated pairs. This leads to an explicit description of the topology of (the real part of) the general fiber of the momentum map, which are in general neither tori nor cylinders; we encounter here a much larger class of topological types than in all other studies, the reason being that the fibers of the momentum map of our systems have in general nothing to do with Abelian varieties.

The compactification of the smooth fibers of the momentum map, of major interest in several studies in this field, is discussed in Paragraph III.3.4. One obvious compactification of such a fiber is given by the symmetric product of the underlying curve; in general however, a smooth compactification such that the integrable vector fields extend in a holomorphic way, does *not* exist.

8

For fixed F and d the algebra $\mathcal{A}_{F,d}$ which is obtained is fixed, but it leads to many (very) different integrable Hamiltonian systems by varying the polynomial φ which dictates the Poisson structure; since all these Poisson structures are compatible we obtain compatible integrable Hamiltonian systems, however they are not multi-Hamiltonian since the integrable vector fields which are generated by using the different Poisson structures are completely different (a multi-Hamiltonian formulation for some of these systems will be discussed in Paragraph VI.3; for the general case see [Van5]). For different choices of $\varphi(x,y)$ all corresponding integrable vector fields are tangent to the fibers of the momentum map and they are related in a quite simple way. For example the relation between the integrable vector fields corresponding to $\varphi = 1$ and arbitrary φ is described by a transfer matrix $\mathcal{T}_1^\varphi$, defined as

$$\left(X_{H_{d-1}}^\varphi, \ldots, X_{H_0}^\varphi\right) = \left(X_{H_{d-1}}^1, \ldots, X_{H_0}^1\right) \mathcal{T}_1^\varphi;$$

we show that this transfer matrix is given by $\mathcal{T}_1^\varphi = \varphi(M, v(M))$ where

$$M = \begin{pmatrix} -u_{d-1} & -u_{d-2} & -u_{d-3} & \cdots & -u_0 \\ 1 & 0 & 0 & \cdots & 0 \\ 0 & 1 & 0 & \cdots & 0 \\ \vdots & \ddots & \ddots & \ddots & \vdots \\ 0 & \cdots & 0 & 1 & 0 \end{pmatrix}.$$

4. Algebraic completely integrable systems

For many integrable Hamiltonian systems the general fiber of the momentum map is isomorphic to an affine part of an Abelian variety; Abelian varieties were intensively studied in algebraic geometry, an important class of them, the Jacobi varieties, being introduced to study algebraic curves. In Chapter IV we will describe the basics of the theory of Abelian varieties in order to provide the reader with a better understanding of the integrable Hamiltonian systems discussed below, in order to be able to use this powerful machinery to perform certain constructions with these systems and finally because we will be able to use some of these integrable Hamiltonian systems to obtain results about Abelian varieties which are difficult to obtain in a direct way.

Abelian varieties are complex tori which can be embedded in projective space, hence are algebraic. The embedding is done by means of the holomorphic sections of a (very ample) line bundle. There is a basic correspondence between divisors and line bundles and under this correspondence the holomorphic sections of the line bundle correspond to meromorphic functions having a pole along the divisor which corresponds to the line bundle. Hence the embedding can be made concrete using the functions with a (certain) pole along some divisor; given an ideal $\mathcal{I} \subset \mathbf{C}[X_1, \ldots, X_n]$ defining an affine part of an Abelian variety for example, the divisor can be taken to be the divisor at infinity and the embedding is provided by *certain* regular functions in the coordinate ring $\mathbf{C}[X_1, \ldots, X_n]/\mathcal{I}$ of this affine part.

The above construction is a very precise one and basically everything about Abelian varieties can be computed. However the construction is not explicit and here is where the theory of integrable Hamiltonian systems comes in. A first question is how to find an ideal which defines an affine part of an Abelian variety. Since in many integrable Hamiltonian systems the general fiber of the momentum map is isomorphic to an affine part of an Abelian variety we find such ideals at once from the integrable algebra defining the integrable Hamiltonian system. Notice that the integrable vector fields (and the Poisson structure) are not used to solve this first question (but one may want to use the vector field to prove that the general fiber *is* an affine part of an Abelian variety). It should be remarked that, before the theory of integrable systems was used, the only known explicit examples of such ideals were for elliptic curves! The second question is how to determine explicitly the *certain* regular functions which provide the embedding of the Abelian variety into projective space. The method of doing this was developed by Adler and van Moerbeke and is based on an idea due to Kowalevski; one searches for the Laurent solutions to one of the integrable vector fields and it turns out that these encode all necessary information about the compactification and allow to determine explicitly the functions which define an embedding of the Abelian variety.

In Chapter IV we will restrict ourselves to the "abstract" theory, leaving the issues discussed in the last paragraph to Chapter V. Everything which is contained in Chapter IV is very classical and well-known to the algebraic geometer. We will basically explain what is needed later on: the correspondence between divisors and line bundles, how line bundles provide embeddings in projective space; also Abelian varieties, in particular Abelian surfaces and Jacobians; *en passant* we give some useful information about algebraic curves (in particular hyperelliptic curves) and about Kummer's classical quartic surface. References on Abelian varieties and Jacobians are given at the beginning of Section IV.1.

We give in Section V.2 the definition of an a.c.i. system and explain in Section V.3 the above described program by which explicit embeddings of Abelian varieties (and associated varieties such as their Kummer varieties) in projective space are obtained. Actually different definitions of an a.c.i. system (on a symplectic or Poisson manifold) have been given; we compare these and extract from it the following definition of an a.c.i. system on an affine Poisson variety (another, more restrictive definition will also be given).

Definition 4.1 Let $(M, \{\cdot, \cdot\}, \mathcal{A})$ be an integrable Hamiltonian system on an affine Poisson variety and $\pi_{\mathcal{A}} : M \to \operatorname{Spec}(\mathcal{A})$ its momentum map. Then $(M, \{\cdot, \cdot\}, \mathcal{A})$ is called an algebraic completely integrable Hamiltonian system or an a.c.i. system if there exists a Zariski open subset $B \subset \operatorname{Spec}(\mathcal{A})$ and a bundle of Abelian groups $\pi : \mathcal{T} \to B$ such that for each $b \in B$ there exists a divisor $\mathcal{D}_b \subset \pi^{-1}(b)$ and an isomorphism $\phi_b : \pi_{\mathcal{A}}^{-1}(b) \to \pi^{-1}(b) \setminus \mathcal{D}_b$, such that the restriction of each vector field in $\operatorname{Ham}(\mathcal{A})$ to $\pi_{\mathcal{A}}^{-1}(b)$ is ϕ-related to a linear vector field on $\pi^{-1}(b)$.

We will also discuss several constructions of a.c.i. systems. The fact that one can take a quotient of an a.c.i. system (by a finite group) is very helpful, e.g. when dealing with morphisms between integrable Hamiltonian systems: if the morphism is invariant for the group action then it can be factorized via the quotient, leading often to an isomorphism of the quotient with the image. There are two very different types of quotients. For one type the group action leaves all fibers of the momentum map invariant and the quotient will have a momentum map whose general fiber is isomorphic to an affine part of an Abelian variety, but having a different polarization type. For the other type the group action interchanges the general fibers of the momentum map and the general fibers of the momentum map of the quotient system will be isomorphic to a general fiber of the original momentum map. In general the action will be such as to give something which interpolates between these two types (see Proposition V.2.5).

The basics about Lax equations are given in Section V.5. Since Lax equations are not central in this text — we will construct Lax equations for some integrable Hamiltonian systems, but we will not use them — we will only give a brief sketch.

5. The Mumford systems

The integrable Hamiltonian systems which are introduced in Chapter III are not a.c.i., except for a very small subclass, which corresponds to the even and the odd Mumford systems; the odd Mumford systems were constructed by Mumford in [Mum5] while the even Mumford systems were introduced by us in [Van2].

In this text we will neither construct the odd Mumford systems as a special case of the systems from Chapter III, nor will we follow Mumford's construction. Instead we will construct its phase space as a natural space of pairs of commuting differential operators (Section VI.2) and we will obtain (in Section VI.3) its multi-Hamiltonian structure (including the commuting Hamiltonians and hence the integrable vector fields) by a Poisson reduction on the loop algebra $\widetilde{\mathfrak{sl}}(2)$. This construction has among other things the advantage of admitting several generalizations, due to its Lie algebraic nature. In fact, we will present in this text a $\mathfrak{gl}(q)$ generalization of the odd Mumford systems, while for further generalizations we refer to [KV] (for matrix differential operators) and [PV1] (for arbitrary simple Lie algebras).

The odd Mumford systems are obtained from this construction by taking $q = 2$ and by fixing some of the Casimirs (corresponding to reducing $\mathfrak{gl}(q)$ to $\mathfrak{sl}(q)$). Since all formulas for the even Mumford systems are very similar to the ones for the odd Mumford systems we believe that the even Mumford systems can be obtained in a similar way (which would lead e.g. also to a Lie algebraic generalization), but despite several efforts, no result in this direction has been obtained at present. Therefore, we derive the even Mumford systems from the systems which were introduced in Chapter III, making special choices for F, φ and d. Since the same construction can be applied to construct the odd Mumford systems in a different way, we have two different interpretations of the multi-Hamiltonian structure of the odd Mumford systems, namely their compatible brackets are one the one hand product brackets, when expressed in terms of linearizing variables, on the other hand they are Lie-Poisson brackets, when expressed in terms of the original phase variables.

The importance of having a multi-Hamiltonian formulation of the Mumford systems comes, in our point of view, from the added flexibility of defining morphisms. Namely, for many (i.e., most) integrable Hamiltonian systems linearizing variables have been found (by various methods, often by trial and error or generalizing from easy examples to more complicated ones), while linearizing variables for the even and the odd Mumford systems (as well as their analogs of Chapter III which are not a.c.i.) are so closely related to the original phase variables of the Mumford systems that the linearizing variables for an a.c.i. system lead automatically to a morphism to one of the Mumford systems. These morphisms turn out to be Poisson, when picking the right Poisson structure for the corresponding Mumford system (this Poisson structure differs from case to case); thus the existence of a multi-Hamiltonian formulation for the Mumford systems explains why the morphisms one finds from linearizing variables usually turn out to be morphisms of integrable Hamiltonian systems.

We discuss the algebraic complete integrability of the even and the odd Mumford systems in Paragraph VI.4.2 and we show how Painlevé analysis for the Mumford systems leads to a natural family of stratifications of hyperelliptic Jacobians. This stratification can e.g. in the even case (when the affine hyperelliptic curve underlying the Jacobian has two points at infinity) be depicted as follows.

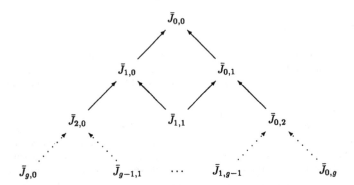

necessarily Here $\bar{J}_{0,0}$ is the entire Jacobian, $\bar{J}_{1,0}$ and $\bar{J}_{0,1}$ are two touching translates of the theta divisor and the lowest row contains $g+1$ points; the arrows are inclusion maps. For the odd case the stratification is much simpler, having precisely one stratum in each dimension. This statification is related to the natural stratification of the Sato Grassmannian but this will not be discussed here (see [Van4]).

Finally, we give in Paragraph VI.5 an algebraic completely integrable analog of the integrable Hamiltonian systems constructed in Chapter III, thereby providing an algebraic completely integrable generalization of the odd and the even Mumford systems. To do this, we fix the Poisson structure on $\mathbf{C}^{2g}$ which corresponds to $\varphi = 1$ and we pick a non-singular curve of genus g in $\mathbf{C}^2$. From an equation $F(x,y) = 0$ of the curve we obtain an algebra $\mathcal{A}$ of functions by expanding $F(\lambda, v(\lambda)) \bmod u(\lambda)$ in terms of g polynomials which are constructed from a basis of the vector space of holomorphic differentials on the curve. $\mathcal{A}$ has dimension g and is involutive, however the elements of $\mathcal{A}$ are not regular (they are rational); thus we need to change our phase space a little bit, which can be done rigorously by constructing a new affine Poisson variety and a dominant injective morphism which maps onto the locus where all elements of $\mathcal{A}$ are regular; we obtain an integrable Hamiltonian system on this affine Poisson variety. Surprisingly enough[4], the process of removing this divisor from $\mathbf{C}^{2g}$ has exactly the effect of removing from the general fiber of the momentum map — which we know to be isomorphic to an affine part of a symmetric product of a curve — the locus which is blown down by the Abel-Jacobi map; thus these fibers become now isomorphic to genuine affine parts of Jacobians. We arrive in this way at a large class of a.c.i. systems, generalizing the Mumford systems.

---

[4] To be more precise, there is an assumption about the curve which is generically satisfied, in particular it is satisfied for many curves of interest; if it is not satisfied then the claimed property continues to hold for at least the fiber over 0.

6. Further examples and applications

In Chapter VII we give some applications of integrable Hamiltonian systems to algebraic geometry and we discuss the geometry — in particular the algebraic complete integrability — of a several two-dimensional examples.

- The genus two Mumford systems and the BvM system

We give in Section VII.2 explicit formulas for the genus two even and odd Mumford systems. These formulas will be used in our study of all examples and applications that follow. In that section we also present the so-called Bechlivanidis-van Moerbeke system (see [BM]) because, as we show, it is isomorphic to the genus two even Mumford system. As a consequence, this proves the algebraic complete integrability of the Bechlivanidis-van Moerbeke system.

We use the genus two even Mumford system to give in Section VII.3 a first example which shows that integrable Hamiltonian systems may be used to study and solve some questions in algebraic geometry, especially in curve theory and the theory of Abelian varieties; the present example has been worked out in collaboration with J. Bertin (see [BV2]). The question considered here is to find an explicit projective realization of a generalized Kummer surface. The Kummer surface is obtained by considering a curve of genus two which has an automorphism of order three, i.e., it has an equation

$$y^2 = x^6 + 2\kappa x^3 + 1.$$

Here $\kappa^2 \neq 1$ is the modular parameter as we will show. The symmetry of order three extends to the Jacobian and the quotient is a singular surface which we call a generalized Kummer surface. It has a 9_4 configuration consisting of 9 invariant theta curves and 9 fixed points, similar to Kummer's classical 16_6 configuration. This 9_4 configuration, which has essentially only one projective realisation, has been considered by Castelnuovo and Segre (see [Cas] and [Seg]). The generalized Kummer surface can be embedded in $\mathbf{P}^4$ as the intersection of a quadric and a cubic hypersurface. The main question addressed is to compute explicit equations for the quadric and cubic hypersurface. To this aim we use the genus two even Mumford system: we identify the Jacobians of curves with an automorphism of order three among the fibers of the momentum map and we apply Painlevé analysis to this system to construct an embedding of these Jacobians in projective space $\mathbf{P}^8$. By using only symmetric sections of the embedding line bundle we obtain an explicit embedding of the generalized Kummer surface in $\mathbf{P}^4$. We show that, in appropriate coordinates which are suggested by the 9_4 configuration, the generalized Kummer surface is given by

$$c^2 x_1 x_2 x_3 + \bar{c}^2 x_4 x_5 x_6 = 0,$$
$$c(x_1 x_2 + x_2 x_3 + x_1 x_3) + \bar{c}(x_4 x_5 + x_5 x_6 + x_4 x_6) = 0,$$
$$x_1 + x_2 + x_3 + x_4 + x_5 + x_6 = 0,$$

where the last equation defines $\mathbf{P}^4$ as a hyperplane of $\mathbf{P}^5$ and $c = 1 + \kappa$, $\bar{c} = 1 - \kappa$. It has been pointed out to us by I. Dolgacev that such equations already appeared in the basic work of Enriques and Severi on hyperelliptic surfaces. They proved that the associated Kummer surface is hyperelliptic with the aid of results of Segre and Castelnuovo. The indirect arguments used there seem not to be complete. Our direct method shows also the precise relationship between the two parameters and can be used in other similar situations.

- The Garnier system and geodesic flow on $SO(4)$

In Sections VII.4 and VII.5 we study two a.c.i. systems which are related to Abelian surfaces of type $(1, 4)$. In its simplest form, the Garnier system is defined by the quartic potential

$$V_{\alpha\beta} = \left(q_1^2 + q_2^2\right)^2 + \alpha q_1^2 + \beta q_2^2 \tag{6.1}$$

on the plane, whose integrability was discovered by Garnier in the beginning of the last century. In fact the Garnier system is a much more general system which contains a lot of integrable Hamiltonian systems; for a derivation of the potentials $V_{\alpha\beta}$ (and their generalizations to higher dimensions) see the Appendix of [Van3].

To prove that the potentials $V_{\alpha\beta}$ define an a.c.i. system we use the result of [BLS] (explained in Section IV.5) which states that the line bundle $\mathcal{L}$ which defines the polarization on a generic Abelian surface of type $(1, 4)$ induces a birational morphism $\phi_{\mathcal{L}} : \mathcal{T}^2 \to \mathbf{P}^3$, whose image is an octic of a certain type; an equation for this octic is given with respect to well-chosen coordinates for $\mathbf{P}^3$ by

$$\lambda_0^2 y_0^2 y_1^2 y_2^2 y_3^2 + \lambda_1^2 (y_0^4 y_1^4 + y_2^4 y_3^4) + \lambda_2^2 (y_1^4 y_3^4 + y_0^4 y_2^4) + \lambda_3^2 (y_0^4 y_3^4 + y_1^4 y_2^4) +$$
$$2\lambda_1 \lambda_2 (y_0^2 y_1^2 + y_2^2 y_3^2)(y_1^2 y_3^2 - y_0^2 y_2^2) + 2\lambda_1 \lambda_3 (y_0^2 y_3^2 - y_1^2 y_2^2)(y_0^2 y_1^2 - y_2^2 y_3^2) +$$
$$2\lambda_2 \lambda_3 (y_1^2 y_2^2 + y_0^2 y_3^2)(y_1^2 y_3^2 + y_0^2 y_2^2) = 0,$$

for some $(\lambda_0 : \lambda_1 : \lambda_2 : \lambda_3) \in \mathbf{P}^3 \setminus S$ where S is some divisor of $\mathbf{P}^3$, which we will determine. Moreover each octic of this type occurs in that way. It will allow us to show that the general fiber of the momentum map of the integrable Hamiltonian system defined by the potential $V_{\alpha\beta}$, $(\alpha \neq \beta)$, is isomorphic to an affine part of an Abelian surface (of polarization type $(1, 4)$) and we show that the integrable vector fields are linearized by these isomorphisms (Proposition VII.4.1). Our proof of algebraic complete integrability is unusual in the sense that we do not use the Laurent solutions to the differential equations nor the Lax equations. We also give another proof which uses a morphism to the genus two odd Mumford system and which gives as a by-product a Lax representation for the potentials $V_{\alpha\beta}$.

We make a detailed study of the moduli space $\mathcal{A}_{(1,4)}$ of Abelian surfaces of type $(1, 4)$ and of some associated moduli spaces (Section VII.4.2) and we use some results from [BLS] to construct a morphism ψ from $\mathcal{A}_{(1,4)}$ to an algebraic cone $\mathcal{M}^3$ of dimension 3 which lives in weighted projective space $\mathbf{P}^{(1,2,2,3,4)}$. The morphism is bijective on the dense subset $\tilde{\mathcal{A}}_{(1,4)}$ of Abelian surfaces for which the above morphism $\phi_{\mathcal{L}}$ is birational and the image is an affine variety $\mathcal{M}^3 \setminus \mathcal{D}$ where $\mathcal{D}$ is some divisor in $\mathcal{M}^3$ which we will also compute. We also define a morphism from $\tilde{\mathcal{A}}_{(1,4)}$ onto the moduli space of two-dimensional Jacobians, or what is the same, the moduli space of smooth curves of genus two: for a given $\mathcal{T}^2 \in \tilde{\mathcal{A}}_{(1,4)}$ the corresponding Jacobi surface, called the canonical Jacobian (of $\mathcal{T}^2$), is defined by the following commutative diagram.

The problem arises to calculate this map explicitly as well as the extra data. We know of no direct algebraic way to do this. Instead we solve this problem (in Section VII.4.4) by relying heavily on the particular coordinates provided by the potentials $V_{\alpha\beta}$. Our geometrical investigations then lead to the following result: for the Abelian surface $\mathcal{T}^2 \in \tilde{\mathcal{A}}_{(1,4)}$ with moduli $(\lambda_0 : \lambda_1 : \lambda_2 : \lambda_3)$ the canonical Jacobian is the Jacobian of the curve

$$y^2 = x(x-1)\left(4\lambda_2^2 x^3 - (\lambda_0^2 + 2\lambda_1^2 + 6\lambda_2^2 + 2\lambda_3^2)x^2 + (\lambda_0^2 - 2\lambda_1^2 + 2\lambda_2^2 + 6\lambda_3^2)x - 4\lambda_3^2\right).$$

In this representation the coordinate x is chosen such that it sends the points of $\mathcal{W}_2$ to $0, 1$ and ∞; $\mathcal{W}_1$ contains the other 3 Weierstrass points on this curve. We obtain this result in two different ways: one way uses the cover $J \to \mathcal{T}^2$ and the other uses the cover $\mathcal{T}^2 \to J$. It would be interesting to be able to calculate this map in a direct way, i.e., without using the potentials $V_{\alpha\beta}$.

In Section VII.5 we show that the geodesic flow on $SO(4)$ corresponding to some special metric is also related to Abelian surfaces of type $(1, 4)$. Here, the general fiber of the momentum map of the corresponding a.c.i. system that we consider, is an affine part of a hyperelliptic Jacobian, where four translates of the theta divisor are missing. These translates all differ by a translation over a half-period and the corresponding group $\mathfrak{T}$ of translations leads to a quotient which is an Abelian surface of type $(1, 4)$. Thus there is an induced map from the base space of this a.c.i. system to the above-mentioned cone $\mathcal{M}^3$ in $\mathbf{P}^{(1,2,2,3,4)}$. Using the geometry of the octic, which is the projective image of the Jacobian by the $\mathfrak{T}$-invariant sections of the line bundle which is defined by the four translates, we will explicitly compute this map. This map solves the moduli problem for this a.c.i. system, as posed in [BV4].

- The Hénon-Heiles hierarchy

The Hénon-Heiles hierarchy consists of a sequence of integrable polynomials on the plane, defined by

$$V_n = \sum_{k=0}^{[n/2]} 2^{n-2k} \binom{n-k}{k} q_1^{2k} q_2^{n-2k}.$$

The geometry of the first non-trivial one $(n = 3)$ was first described by Adler and van Moerbeke (see [AM9]) and the second one $(n = 4)$ by us in [Van2]. We reconsider them along with the higher potentials from the point of view of morphisms. It is easy to see that these integrable Hamiltonian systems have an automorphism of order two, thus leading to a quotient system. We show that for $n = 3$ and $n = 4$ the quotient system is isomorphic to a trivial subsystem of one of the genus two Mumford systems, thereby proving its algebraic complete integrability. For the potential V_3 its algebraic complete integrability follows and it is seen that the general fiber of its momentum map is isomorphic to an affine part of an Abelian surfaces of type $(1, 2)$, confirming the results in [AM9]. For V_4 its algebraic complete integrability does not follow: the general fiber cannot be compactified into an Abelian surface, although it is an unramified cover of an affine part of an Abelian surface; a first example of an integrable Hamiltonian system whose fibers of the momentum map are of this type was discovered by Bechlivianidis and van Moerbeke (see [BM]). For the higher potentials the quotient system is not isomorphic to a subsystem of one of the Mumford systems but of one of the hyperelliptic systems considered in Chapter III which are not algebraic completely integrable (the genus of the hyperelliptic curve does not agree with the dimension of the fibers of the momentum map). We use the morphism to describe the general fiber of the momentum map for the potential V_n as a $2 : 1$ unramified cover of a specific stratum of a hyperelliptic Jacobian.

- The periodic three body Toda lattice

From all integrable Hamiltonian systems, the Toda lattice (and its generalizations) is the one which has been most intensively studied in the literature. We restrict ourselves here to the periodic $\mathfrak{sl}(3, \mathbf{C})$ Toda lattice. We will present in Paragraph VII.7.1 the Toda lattice in some different forms, each of which has its own virtue for the applications which we have in mind. A first one is to show its algebraic complete integrability by using a morphism to the Bechlivanidis-van Moerbeke system (which we have shown to be isomorphic to the genus two even Mumford systems). This morphism is neither injective nor surjective, even not surjective to a trivial subsystem of the Bechlivanidis-van Moerbeke system. This can for example be detected by using the invariant polynomial for their Poisson structures, as defined in Chapter II. It follows that some more work needs to be done to prove that the Toda lattice is algebraic completely integrable. This is done by removing a divisor from the phase space of the Toda lattice on the one hand and removing a divisor from a trivial subsystem of the Bechlivanidis-van Moerbeke system on the other hand. The restriction of the morphism to these subsystems is an isomorphism (of integrable Hamiltonian systems) and we deduce from it that the Toda lattice is a.c.i.. This is done in Paragraph VII.7.2.

Finally it is easy to see that the Toda lattice has an automorphism of order three. Thus we can construct a quotient system, whose general fiber is isomorphic to an affine part of an Abelian surface of type $(1, 3)$. We will give an explicit description of the corresponding phase space and the equations which define the Abelian surfaces are obtained from it at once. One notices that these affine parts are not given as complete intersections.

Chapter II

Integrable Hamiltonian systems
on affine Poisson varieties

1. Introduction

In this chapter we give the basic definitions and properties of integrable Hamiltonian systems on affine Poisson varieties and their morphisms. In Section 2 we define the notion of a Poisson bracket (or Poisson structure) on an affine algebraic variety. The Poisson bracket is precisely what is needed to define Hamiltonian mechanics on a space, as is well-known from the theory of symplectic and Poisson manifolds. We shortly describe the simplest Poisson structures (i.e., constant, linear, affine and quadratic Poisson structures; also general Poisson structures on $\mathbf{C}^2$ and $\mathbf{C}^3$) and describe two natural decompositions of affine Poisson varieties, one is given by the algebra of Casimirs, the other comes from the notion of rank of a Poisson structure (at a point). We also describe several ways to build new affine Poisson varieties from old ones.

Morphisms of affine Poisson variety are regular maps which preserve the Poisson bracket. Isomorphisms preserve the rank at each point, leading to a polynomial invariant for affine Poisson varieties. This invariant permits us on the one hand to distinguish many different affine Poisson varieties, on the other hand it allows us to display in a structured way the basic characteristics of the Poisson structure. It will be computed for many different examples and a refinement of this invariant is also discussed.

In Section 3 we turn to integrable Hamiltonian systems. We motivate our definition by several propositions and (counter-)examples. The notions of super-integrability, compatibility and integrable multi-Hamiltonian systems fit very well into the picture and most of our propositions are easily adapted to the case that the integrable Hamiltonian systems under discussion have one of these extra structures. The notion of momentum map leads to a decomposition of the variety, as the one given by the Casimirs (however it is much finer).

We also define morphisms of integrable Hamiltonian systems; they are Poisson morphisms which preserve the algebra of functions in involution. It allows one to state precisely the relation between different integrable Hamiltonian systems, for example between new systems and the old ones from which they were constructed. Our discussion is parallel to the one of affine Poisson varieties (up to some modifications). Some really interesting examples of integrable Hamiltonian systems will be given in later chapters.

The final section (Section 4) is devoted to a generalization of our definitions to the case of other spaces. We draw special attention to the case of real Poisson manifolds. The main difference is that on the one hand the algebras we work with in the case of an affine Poisson variety are in general not finitely generated so that many constructions do not apply (e.g., the polynomial invariant), on the other hand many local constructions (e.g., Darboux coordinates, action-angle variables) which cannot be performed for affine Poisson varieties, play a dominant role in the study of some other Poisson spaces, including Poisson manifolds.

Apart from Section 4 we will in this chapter always work over the field of complex numbers.

2. Affine Poisson varieties and their morphisms

2.1. Affine Poisson varieties

Phase space will always be an affine variety in the sense of [Har], i.e., an irreducible closed subset of $\mathbf{C}^n$ (closed for the Zariski topology). Such a variety $M \subset \mathbf{C}^n$ is the zero locus of a prime ideal $\mathcal{I}_M$ of $\mathbf{C}[x_1, \ldots, x_n]$, and its ring (or $\mathbf{C}$-algebra) of regular functions is denoted, resp. defined by

$$\mathcal{O}(M) = \frac{\mathbf{C}[x_1, \ldots, x_n]}{\mathcal{I}_M}.$$

$\mathcal{O}(M)$ is an integral domain (it has no zero divisors) and it is finitely generated; M can be reconstructed, up to isomorphism, from $\mathcal{O}(M)$ as $\operatorname{Specm} \mathcal{O}(M)$, the set of closed points in $\operatorname{Spec} \mathcal{O}(M)$.

The extra structure which we use to describe Hamiltonian systems on M is given by a Poisson bracket on its algebra of regular functions.

Definition 2.1 Let M be an affine variety. A *Poisson bracket* or *Poisson structure* on M is a Lie algebra structure $\{\cdot, \cdot\}$ on $\mathcal{O}(M)$, which is a bi-derivation, i.e., for any $f \in \mathcal{O}(M)$ the $\mathbf{C}$-linear map

$$X_f : \mathcal{O}(M) \to \mathcal{O}(M)$$
$$g \mapsto \{g, f\}$$

is a derivation (satisfies the Leibniz rule),

$$X_f(gh) = (X_f g)h + g X_f h \tag{2.1}$$

for all $g, h \in \mathcal{O}(M)$. The derivation X_f is called the *Hamiltonian derivation* associated to the *Hamiltonian* f and we write $\operatorname{Ham}(M, \{\cdot, \cdot\})$ for the (vector) space

$$\{X_f = \{\cdot, f\} \mid f \in \mathcal{O}(M)\}$$

of Hamiltonian derivations. A function $f \in \mathcal{O}(M)$ whose Hamiltonian vector field is zero, $X_f = 0$, is called a *Casimir function* or a *Casimir* and we denote

$$\operatorname{Cas}(M, \{\cdot, \cdot\}) = \{f \in \mathcal{O}(M) \mid X_f = 0\}$$

for the (vector) space of Casimirs; it is the center of the Lie bracket $\{\cdot, \cdot\}$ hence it is a Lie ideal of $(\mathcal{O}(M), \{\cdot, \cdot\})$. When no confusion can arise, either argument in $\operatorname{Ham}(M, \{\cdot, \cdot\})$ and $\operatorname{Cas}(M, \{\cdot, \cdot\})$ is omitted.

Remarks 2.2
1. X_f being a derivation may be refrased in a geometric way by saying that it is a global section of $\mathcal{T}_M$, the tangent sheaf to M, i.e., $X_f \in H^0(M, \mathcal{T}_M)$ (for the definition of the sheaf of differentials and the tangent sheaf to an algebraic variety see [Har] Section II.8). For this reason we usually call the elements X_f of $\operatorname{Ham}(M, \{\cdot, \cdot\})$ *Hamiltonian vector fields*.

2. Using the above mentionned correspondence between an affine variety and its algebra of regular functions we have that affine Poisson varieties correspond to finitely generated Poisson algebras without zero divisors.

3. Turning the above definition upside down one gets at the following, equivalent definition of a Poisson bracket. Let us denote the vector space $\mathrm{Hom}(\bigwedge^n \mathcal{O}(M), \mathcal{O}(M))$ by $C^n(M)$ and its subspace of skew-symmetric n-derivations by $\mathrm{Der}^n(M)$. For every $p, q \geq 0$ a bilinear map

$$[\cdot,\cdot] : C^p(M) \times C^q(M) \to C^{p+q-1}(M)$$

is defined for $P \in C^p(M)$, $Q \in C^q(M)$ and for $f_1, \ldots, f_{p+q-1} \in \mathcal{O}(M)$ by

$$
\begin{aligned}
[P,Q](f_1, &\ldots, f_{p+q-1}) \\
&= \sum_{\sigma \in S_{q,p-1}} (-1)^{pq+q} \epsilon(\sigma) P\left(Q(f_{\sigma(1)}, \ldots, f_{\sigma(q)}), f_{\sigma(q+1)}, \ldots, f_{\sigma(p+q-1)}\right) \\
&+ \sum_{\sigma \in S_{p,q-1}} (-1)^q \epsilon(\sigma) Q\left(P(f_{\sigma(1)}, \ldots, f_{\sigma(p)}), f_{\sigma(p+1)}, \ldots, f_{\sigma(p+q-1)}\right)
\end{aligned}
$$

where $S_{p,q}$ denotes the set of (p,q) shuffles (permutations σ of $\{1, \ldots, p+q\}$ such that $\sigma(1) < \ldots < \sigma(p)$ and $\sigma(p+1) < \ldots < \sigma(p+q)$; $\epsilon(\sigma)$ is the sign of σ). It is easy to see that if $P \in \mathrm{Der}^p(M)$ and $Q \in \mathrm{Der}^q(M)$ then $[P,Q] \in \mathrm{Der}^{p+q-1}(M)$. Thus $[\cdot,\cdot]$ restricts to a bracket

$$[\cdot,\cdot]_S : \mathrm{Der}^p(M) \times \mathrm{Der}^q(M) \to \mathrm{Der}^{p+q-1}(M),$$

called the *Schouten bracket*. For $P \in \mathrm{Der}^2(M)$ we have that

$$[P,P]_S(f,g,h) = 2(P(P(f,g),h) + P(P(g,h),f) + P(P(h,f),g)),$$

so that Poisson structures can also be defined as skew-symmetric bi-derivations P such that $[P,P]_S = 0$. This point of view leads to the following interesting interpretations. If $P \in \mathrm{Der}^2(M)$ defines a Poisson structure then the (graded) Jacobi identity for $[\cdot,\cdot]_S$ implies that $\mathrm{Der}^*(M)$ becomes a complex when the coboundary operator

$$\delta_P : \mathrm{Der}^q(M) \to \mathrm{Der}^{q+1}(M)$$

is defined for $Q \in \mathrm{Der}^q(M)$ by $\delta_P(Q) = [P,Q]_S$. The corresponding cohomology is called *Poisson cohomology*. One observes that the Casimirs are precisely the 0-cocycles and that the Hamiltonian vector fields are the 1-coboundaries. For $X \in \mathrm{Der}^1(M)$, $\delta_P X = -\mathcal{L}_X P$, where $\mathcal{L}_X$ is the *Lie derivative* of P with respect to X, hence the 1-cocycles are the vector fields that preserve the Poisson structure P (such vector fields are called *Poisson vector fields*). A similar interpretation of the 2-cocycles and the 2-coboundaries will be given at the end of this section.

The following properties follow at once from Definition 2.1.

Proposition 2.3 *Let $(M, \{\cdot,\cdot\})$ be an affine Poisson variety.*

(1) $\mathrm{Ham}(M)$ *is a Lie subalgebra of* $\mathrm{Der}^1(M)$ *(with the commutator* $[\cdot,\cdot]$ *as Lie bracket);* $\mathrm{Ham}(M)$ *is however in general not an* $\mathcal{O}(M)$-*module, as opposed to* $\mathrm{Der}^1(M)$;

(2) $\mathrm{Cas}(M)$ *is a subalgebra of* $\mathcal{O}(M)$;

(3) *The adjoint map* $\mathrm{ad} : \mathcal{O}(M) \to \mathrm{Ham}(M)$ *which is defined by* $f \mapsto -X_f$ *is a Lie algebra homomorphism;*

(4) *For all* $f, g \in \mathcal{O}(M)$, $\mathrm{ad}(fg) = f\,\mathrm{ad}(g) + \mathrm{ad}(f)g$; *equivalently,* $X_{fg} = fX_g + gX_f$;

(5) *There is a short exact sequence of Lie algebras*

$$0 \longrightarrow \mathrm{Cas}(M) \longrightarrow \mathcal{O}(M) \xrightarrow{\ \mathrm{ad}\ } \mathrm{Ham}(M) \longrightarrow 0$$

2. Affine Poisson varieties and their morphisms

By (2.1) the Hamiltonian vector field X_f is completely determined by its action on any system of generators $g_1, \ldots, g_s$ of $\mathcal{O}(M)$. It leads to a system of first order polynomial differential equations on $\mathcal{O}(M)$, namely

$$\dot{g}_i = X_f g_i = \{g_i, f\}, \qquad (i = 1, \ldots, s), \tag{2.2}$$

where $\dot{g}_i$ is a convenient notation for $X_f g_i$ when a particular choice of $f \in \mathcal{O}(M)$ has been fixed. Similarly a Poisson structure $\{\cdot, \cdot\}$ is, in view of the Leibniz rule, completely described in terms of the system of generators $g_1, \ldots, g_s$ by the *Poisson matrix*

$$g = (\{g_i, g_j\})_{1 \leq i, j \leq s},$$

whose entries g_{ij} are polynomials in $g_1, \ldots, g_s$. Morevoer, the Jacobi identity will be satisfied for any three functions in $\mathcal{O}(M)$ as soon as it is satisfied for all triples of generators g_i, g_j, g_k of $\mathcal{O}(M)$, with $i < j < k$.

Every vector field, in particular every Hamiltonian vector field, can be (non-uniquely) extended to a vector field on the ambient affine space of M. Namely, suppose that $M \subset \mathbf{C}^s$, let $\bar{g}_1, \ldots, \bar{g}_s$ be the standard coordinates of $\mathbf{C}^s$ and denote the corresponding generators of $\mathcal{O}(M)$ by $g_1, \ldots, g_s$. The vector field determined by $\dot{g}_i = F_i(g_1, \ldots, g_s)$ then leads to a vector field on $\mathbf{C}^s$ determined by $\dot{\bar{g}}_i = F_i(\bar{g}_1, \ldots, \bar{g}_s)$. Conversely, a system $\dot{\bar{g}}_i = F_i(\bar{g}_1, \ldots, \bar{g}_s)$ of first order polynomial differential equations on $\mathbf{C}^s$ defines a derivation on $\mathcal{O}(M) = \mathbf{C}[g_1, \ldots, g_s]/\mathcal{I}_M$ precisely when $\dot{f} \in \mathcal{I}_M$ for every $f \in \mathcal{I}_M \subset \mathbf{C}[\bar{g}_1, \ldots, \bar{g}_s]$, i.e., when the vector field is tangent to M as a subvariety of $\mathbf{C}^s$. Similarly, a Poisson bracket on $\mathcal{O}(M)$ defines a skew-symmetric bi-derivation on $\mathbf{C}^s$, but the latter is in general not a Poisson structure because the Jacobi identity need not hold. Explicitly, let $\bar{g}_{ij} = -\bar{g}_{ji}$ denote any extensions of g_{ij} to $\mathbf{C}[\bar{g}_1 \ldots, \bar{g}_s]$ $(1 \leq i, j \leq k)$ and define a skew-symmetric bi-derivation $\bar{P}$ of $\mathbf{C}[\bar{g}_1, \ldots, \bar{g}_s]$ by

$$\bar{P} = \sum_{i < j} \bar{g}_{ij} \frac{\partial}{\partial \bar{g}_i} \wedge \frac{\partial}{\partial \bar{g}_j}. \tag{2.3}$$

Then the Jacobi identity for $\bar{P}$ is tantamount to the following system of partial differential equations.

$$\sum_{l=1}^{s} \left(\bar{g}_{il} \frac{\partial}{\partial \bar{g}_l} \bar{g}_{jk} + \bar{g}_{jl} \frac{\partial}{\partial \bar{g}_l} \bar{g}_{ki} + \bar{g}_{kl} \frac{\partial}{\partial \bar{g}_l} \bar{g}_{ij} \right) \in \mathcal{I}_M, \qquad (1 \leq i < j < k \leq s), \tag{2.4}$$

showing that the Poisson matrix $\bar{g} = (\bar{g}_{ij})$ does not define a Poisson bracket on $\mathbf{C}[\bar{g}_1, \ldots, \bar{g}_s]$ in general.

We will often define a Poisson bracket by specifying its Poisson matrix (in terms of a system of generators); necessary and sufficient conditions for a matrix to define a Poisson structure are given by the following proposition.

Proposition 2.4 *Let $M = \mathbf{C}[\bar{g}_1, \ldots, \bar{g}_s]/\mathcal{I}$ be an affine variety and let $g_1, \ldots, g_s$ denote the corresponding system of generators of $\mathcal{O}(M)$. Let g be a skew-symmetric $s \times s$ matrix, with entries in $\mathcal{O}(M)$, and let $\bar{g}$ be any skew-symmetric matrix, whose entries are extensions of the corresponding elements in g to $\mathbf{C}[\bar{g}_1, \ldots, \bar{g}_s]$. Then g is the Poisson matrix (in terms of $g_1, \ldots, g_s$) of a Poisson bracket on M if an only if*

(1) *the Jacobi identity (2.4) is satisfied for all triples $(\bar{g}_i, \bar{g}_j, \bar{g}_k)$ (with $i < j < k$);*
(2) *for any $G \in \mathcal{I}_M \subset \mathbf{C}[\bar{g}_1, \ldots, \bar{g}_s]$ one has*

$$\sum_{i=1}^{s} \frac{\partial G}{\partial \bar{g}_i} \bar{g}_{ij} \in \mathcal{I}_M, \qquad (j = 1, \ldots, s). \tag{2.5}$$

Proof

We have seen above that (1) is necessary; if $G \in \mathcal{I}_M$ then for any g_j one has $\{g_j, G\} = 0$ (in $\mathcal{O}(M)$), which implies (2.5) upon using (2.3). Let us show that (1) and (2) are also sufficient. Define a skew-symmetric bi-derivation $\{\cdot, \cdot\}'$ on $\mathbf{C}[\bar{g}_1, \ldots, \bar{g}_s]$ by (2.3). It is not necessarily a Poisson bracket since (1) only guarantees that the Jacobi identity is satisfied in $\mathcal{O}(M)$, i.e., its right hand side belongs to $\mathcal{I}_M$. Condition (2) implies that for all $F_1, F_2 \in \mathbf{C}[\bar{g}_1, \ldots, \bar{g}_s]$ and for all $G_1, G_2 \in \mathcal{I}_M$ we have

$$\{F_1 + G_1, F_2 + G_2\}' = \{F_1, F_2\}' + G_3$$

for some $G_3 \in \mathcal{I}_M$ hence we have an induced bracket $\{\cdot, \cdot\}$ on $\mathcal{O}(M)$. It satisfies the Leibniz rule because the original bracket does; because of condition (1) it satisfies the Jacobi identity for a system of generators hence for all regular functions. Hence it defines a Poisson bracket on $\mathcal{O}(M)$. ∎

Remark 2.5 It is natural to ask if for every affine Poisson variety M one can find a system of generators $g_1, \ldots, g_s$ of $\mathcal{O}(M)$ (realizing M as a subvariety of $\mathbf{C}^s$) and extensions of their Poisson brackets to $\mathbf{C}[\bar{g}_1, \ldots, \bar{g}_s]$, such that the bi-derivation (2.3) is a Poisson bracket on $\mathbf{C}^s$. The answer to this question is at present unknown.

Definition 2.6 Let $(M, \{\cdot, \cdot\})$ be an affine Poisson variety and let $x \in M$. We define the *rank* of the Poisson structure *at* x, denoted $\mathrm{Rk}_x\{\cdot, \cdot\}$, as the even number defined by either one of the following:

(i) The rank at x of any Poisson matrix representing $\{\cdot, \cdot\}$;
(ii) The number of linearly independent Hamiltonian derivations at x.

The *rank* of the Poisson structure is defined as $\max\{\mathrm{Rk}_x\{\cdot, \cdot\} \mid x \in M\}$ and is denoted by $\mathrm{Rk}\{\cdot, \cdot\}$ and the co-rank $\dim M - \mathrm{Rk}\{\cdot, \cdot\}$ is denoted by $\mathrm{CoRk}\{\cdot, \cdot\}$. The Poisson structure (or variety) is called *regular* if it has constant rank and *trivial* if its rank is zero.

The equivalence of the two definitions is easily established; since *(ii)* is intrinsic we see that *(i)* does not depend on the chosen system of generators of $\mathcal{O}(M)$. Notice that the rank is also defined at the singular points of M and that, in view of *(i)* it is maximal on a Zariski open subset of M (see also Paragraph 2.4 below). Moreover, since at the general point of M the number of independent Hamiltonian derivations is bounded by $\dim M$ we have that $\mathrm{Rk}\{\cdot, \cdot\} \leq \dim M$. We give some first examples of affine Poisson varieties.

2. Affine Poisson varieties and their morphisms

Example 2.7 Any constant skew-symmetric $n \times n$ matrix is the matrix of a Poisson structure on $\mathbb{C}^n$, in terms of its standard coordinates, as follows from (2.4). We refer to such a Poisson structure as a *constant Poisson structure*. By the classification theorem for skew-symmetric bilinear forms the Poisson matrix takes the standard form

$$\begin{pmatrix} 0 & I_r & 0 \\ -I_r & 0 & 0 \\ 0 & 0 & 0 \end{pmatrix}$$

after a linear change of coordinates. Here $2r$ is the rank of the Poisson matrix. This structure is often called the *canonical* or *standard Poisson structure* of rank $2r$ on $\mathbb{C}^n$. Since in the case of a constant bracket (2.5) is of degree $\deg G - 1$ we see that a necessary condition for such a bracket to define a bracket on an affine subvariety M of $\mathbb{C}^n$ is that all elements of minimal degree in $\mathcal{I}_M$ are Casimirs.

Example 2.8 Let $\mathfrak{g}$ be any finite-dimensional (complex) Lie algebra, with Lie bracket $[\cdot\,,\cdot]$. By Proposition 2.4, the Lie bracket extends to the symmetric algebra $\operatorname{Sym} \mathfrak{g}$ of $\mathfrak{g}$, making $\operatorname{Sym} \mathfrak{g}$ into a Poisson algebra. Since $\operatorname{Sym} \mathfrak{g} \cong \mathcal{O}(\mathfrak{g}^*)$ the dual space $\mathfrak{g}^*$ of $\mathfrak{g}$ carries a natural Poisson structure. For $f, g \in \mathcal{O}(\mathfrak{g}^*)$ it is given at $\xi \in \mathfrak{g}^*$ by

$$\{f, g\}(\xi) = \xi\left(\left[\widehat{df(\xi)}, \widehat{dg(\xi)} \right] \right), \tag{2.6}$$

the hat denoting the natural pairing $(\mathfrak{g}^*)^* \to \mathfrak{g}$. This Poisson structure is known as the *canonical Poisson structure* or the *Lie-Poisson structure* of $\mathfrak{g}^*$. The rank at the origin is always zero, so it is never regular (unless $\mathfrak{g}$ is commutative so that $\{\cdot\,,\cdot\}$ is trivial). Notice that the resulting Poisson algebra is characterized by having independent generators with respect to which the Poisson matrix is linear.

Choosing a non-degenerate bilinear form $\langle \cdot\,,\cdot \rangle$ on $\mathfrak{g}$ we have an isomorphism $\chi : \operatorname{Sym} \mathfrak{g} \cong \mathcal{O}(\mathfrak{g})$ defined by

$$\chi\left(\bigotimes_{\text{sym}} e_i \right) = \prod \langle e_i, \cdot \rangle.$$

It allows us to transfer the Poisson structure to $\mathfrak{g}$; it is easy to see that it is given, for $f, g \in \mathcal{O}(\mathfrak{g})$ at $x \in \mathfrak{g}$ by

$$\{f, g\}(x) = \langle x, [\nabla f(x), \nabla g(x)] \rangle,$$

where the gradient $\nabla f(x)$ of $f \in \mathcal{O}(\mathfrak{g})$ at x is defined by

$$\forall y \in \mathfrak{g} \qquad \langle \nabla f(x), y \rangle = \frac{d}{dt}_{|t=0} f(x + ty).$$

Example 2.9 For quadratic Poisson structures (on $\mathbb{C}^n$) a general theory is not known. A 3-dimensional family of quadratic structures on $\mathbb{C}^4$ was given by Sklyanin (see [Skl1]). Let x_0, x_1, x_2 and x_3 be linear coordinates on $\mathbb{C}^4$ and consider the following matrix:

$$\begin{pmatrix} 0 & b_1 x_2 x_3 & b_2 x_1 x_3 & b_3 x_1 x_2 \\ -b_1 x_2 x_3 & 0 & a_3 x_0 x_3 & a_2 x_0 x_2 \\ -b_2 x_1 x_3 & -a_3 x_0 x_3 & 0 & a_1 x_0 x_1 \\ -b_3 x_1 x_2 & -a_2 x_0 x_2 & -a_1 x_0 x_1 & 0 \end{pmatrix}.$$

By Proposition 2.4 four checks of the Jacobi identity suffice to show that this is a Poisson matrix and they are all satisfied if and only if

$$a_1 b_1 - a_2 b_2 + a_3 b_3 = 0;$$

notice that this is also equivalent to the vanishing of the determinant of the Poisson matrix (for all values of the x_i). This gives a 5-dimensional family of quadratic Poisson structures on $\mathbf{C}^4$. Except for the trivial structure they are all of rank two and two Casimirs are given by $a_1 x_1^2 - a_2 x_2^2 + a_3 x_3^2$ and $a_1 x_0^2 - b_3 x_2^2 + b_2 x_3^2$. The Poisson structures given by Sklyanin correspond to

$$a_1 = -a_2 = a_3,$$
$$b_1 + b_2 + b_3 = 0.$$

We will see in Example 2.52 that the 5-dimensional family given above is more general (up to isomorphism) than the family given by Sklyanin.

Example 2.10 It is easy to describe all Poisson structures on $\mathbf{C}^2$ since in this case the Jacobi identity is satisfied for any skew-symmetric bracket. Let x and y be coordinate functions then any polynomial φ in two variables defines a Poisson bracket on $\mathbf{C}^2$ by $\{x, y\} = \varphi(x, y)$ and conversely every Poisson bracket is of this form. It is regular if and only if φ is constant, otherwise the rank drops to zero along the algebraic curve $\varphi(x, y) = 0$.

Example 2.11 For $\mathbf{C}^3$ the Jacobi identity is not trivially satisfied. Indeed the matrix

$$\begin{pmatrix} 0 & H & -G \\ -H & 0 & F \\ G & -F & 0 \end{pmatrix}$$

will be a Poisson matrix if and only if

$$(\nabla \times \vec{E}) \cdot \vec{E} = 0, \tag{2.7}$$

where $\vec{E} = (F, G, H)$. For χ and φ arbitrary polynomials $\vec{E} = \chi \nabla \varphi$ is a solution to (2.7), leading to a large number of Poisson structures on $\mathbf{C}^3$; explicitly for such $\vec{E}$ the Poisson matrix is given by

$$\chi \begin{pmatrix} 0 & \frac{\partial \varphi}{\partial z} & -\frac{\partial \varphi}{\partial y} \\ -\frac{\partial \varphi}{\partial z} & 0 & \frac{\partial \varphi}{\partial x} \\ \frac{\partial \varphi}{\partial y} & -\frac{\partial \varphi}{\partial x} & 0 \end{pmatrix},$$

and φ is seen to be a Casimir of this Poisson structure. The rank is two except at the zero locus of χ and at the points where the gradient of φ vanishes. Notice that not all solutions to (2.7) are of this form, for example the Poisson matrix

$$\begin{pmatrix} 0 & 0 & x \\ 0 & 0 & y \\ -x & -y & 0 \end{pmatrix}$$

is not of the form $\chi \nabla \varphi$ (with χ and φ regular).

2. Affine Poisson varieties and their morphisms

In many cases, especially in the theory of integrable systems, the space comes naturally equipped with several Poisson structures, moreover they are often compatible in the following sense.

Definition 2.12 Let M be an affine variety and let $\{\cdot,\cdot\}_1$ and $\{\cdot,\cdot\}_2$ two Poisson brackets on M. Then we say that they are *compatible* if for any $a, b \in \mathbf{C}$ the linear combination

$$\{\cdot,\cdot\}_{a,b} = a\{\cdot,\cdot\}_1 + b\{\cdot,\cdot\}_2$$

is a Poisson bracket on M. Similarly n Poisson brackets on M are said to be *compatible* if any linear combination of them is a Poisson bracket on M.

Remark 2.2 implies that $\{\cdot,\cdot\}_1$ and $\{\cdot,\cdot\}_2$ are compatible if and only if $[a\{\cdot,\cdot\}_1 + b\{\cdot,\cdot\}_2, a\{\cdot,\cdot\}_1 + b\{\cdot,\cdot\}_2]_S = 0$, for all $a, b \in \mathbf{C}$, which is equivalent to $[\{\cdot,\cdot\}_1, \{\cdot,\cdot\}_2]_S = 0$. This means that $\{\cdot,\cdot\}_2$ defines a 2-cocycle in the Poisson cohomology of $\{\cdot,\cdot\}_1$. It is from this point of view a natural question to ask whether the cohomology class defined by $\{\cdot,\cdot\}_2$ is trivial, i.e., whether there exist a vector field X on M such that $L_X\{\cdot,\cdot\}_1 = \{\cdot,\cdot\}_2$. Such a vector field is said to have the *deformation property* with respect to $\{\cdot,\cdot\}_1$. Notice that if P and $L_X P$ are both Poisson structures then they are automatically compatible Poisson structures since

$$[P, L_X P]_S = -[P, \delta_P X] = -\delta_P^2 X = 0.$$

Explicitly, the condition $[\{\cdot,\cdot\}_1, \{\cdot,\cdot\}_2]_S = 0$ takes the following form

$$\begin{aligned}
&\{\{f,g\}_1, h\}_2 + \{\{g,h\}_1, f\}_2 + \{\{h,f\}_1, g\}_2 + \\
&\{\{f,g\}_2, h\}_1 + \{\{g,h\}_2, f\}_1 + \{\{h,f\}_2, g\}_1 = 0,
\end{aligned} \tag{2.8}$$

for all f, g and h in $\mathcal{O}(M)$ (or, equivalently, for a system of generators). It is also clear from the Schouten bracket that n brackets are compatible if and only if they are pairwise compatible.

Example 2.13 Constant Poisson structures are always compatible.

Example 2.14 Examples 2.7 and 2.8 describe all possible constant and linear Poisson structures on $\mathbf{C}^n$, i.e., Poisson structures whose Poisson matrix (with respect to some, hence with respect to any system of linear coordinates for $\mathbf{C}^n$) consists only of constant resp. linear elements. One may wonder about their combination, that is, Poisson brackets on $\mathbf{C}^n$ whose Poisson matrices (as above) have entries of degree at most one; let us call them *affine Poisson brackets* on $\mathbf{C}^n$. Obviously both the constant and linear part of such Poisson structures are Poisson structures, hence a constant Poisson structure on $\mathbf{C}^n$ which is *compatible* with a Lie-Poisson structure on $\mathbf{C}^n$ defines (by taking their sum) an affine Poisson structure on $\mathbf{C}^n$ and any affine Poisson structure on $\mathbf{C}^n$ is of this form. These affine structures are known as *modified canonical Poisson structures* or *modified Lie-Poisson structures*. If we denote the Poisson bracket determined by the linear part by $\{\cdot,\cdot\}_1$ and the one determined by the constant part by $\{\cdot,\cdot\}_0$ then we see that condition (2.8) reduces in terms of linear coordinates $x_1, \ldots, x_n$ for $\mathbf{C}^n$ to

$$\{\{x_i, x_j\}_1, x_k\}_0 + \{\{x_j, x_k\}_1, x_i\}_0 + \{\{x_k, x_i\}_1, x_j\}_0 = 0. \tag{2.9}$$

Using the hat-notation from Example 2.8, we look at $\mathbf{C}^n$ as the dual of a Lie algebra $\mathfrak{g}$ with basis $\hat{x}_i$ and Lie bracket determined by

$$[\hat{x}_i, \hat{x}_j] = \widehat{\{x_i, x_j\}}_1.$$

Then the second bracket (the constant bracket) determines a linear map

$$C : \mathfrak{g} \wedge \mathfrak{g} \to \mathbf{C} : (\hat{x}_i, \hat{x}_j) \mapsto \{x_i, x_j\}_0.$$

With this notation

$$\{\{x_i, x_j\}_1, x_k\}_0 = C(\{\widehat{x_i, x_j}\}_1, \hat{x}_k) = C([\hat{x}_i, \hat{x}_j], \hat{x}_k]),$$

hence (2.9) becomes

$$C([\hat{x}_i, \hat{x}_j], \hat{x}_k) + C([\hat{x}_j, \hat{x}_k], \hat{x}_i) + C([\hat{x}_k, \hat{x}_i], \hat{x}_j) = 0.$$

This formula expresses that C is a 2-cocycle in the cohomology[5] of $\mathfrak{g}$ associated with the trivial representation of $\mathfrak{g}$ on $\mathbf{C}$, giving another way to describe affine Poisson structures on a vector space. As an application of this point of view, recall that if $\mathfrak{g}$ is semisimple the first and second cohomology groups are trivial; then C is a coboundary, $C = \partial C'$, written out, $C(\hat{x}_i, \hat{x}_j) = C'([\hat{x}_i, \hat{x}_j])$ and we see that the affine Poisson bracket is nothing but the original Lie-Poisson bracket with x_i replaced by $x_i + C'(\hat{x}_i)$, i.e., both brackets are the same up to an *affine* change of variables.

2.2. Morphisms of affine Poisson varieties

We recall that a map $\phi : M_1 \to M_2$ of affine varieties is called a *regular map* or a *morphism* if $\phi^* \mathcal{O}(M_2) \subset \mathcal{O}(M_1)$, where $\phi^*(f) = f \circ \phi$ for functions $f \in \mathcal{O}(M_2)$; thus ϕ induces and is uniquely defined by an algebra homomorphism $\phi^* : \mathcal{O}(M_2) \to \mathcal{O}(M_1)$. A regular map which has a regular inverse is called a *biregular map* or an *isomorphism*.

Definition 2.15 Let $(M_1, \{\cdot, \cdot\}_1)$ and $(M_2, \{\cdot, \cdot\}_2)$ be two affine Poisson varieties, then a map $\phi : M_1 \to M_2$ is called a *Poisson morphism* or a *morphism of affine Poisson varieties* if

(1) ϕ is a morphism, $\phi^* \mathcal{O}(M_2) \subset \mathcal{O}(M_1)$;
(2) For all $f, g \in \mathcal{O}(M_2)$, $\phi^*\{f, g\}_2 = \{\phi^* f, \phi^* g\}_1$.

Both conditions are conveniently summarized by the commutativity of the following diagram:

$$
\begin{array}{ccc}
\mathcal{O}(M_2) \times \mathcal{O}(M_2) & \xrightarrow{\{\cdot,\cdot\}_2} & \mathcal{O}(M_2) \\
\Big\downarrow{\scriptstyle \phi^* \times \phi^*} & & \Big\downarrow{\scriptstyle \phi^*} \\
\mathcal{O}(M_1) \times \mathcal{O}(M_1) & \xrightarrow[\{\cdot,\cdot\}_1]{} & \mathcal{O}(M_1)
\end{array}
$$

---

[5] For an introduction to the cohomology of Lie algebras see e.g. Appendix 5 in [LM3].

The standard adjectives which are used for morphisms of affine varieties (e.g., *injective, dominant, finite, ...*) may also be used for Poisson morphisms. A biregular map is a Poisson morphism if and only if its inverse is a Poisson morphism; we call such a map a *(Poisson) isomorphism*. When M_1 is an affine subvariety of M_2 and the inclusion map $(M_1, \{\cdot, \cdot\}_1) \to (M_2, \{\cdot, \cdot\}_2)$ is a Poisson morphism then M_1 is called an *affine Poisson subvariety* of M_2.

The image of an affine variety by a morphism needs not be an affine variety; consider for example the image of the map $(x, y) \mapsto (x, xy)$, defined on $\mathbf{C}^2$. If the image of a Poisson morphism is an affine subvariety (i.e., a (Zariski) closed subset) of the target space[6] then there is an induced Poisson structure on it, making it into an affine Poisson subvariety, as shown in the following proposition.

Proposition 2.16 *If $\phi : M_1 \to M_2$ is a morphism of affine Poisson varieties and the image $\phi(M_1)$ is an affine subvariety of M_2 then $\phi(M_1)$ has a unique structure of an affine Poisson variety such that ϕ can be factorized as the composition of a surjective and an injective Poisson morphism, as in the following diagram:*

Proof

By assumption ϕ can be factorized as $\phi = \imath \circ \bar{\phi}$, where $\bar{\phi} : M_1 \to \phi(M_1)$ is a surjective morphism (of affine varieties) and $\imath : \phi(M_1) \to M_2$ is an inclusion map. Since $\bar{\phi}$ is surjective, $\bar{\phi}^* : \mathcal{O}(\phi(M_1)) \to \mathcal{O}(M_1)$ is injective; since $\imath : \phi(M_1) \to M_2$ is an inclusion map, $\imath^* : \mathcal{O}(M_2) \to \mathcal{O}(\phi(M_1))$ is surjective.

This leads to the following definition of a Poisson bracket on $\phi(M_1)$. Let $f, g \in \mathcal{O}(\phi(M_1))$, then by surjectivity of $\imath^*$ there exist f' and g' in $\mathcal{O}(M_2)$ such that $f = \imath^* f'$ and $g = \imath^* g'$. We define

$$\{f, g\} = \imath^* \{f', g'\}_2, \tag{2.10}$$

and verify that it is independent of the choices made for f' and g'; since $\bar{\phi}^*$ is injective it suffices to show that $\bar{\phi}^* \imath^* \{f', g'\}_2$ depends only on f and g. Using the fact that ϕ is a Poisson morphism we find

$$\bar{\phi}^* \imath^* \{f', g'\}_2 = \{\bar{\phi}^* f, \bar{\phi}^* g\}_1. \tag{2.11}$$

It follows from (2.10) that $\{\cdot, \cdot\}$ satisfies the Jacobi identity and that $\imath$ is a Poisson morphism. Now (2.11) can also be written as

$$\bar{\phi}^* \{f, g\} = \{\bar{\phi}^* f, \bar{\phi}^* g\}_1,$$

so that $\bar{\phi}$ is a Poisson morphism. ∎

We now show how the rank of the Poisson structure at a point and at its image by a Poisson morphism are related; it implies equality for an isomorphism.

---

[6] Examples include closed, proper and finite morphisms, see [Har] Ch. II ¶4.

Proposition 2.17 *Let* $(M_i, \{\cdot, \cdot\}_i)$, $(i = 1, 2)$ *be two affine Poisson varieties and let* $\phi :$
$M_1 \to M_2$ *a Poisson morphism. Then* $\mathrm{Rk}_x\{\cdot, \cdot\}_1 \geq \mathrm{Rk}_{\phi(x)}\{\cdot, \cdot\}_2$ *for any* $x \in M_1$.

Proof

Let $g_1, \ldots, g_s$ be a system of generators of $\mathcal{O}(M_2)$. Then $h_i = \phi^* g_i$, $(i = 1, \ldots, s)$ can be completed into a system of generators $h_1, \ldots, h_{s+t}$ of $\mathcal{O}(M_1)$. By definition of the rank we have

$$\begin{aligned}
\mathrm{Rk}_x\{\cdot, \cdot\}_1 &= \mathrm{Rk}\left(\{h_i, h_j\}(x)\right)_{1 \leq i, j \leq s+t} \\
&\geq \mathrm{Rk}\left(\{h_i, h_j\}(x)\right)_{1 \leq i, j \leq s} \\
&= \mathrm{Rk}_{\phi(x)}\{\cdot, \cdot\}_2.
\end{aligned}$$

■

The proposition implies that in general an affine subvariety of an affine Poisson variety does not carry a Poisson structure which makes it into an affine Poisson subvariety. Necessary and sufficient conditions for this to happen will be given in Proposition 2.18.

2.3. Constructions of affine Poisson varieties

There are four (known) basic constructions of Poisson brackets on finite dimensional spaces (here taken to be affine varieties). The first one is that of the canonical Lie-Poisson structure (Example 2.8). In the most important examples, at least from the point of view of integrable Hamiltonian systems, the relevant Lie bracket is an R-bracket, see Section V.5. The second one consists of the canonical Poisson structure associated to a symplectic structure (see Example 4.2 below), the prime example being here the one of cotangent bundles. Notice that in the first case the Poisson structure is never regular while in the second case it is always regular. Both these are very classical, as opposed to the other two which will not be discussed in detail here. The first of these two deals with Poisson structures on Lie groups, in particular Lie-Poisson groups; an excellent account of this is given in Semenov's paper in [BCK1]. The last one consists of the construction of higher order brackets, starting from a Lie bracket, also within the R-matrix approach (see [LP]).

Apart from these four basic constructions there are also several constructions which allow one to build new Poisson brackets from given ones. We will discuss these here in the context of affine varieties, in fact we will show how the standard constructions by which new affine varieties are built from given ones, have their equivalents for affine Poisson varieties.

We think here of the following:

(1) the restriction to an affine subvariety;
(2) the product of two affine Poisson varieties;
(3) the quotient and fixed point set of an affine Poisson variety under the action of a finite or reductive group;
(4) removing a divisor.

They are considered next in the above order: we start by giving a precise condition for a Poisson structure to restrict to an affine subvariety (for an important example, see Proposition 2.38 below).

Proposition 2.18 *Let* $(M, \{\cdot, \cdot\})$ *be an affine Poisson variety and suppose that* N *is an affine subvariety of* M. *Then the following are equivalent:*

    (i) There exists a Poisson structure on N with respect to which N is an affine Poisson subvariety of M;

    (ii) The ideal of N is a Poisson ideal of $\mathcal{O}(M)$;

    (iii) The restriction of every Hamiltonian vector field X_f, $f \in \mathcal{O}(M)$ to N is tangent to N.

Proof

    The equivalence of *(ii)* and *(iii)* is immediate from the definition of a Poisson ideal: an ideal of a Poisson algebra is called a *Poisson ideal* if it has the additional property of being a Lie ideal; thus, $\mathcal{I}_N$, the ideal of regular functions vanishing on N, is a Poisson ideal if and only if $\{\mathcal{I}_N, \mathcal{O}(M)\} \subset \mathcal{I}_N$ which is in turn equivalent to

$$X_f g = \{g, f\} \text{ vanishes at all points of } N,$$

for every Hamiltonian vector field X_f, $f \in \mathcal{O}(M)$ and every $g \in \mathcal{I}_N$.

    If $\mathcal{I}_N$ is a Poisson ideal of $\mathcal{O}(M)$ then for any $f, g \in \mathcal{O}(M)$ and $n \in N$

$$\{f + \mathcal{I}_N, g + \mathcal{I}_N\}(n) = \{f, g\}(n), \tag{2.12}$$

and we can define $\{\cdot, \cdot\}_N$ at $n \in N$ by $\{\imath^* f, \imath^* g\}_N(n) = \{f, g\}(n)$, where $\imath : N \to M$ is the inclusion map; clearly the latter becomes a Poisson morphism. This shows that *(ii)* implies *(i)*. If $\imath : N \to M$ is a Poisson morphism then (2.12) holds, in particular $\{\mathcal{I}_N, g\}$ vanishes on N for any $g \in \mathcal{O}(M)$. Since $\mathcal{I}_N$ is a prime ideal it follows that $\{\mathcal{I}_N, g\} \subset \mathcal{I}_N$ for any $g \in \mathcal{O}(M)$, so *(i)* implies *(ii)*. ∎

Example 2.19 Suppose that $\mathfrak{h}$ is an ideal of a Lie algebra $\mathfrak{g}$ and denote by $(\mathfrak{h})$ the ideal of $\operatorname{Sym}\mathfrak{g}$ generated by $\mathfrak{h}$. For $h \in \mathfrak{h}$ and for $g_1, \ldots, g_m \in \mathfrak{g}$ a direct application of the Leibniz rule shows that every term in

$$\{h, g_1 \cdot g_2 \cdots g_m\}$$

belongs to $(\mathfrak{h})$, where $\{\cdot, \cdot\}$ denotes the canonical Poisson structure on $\operatorname{Sym}\mathfrak{g}$. This shows that $(\mathfrak{h})$ is a Poisson ideal of $\operatorname{Sym}\mathfrak{g}$. Under the canonical isomorphism $\operatorname{Sym}\mathfrak{g} \cong \mathcal{O}(\mathfrak{g}^*)$ the ideal $(\mathfrak{h})$ corresponds to the ideal of functions vanishing on $\mathfrak{h}$. Therefore, Proposition 2.18 implies that the subspace of $\mathfrak{g}^*$ which consists of the elements of $\mathfrak{g}^*$ that vanish on $\mathfrak{h}$ is an affine Poisson subvariety of $\mathfrak{g}^*$ with its Lie-Poisson structure.

Remark 2.20 There are, of course, other ways in which a subvariety may inherit — in one way or another — a Poisson structure from its ambient affine Poisson variety. Think for example of a proper symplectic subvariety of a symplectic manifold, which is never a Poisson subvariety, but still carries a natural Poisson structure "inherited" by the symplectic 2-form on its ambient manifold. This will be discussed later in this paragraph.

Second, we consider products of affine Poisson varieties.

Proposition 2.21 *Let $(M_1, \{\cdot,\cdot\}_1)$ and $(M_2, \{\cdot,\cdot\}_2)$ be two affine Poisson varieties, then the product $M_1 \times M_2$ has a natural structure of an affine Poisson variety such that the projection maps $\pi_i : M_1 \times M_2 \to M_i$ are Poisson morphisms. Moreover it has rank $\mathrm{Rk}\{\cdot,\cdot\}_1 + \mathrm{Rk}\{\cdot,\cdot\}_2$, the rank at $(m_1, m_2) \in M_1 \times M_2$ being given by $\mathrm{Rk}_{m_1}\{\cdot,\cdot\}_1 + \mathrm{Rk}_{m_2}\{\cdot,\cdot\}_2$.*

Proof

The algebra of regular functions on the product $M_1 \times M_2$ is given by

$$\mathcal{O}(M_1 \times M_2) = \pi_1^* \mathcal{O}(M_1) \otimes \pi_2^* \mathcal{O}(M_2), \qquad (2.13)$$

hence the construction amounts to making the tensor product of two Poisson algebras into a Poisson algebra. Formula (2.13) implies that $\mathcal{O}(M_1 \times M_2)$ is generated by the functions $\pi_1^* f_1, \ldots, \pi_1^* f_n, \pi_2^* g_1, \ldots, \pi_2^* g_m$, where $f_1, \ldots, f_n$ is an arbitrary system of generators of $\mathcal{O}(M_1)$ and $g_1, \ldots, g_m$ for $\mathcal{O}(M_2)$. In order for π_1 and π_2 to be Poisson morphisms it is necessary and sufficient to define $\{\pi_1^* f_i, \pi_1^* f_j\} = \pi_1^* \{f_i, f_j\}_1$ and $\{\pi_2^* g_i, \pi_2^* g_j\} = \pi_2^* \{g_i, g_j\}_2$ for all i and j. A *natural* choice for the remaining brackets $\{\pi_1^* f_i, \pi_2^* g_j\}$ is to make them all zero: with this choice the Jacobi identity is surely satisfied. The Poisson matrix of $\{\cdot,\cdot\}$ with respect to the generators $\pi_1^* f_1, \ldots, \pi_2^* g_m$ has a block form, hence $\mathrm{Rk}_{(m_1, m_2)}\{\cdot,\cdot\} = \mathrm{Rk}_{m_1}\{\cdot,\cdot\}_1 + \mathrm{Rk}_{m_2}\{\cdot,\cdot\}_2$ for any $(m_1, m_2) \in M_1 \times M_2$. It is also easy to see that the algebra of Casimirs of $(M_1 \times M_2, \{\cdot,\cdot\})$ is given by $\pi_1^* \mathrm{Cas}(M_1) \otimes \pi_2^* \mathrm{Cas}(M_2)$. ∎

Definition 2.22 The Poisson bracket on $M_1 \times M_2$ given by Proposition 2.21 is called the *product bracket* and is denoted by $\{\cdot,\cdot\}_{M_1 \times M_2}$.

Example 2.23 Suppose that $(G, \{\cdot,\cdot\})$ is an affine Poisson variety and that G is an algebraic group with multiplication $\chi : G \times G \to G$. Then $(G, \{\cdot,\cdot\})$ is called an *affine Lie-Poisson group* if χ is a Poisson morphism, the Poisson bracket on $G \times G$ being the product bracket.

Example 2.24 A related construction appears when having a family of affine Poisson varieties which depend on a (or several) parameter(s). More precisely we assume that there is given a dominant morphism $\pi : P \to N$ of affine varieties (N being the parameter space) and a Poisson bracket $\{\cdot,\cdot\}_n$ on each non-empty fiber $\pi^{-1}(n)$. Define for $f, g \in \mathcal{O}(P)$ and $p \in P$

$$\{f, g\}(p) = \left\{ f_{|\pi^{-1}(\pi(p))}, g_{|\pi^{-1}(\pi(p))} \right\}_{\pi(p)} (p). \qquad (2.14)$$

If for any $f, g \in \mathcal{O}(P)$ one has $\{f, g\} \in \mathcal{O}(P)$ (roughly speaking, if the brackets $\{\cdot,\cdot\}_n$ vary regularly with $n \in N$) then $\{\cdot,\cdot\}$ defines a Poisson bracket on P and for any $n \in N$ the bracket $\{\cdot,\cdot\}_n$ makes every irreducible component of $\pi^{-1}(n)$ into an affine Poisson subvariety of $(P, \{\cdot,\cdot\})$. As a special case, let $P = M \times N$, where M is an affine Poisson variety and take π as projection on the second factor. Clearly this leads to a Poisson structure on each of the fibers of π which varies regularly with $n \in N$. The resulting Poisson structure on P coincides with the product bracket on $M \times N$, where N is given the trivial (zero) bracket.

Third, we consider the Poisson structure on the fixed point set and on the quotient space M/G where G is a finite group or a reductive algebraic group which is equipped with a Poisson structure (notice that it needs not be an affine Lie-Poisson group).

2. Affine Poisson varieties and their morphisms

Before doing this we recall a few facts about group actions on affine varieties. All groups considered here will be either finite or reductive; moreover, when we want to consider Poisson structures on reductive groups we will assume that they are affine varieties, so as to stay in the category of affine Poisson varieties. A (finite or algebraic) group G is said to act on an affine variety $M \subset \mathbf{C}^n$ if the action is the restriction to M of a representation of G on $\mathbf{C}^n$. When G is finite every representation on $\mathbf{C}^n$ is completely reducible, i.e., if the action of G leaves invariant some subspace of $\mathbf{C}^n$ then it leaves invariant a complementary subspace. For infinite groups the above property characterizes reductive groups (since we are working over $\mathbf{C}$). We recall also that there is an induced action of G on $\mathcal{O}(M)$, given for $g \in G$, $f \in \mathcal{O}(M)$ and $x \in M$ by $g^* f(x) = f(g^{-1}x)$.

If G is finite then we say that the action of G on M is a *Poisson action* if for every $g \in G$ the isomorphism $M \to M$ defined by $m \mapsto gm$ is Poisson. If G is infinite, say G is an affine algebraic group, it may itself carry a Poisson structure and the proper generalization of the above notion of Poisson action is that the map $G \times M \to M$ is a Poisson map, where $G \times M$ is given the product Poisson structure.

Proposition 2.25 *Let $(M, \{\cdot, \cdot\}_M)$ and affine Poisson variety and let G be an affine algebraic group acting on M.*

(1) *If there is a Poisson structure on G for which the action is a Poisson action, then the algebra $\mathcal{O}(M)^G$ of G-invariant functions is a Poisson subalgebra of $\mathcal{O}(M)$.*

(2) *If G is moreover reductive or finite, then $\mathcal{O}(M)^G$ is finitely generated, hence corresponds to an affine Poisson variety M/G, leading to the following commutative diagram of Poisson morphisms (π_2 is projection onto the second component).*

$$
\begin{array}{ccc}
G \times M & \xrightarrow{\;\;\chi\;\;} & M \\
\downarrow{\scriptstyle \pi_2} & & \downarrow{\scriptstyle \pi} \\
M & \xrightarrow[\pi]{} & M/G
\end{array}
\qquad (2.15)
$$

Proof

Clearly $f \in \mathcal{O}(M)$ is G-invariant if and only if the following diagram is commutative.

$$
\begin{array}{ccc}
G \times M & \xrightarrow{\;\;\chi\;\;} & M \\
\downarrow{\scriptstyle \pi_2} & & \downarrow{\scriptstyle f} \\
M & \xrightarrow[f]{} & \mathbf{C}
\end{array}
$$

Thus, if $f, g \in \mathcal{O}(M)^G$ and χ is Poisson then

$$
\chi^* \{f, g\}_M = \{\chi^* f, \chi^* g\}_{G \times M} = \{\pi_2^* f, \pi_2^* g\}_{G \times M} = \pi_2^* \{f, g\}_M,
$$

and we see that the bracket of any two G-invariant functions is G-invariant. Therefore the subalgebra $\mathcal{O}(M)^G$ of $\mathcal{O}(M)$ is also a Lie subalgebra of $\mathcal{O}(M)$, i.e., it is a *Poisson subalgebra* of $\mathcal{O}(M)$.

Assume now that G is reductive or finite. Then $\mathcal{O}(M)^G$ is finitely generated (see e.g. [Mum1] or [Spr]) hence is the algebra of regular functions on an affine variety, denoted M/G and called the (categorical) quotient; M/G is naturally identified with the orbit space of an open subset of M. The natural projection map $M \to M/G$ is regular and yields the commutative diagram (2.15). Granted (1) this proves (2). ∎

We next consider a generalization of the above proposition. We consider again a group G (assumed finite or reductive) acting on M and leaving some subvariety N of M invariant. We will show that N may inherit a Poisson structure from M, even if N is not a Poisson subariety of M. Let us denote by $\mathcal{O}(M, N)^G$, the algebra of regular functions on M that restrict to G-invariant functions on N, and by $\rho : \mathcal{O}(M, N)^G \to \mathcal{O}(N)^G$ the natural map induced by the inclusion $N \hookrightarrow M$.

Definition 2.26 Let $(M, \{\cdot, \cdot\})$ be an affine Poisson variety, $\chi : G \times M \to M$ a Poisson action and N a subvariety of M which is G-stable. Then the triple (M, G, N) is called *Poisson-reducible* if $\mathcal{O}(M, N)^G$ is a Poisson subalgebra of $\mathcal{O}(M)$ and if there exists a Poisson bracket on $\mathcal{O}(N)^G$ such that

$$\{\rho(F_1), \rho(F_2)\}_{\mathcal{O}(N)^G} = \rho\{F_1, F_2\} \tag{2.16}$$

holds for all F_1, $F_2 \in \mathcal{O}(M, N)^G$.

Formula (2.16) says that in order to compute the Poisson bracket of two G-invariant functions on N one computes the Poisson bracket of any extensions to M and then restricts the result to N. Notice also that (2.16) uniquely defines a bracket on $\mathcal{O}(N)^G$ (if it exists) since ρ is surjective. In the following proposition we give necessary and sufficient conditions for (M, G, N) to be Poisson-reducible (for a proof see [PV1]).

Proposition 2.27 Let $(M, \{\cdot, \cdot\})$ be an affine Poisson variety, $\chi : G \times M \to M$ a Poisson action and N a subvariety of M which is G-stable. Then (M, G, N) is Poisson-reducible if and only if

$$\rho\{\mathcal{O}(M, N)^G, I(N)\} = 0; \tag{2.17}$$

it is implicit in this condition that its left hand side makes sense.

In a slightly different vein a Poisson structure is also inherited by the fixed point variety of a group action. This fact was first shown in [FV] in the case of a Poisson involution and is generalized in Proposition 2.29 below. First we need a lemma about the ideal of a fixed point variety.

Lemma 2.28 Let G be a finite or reductive group acting on an affine Poisson variety and let N be its fixed point variety. The ideal $\mathcal{I}_N$ is generated by functions f_j for which there exist $g_j \in G$ and $\xi_j \in \mathbf{C} \setminus \{1\}$ such that $g_j^* f_j = \xi_j f_j$.

Proof

Since G is finite or reductive the representation space $\mathbf{C}^n$ which contains M decomposes as a direct sum of spaces $V_0, \ldots, V_s$ such that V_0 is the fixed point set of the action and such that the action of G on the other V_i is irreducible. Then $V_0 \cap M = N$ and $\mathcal{I}_N$ is generated by $\pi_i^* V_i^*$) where π_i is the natural projection $\mathbf{C}^n \to V_i$ and $i = 1, \ldots, s$. Let i be fixed (between 1 and s) and take any non-zero element $\phi \in V_i^*$. Since the action of G on V_i is irreducible there exists $g \in G$ such that $g^*\phi \neq \phi$. Since G is reductive (or finite), G is generated by its semi-simple elements (see [Hum] p. 162). Therefore, let g be a semi-simple element for which $g^*\phi \neq \phi$. Then we have a linear basis $\phi_1, \ldots, \phi_v$ of V_i^*, where $g^*\phi_j = \xi_j \phi_j$ for $j = 1, \ldots, v$ with $\xi_j = 0$ for $j = 1, \ldots, u$ and $\xi_j \neq 0$ for $j = u+1, \ldots, v$. Since $g^*\phi \neq \phi$ it follows that $u < v$ and we have at least one function ψ for which $g^*\psi = \xi\psi$, with $\xi \neq 1$. Consider now the subspace W_i^* of V_i^* which is the span of all functions $f_j \in V_i^*$ for which there exists $g_j \in G$ and $\xi_j \neq 1$ such that $g_j^* f_j = \xi_j f_j$. We have already established that W_i^* contains a non-zero element. Therefore it suffice to verify that W_i^* is invariant to conclude that $W_i^* = V_i^*$. Let $f = \sum_{j=1}^n c_j f_j \in W_i^*$, with f_j as above, and let $g \in G$. We need to show that $\sum_{j=1}^n c_j g^* f_j \in W_i^*$. This follows at once from

$$(g^{-1}g_j g)^* g^* f_j = g^* g_j^* f_j = \xi_j g^* f_j,$$

since $\xi_j \neq 1$. ∎

Proposition 2.29 *Suppose that G is a finite or reductive group acting on an affine Poisson variety $(M, \{\cdot, \cdot\})$. We assume that for every $g \in G$ the isomorphism $\Phi_g : M \to M$ which corresponds to the action of g is a Poisson map. Let N be the subvariety of M consisting of the fixed points of Φ and denote the inclusion map $N \hookrightarrow M$ by $\imath$. Then N carries a (unique) Poisson structure $\{\cdot, \cdot\}_N$ such that*

$$\imath^*\{F_1, F_2\} = \{\imath^* F_1, \imath^* F_2\}_N$$

for all $F_1, F_2 \in \mathcal{O}(M)$ that are G-invariant.

Proof

For $f_1, f_2 \in \mathcal{O}(N)$ we choose $F_1, F_2 \in \mathcal{O}(M)$ such that $f_1 = \imath^* F_1$ and $f_2 = \imath^* F_2$. Since G is finite or reductive we may assume that F_1 and F_2 are G-invariant. We define

$$\{f_1, f_2\}_N = \imath^*\{F_1, F_2\} \qquad (2.18)$$

and show that this definition is independent of the choice of F_1 and F_2. To do this it is sufficient to show that if F_1 and F_2 are G-invariant, with $\imath^* F_1 = 0$, then $\imath^*\{F_1, F_2\} = 0$. We will actually show that if F_2 is G-invariant and $F_1 \in \mathcal{I}_N$ then $\imath^*\{F_1, F_2\} = 0$. Let us denote by $f_1, \ldots, f_t$ a system of generators of $\mathcal{I}_N$ as given by the previous lemma. If F is G-invariant then the fact that Φ_g is a Poisson map implies

$$\imath^*\{f_j, F\} = \imath^* \Phi_g^*\{f_j, F\} = \imath^*\{\Phi_g^* f_j, \Phi_g^* F\} = \xi_j \imath^*\{f_j, F\},$$

showing our claim, since $\xi_j \neq 1$. Note also that the bracket of any two G-invariant functions is G-invariant. In view of this and because (2.18) is independent of the choice of F_1 and F_2 we have for any $f_1, f_2, f_3 \in \mathcal{O}(N)$ that

$$\{\{f_1, f_2\}_N, f_3\}_N = \imath^*\{\{F_1, F_2\}, \bar{F}_3\},$$

leading at once to the Jacobi identity for $\{\cdot, \cdot\}_N$. Similarly the fact that $\{\cdot, \cdot\}_N$ is an anti-symmetric biderivation follows. ∎

We next give a few examples of Proposition 2.25 in the case of finite group actions on $\mathbf{C}^n$. For examples which involve larger groups see Section VI.3.

Example 2.30 Consider the following automorphism of $\mathbf{C}^2$,

$$\imath_1(p_1, p_2) = (-p_1, p_2),$$

which corresponds to a diagonal action of $\mathbf{Z}_2$ on $\mathbf{C}^2$ which has a line of fixed points. Let us compute the algebras of invariants for the induced action and derive from it the Poisson structures on $\mathbf{C}^2$ which descend to the quotient. If we denote the standard coordinates on $\mathbf{C}^2$ by x_1 and x_2 then the invariant functions are the polynomials whose terms are even in x_1, hence the quotient $\mathbf{C}^2/\imath_1$ is again $\mathbf{C}^2$ and the projection map is given by $(p_1, p_2) \mapsto (q_1, q_2) = (p_1^2, p_2)$. The map $\imath_1$ is a Poisson morphism if and only if $\{-x_1, x_2\} = \imath_1^*\{x_1, x_2\}$. If we denote $\{x_1, x_2\} = F(x_1, x_2)$ then this condition means that $-F(x_1, x_2) = F(-x_1, x_2)$ i.e., F is odd in x_1 and it follows that F can be factorized as x_1 times an invariant polynomial. Then the Poisson structure on the quotient is given by $\{y_1, y_2\}_0 = 2y_1 G(y_1, y_2)$ where G is defined by $F(x_1, x_2) = x_1 G(x_1^2, x_2)$. Notice that these Poisson structures on $\mathbf{C}^2$ and $\mathbf{C}^2/\imath_1$ have a line where the rank is zero: if non-trivial they are never regular.

Example 2.31 The only other possibility (up to isomorphism) for $\mathbf{Z}_2$ to act non-trivially on $\mathbf{C}^2$ corresponds to the following automorphism of $\mathbf{C}^2$:

$$\imath_2(p_1, p_2) = (-p_1, -p_2).$$

In the notations of the previous example, a polynomial function is now invariant if and only if it consists only of terms which are of even total degree in x_1 and x_2. Therefore the algebra of invariants is generated by $y_1 = x_1^2$, $y_2 = x_1 x_2$, $y_3 = x_2^2$, with the single relation $y_2^2 = y_1 y_3$ and the quotient space is a quadratic cone. The projection map is then given by $(p_1, p_2) \mapsto (q_1, q_2, q_3) = (p_1^2, p_1 p_2, p_2^2)$ and $\imath_2$ is a Poisson morphism if and only if the polynomial F which defines the bracket, $\{x_1, x_2\} = F(x_1, y_1)$ is even, $F(x_1, x_2) = F(-x_1, -x_2)$. In this case there exists a polynomial $G(y_1, y_2, y_3)$ for which $G(x_1^2, x_1 x_2, x_2^2) = F(x_1, x_2)$. In terms of the generators y_1, y_2 and y_3 the Poisson structure on the quotient is then described by the following Poisson matrix:

$$2G(y_1, y_2, y_3) \begin{pmatrix} 0 & y_1 & 2y_2 \\ -y_1 & 0 & y_3 \\ -2y_2 & -y_3 & 0 \end{pmatrix}.$$

Even if the original Poisson structure is regular (e.g., if $F = 1$, in which case $(\mathbf{C}^2/\imath_2, \{\cdot, \cdot\}_0)$ is an affine Poisson subvariety of $\mathfrak{sl}(2)^*$, with its standard Lie-Poisson structure) the quotient Poisson structure (if non-trivial) is never regular: it always has rank zero at the vertex of the cone.

Example 2.32 The two preceding examples are easily generalized, giving all possible effective actions of a cyclic group on $\mathbf{C}^2$. Namely let

$$\imath_3(p_1, p_2) = (\epsilon p_1, \delta p_2), \qquad \epsilon^p = \delta^q = 1.$$

Here p and q are assumed to be the smallest integers satisfying $\epsilon^p = \delta^q = 1$. The map $\imath_3$ corresponds to an effective action of the cyclic group of order $l = \text{l.c.m.}(p, q)$ and by the remarks made above every such action is of this form. Suppose first that p and q are not coprime, let d denote their g.c.d. and $p = p'd$, $q = q'd$. Then $y_1 = x_1^p$ and $y_3 = x_2^q$ are invariant functions and, since $\epsilon^{p'}$ and $\delta^{q'}$ are primitive d-th roots of unity, we may suppose that x_1 and x_2 are chosen such that $y_2 = x_1^{p'} x_2^{q'}$ is also invariant. Then the quotient is a cone in $\mathbb{C}^3$ given by $y_1 y_3 = y_2^d$. The bracket $\{x_1, x_2\} = F(x_1, x_2)$ defines a Poisson bracket for which t_3 is a Poisson action if and only if $F(\epsilon x_1, \delta x_2) = \epsilon \delta F(x_1, x_2)$. If $\epsilon \delta \neq 1$ then $F(x_1, x_2)$ is $x_1 x_2$ times an invariant polynomial and we may define $G(y_1, y_2, y_3)$ by $x_1 x_2 G(x_1^p, x_1^{p'} x_2^{q'}, x_2^q) = F(x_1, x_2)$. It is easy to check that the Poisson structure on the quotient is then described by

$$G(y_1, y_2, y_3) \begin{pmatrix} 0 & p y_1 y_2 & p q y_1 y_3 \\ -p y_1 y_2 & 0 & q y_2 y_3 \\ -p q y_1 y_3 & -q y_2 y_3 & 0 \end{pmatrix}.$$

If on the other hand $\epsilon \delta = 1$ then $p = q$ and $v = x_1 x_2$. Then $F(x_1, x_2)$ is invariant and we may define $G(y_1, y_2, y_3)$ by $G(x_1^p, x_1 x_2, x_2^p) = F(x_1, x_2)$. The Poisson structure is in this case described by

$$p G(y_1, y_2, y_3) \begin{pmatrix} 0 & y_1 & p y_2^{p-1} \\ -y_1 & 0 & y_3 \\ -p y_2^{p-1} & -y_3 & 0 \end{pmatrix}.$$

Finally, if p and q are coprime then x_1^p and x_2^q generate the algebra of invariants, hence this algebra is a polynomial algebra. As above F is divisible by $x_1 x_2$ and we define $G(u, v)$ by $G(x_1^p, x_2^q) = F(x_1, x_2)$ and find that the Poisson structure on the quotient is determined by $\{y_1, y_2\}_0 = p q y_1 y_2 G(y_1, y_2)$.

Example 2.33 The cyclic permutation

$$t_4(p_1, p_2, p_3) = (p_2, p_3, p_1)$$

gives an action of $\mathbb{Z}_3$ on $\mathbb{C}^3$ which will also appear later on (see Section VII.7). It is not in diagonal form; in order to diagonalize the induced action on $\mathbb{C}[x_1, x_2, x_3]$ let ϵ be a primitive cubic root of unity and define

$$\begin{aligned} u_1 &= x_1 + x_2 + x_3, \\ u_2 &= x_1 + \epsilon x_2 + \epsilon^2 x_3, \\ u_3 &= x_1 + \epsilon^2 x_2 + \epsilon x_3. \end{aligned} \tag{2.19}$$

Then the action of $1 \in \mathbb{Z}_3$ on $\mathbb{C}[u_1, u_2, u_3]$ is given by $\Psi_1^* u_1 = u_1$, $\Psi_1^* u_2 = \epsilon u_2$ and $\Psi_1^* u_3 = \epsilon^2 u_3$. The algebra of invariants is now generated by u_1, $v = u_2^3$, $w = u_3^3$ and $t = u_2 u_3$ with the single relation $t^3 = vw$, showing that the quotient $\mathbb{C}^3/t_4$ is a cylinder over a cubic surface. In terms of the new coordinates the projection map is given by $(q_1, q_2, q_3) \mapsto (q_1, q_2^3, q_3^3, q_2 q_3)$. Assign to u_i a weight $i - 1$ and let χ and φ be polynomials in u_1, u_2 and u_3, all of whose terms have the same weight modulo 3. According to Example 2.11 these lead to a Poisson structure on $\mathbb{C}^3$; the above action of $\mathbb{Z}_3$ will be a Poisson action if and only if all terms in the product $\chi\varphi$ have weight 0 modulo 3; equivalently χ and φ must both be weight homogeneous and the sum of their weights must be a multiple of 3. The resulting Poisson matrix for the quotient is easily written down.

Example 2.34 As a final example let us consider the natural action of the symmetric group S_d on $\mathbf{C}^d$. It is well-known that the algebra of invariant functions for this action is freely generated by the elementary symmetric functions, in particular the invariant functions constitute a polynomial algebra and the quotient is just $\mathbf{C}^d$, the projection map being

$$(p_1, \ldots, p_d) \mapsto (p_1 + p_2 + \ldots + p_d, \ldots, p_1 p_2 \cdots p_d).$$

By a transformation similar to (2.19) the action can be partly diagonalized. The Poisson brackets which descend to the quotient are the ones for which the bracket of any two symmetric polynomials is a symmetric polynomial.

The fourth and final construction is that of removing a divisor from an affine Poisson variety. We show that the resulting space still has the structure of an affine Poisson variety.

Proposition 2.35 Let $(M, \{\cdot, \cdot\}_M)$ be an affine Poisson variety and let $f \in \mathcal{O}(M)$ be a regular function which is not constant. Then there exists an affine Poisson variety $(N, \{\cdot, \cdot\}_N)$ and a Poisson morphism $N \to M$ which is dominant, having the complement (in M) of the zero locus of f as image.

Proof
Consider a new variable t and define an affine variety N by its ring $\mathcal{O}(N)$ of regular functions as follows:

$$\mathcal{O}(N) = \frac{\mathcal{O}(M)[t]}{\mathrm{idl}(ft - 1)}.$$

Let us denote the canonical projection by $\pi : N \to M$. Clearly π is dominant and its image is the complement of the zero divisor of f. If π is to be a Poisson morphism then we are forced to define $\{g_i, g_j\}_N = \{g_i, g_j\}$ for a system of generators $g_1, \ldots, g_k$ of $\mathcal{O}(M)$ (we made a notational identification between $\pi^* g_i$ and g_i). In view of Proposition 2.4 the only way to extend this to a Poisson structure on N is upon using $\{ft - 1, g_i\} = 0$ (for $i = 1, \ldots, k$). Thus one needs to add the brackets

$$\{g_i, t\}_N = -t^2 \{g_i, f\}_N. \tag{2.20}$$

By the same proposition it now suffices to check the Jacobi identity on the system of generators $g_1, \ldots, g_k, t$; since we know it is valid for $g_1, \ldots, g_k$ the following easily established identity suffices:

$$\{\{g_i, g_j\}, t\}_N + \{\{g_j, t\}, g_i\}_N + \{\{t, g_i\}, g_j\}_N = 0$$

(one uses $\{\{g_i, g_j\}, t\}_N = -t^2 \{\{g_i, g_j\}, f\}_N$, which is an immediate consequence of (2.20)). This gives the desired Poisson bracket on N. Note that if $f \in \mathrm{Cas}(M)$ then

$$\mathrm{Cas}(N) = \frac{\mathrm{Cas}(M)[t]}{\mathrm{idl}(ft - 1)}.$$

∎

Remark 2.36 Another way to state the above result is that the algebra of rational functions on an affine Poisson variety with poles only at some fixed divisor (which need not be irreducible) is also a (finitely generated) Poisson algebra. Clearly, the field of rational functions of an affine Poisson variety can in a similar way be turned into a Poisson algebra.

2.4. Decompositions and invariants of affine Poisson varieties

There are three natural decompositions of an affine Poisson variety, two of which constitute of *algebraic* varieties. The first one discussed here is a level decomposition by the Casimirs, the second is a decomposition according to rank and the last one — the non-algebraic one — is the decomposition into symplectic leaves. Due to its non-algebraic nature, the latter will only indirectly (via the other decompositions) be used in this book and is discussed in Section 4.

- **The Casimir decomposition**

The decomposition of an affine Poisson variety which is naturally associated to its algebra of Casimirs applies (and will be applied) equally for other subalgebras of $\mathcal{O}(M)$, so let us introduce it for an arbitrary subalgebra $\mathcal{A}$ (containing 1) of $\mathcal{O}(M)$. To each $m \in M$ we may associate an algebra homomorphism $\chi_m : \mathcal{A} \to \mathbf{C}$ by $f \mapsto \chi_m(f) = f(m)$, allowing us to associate to $m \in M$ the ideal

$$\{f - \chi_m(f) \mid f \in \mathcal{A}\}$$

of $\mathcal{A}$, which is a point in $\operatorname{Spec} \mathcal{A}$, the spectrum of $\mathcal{A}$; we will denote this natural map $M \to \operatorname{Spec} \mathcal{A}$ by $\pi_{\mathcal{A}}$. Another way to see how this map comes out is as follows. The inclusion map $\imath : \mathcal{A} \subset M$ allows one to associate to a prime ideal $\mathcal{I}$ of $\mathcal{O}(M)$ the prime ideal $\mathcal{I} \cap \mathcal{A}$ of $\mathcal{A}$, hence leading to a morphism

$$\imath^* : \operatorname{Spec} \mathcal{O}(M) \to \operatorname{Spec} \mathcal{A}$$

of affine schemes. The space $\operatorname{Spec} \mathcal{O}(M)$ contains M as the set of its *closed points* — this set may also be seen as the *maximal spectrum*, the space of all maximal ideals, of $\mathcal{O}(M)$ — and our map $\pi_{\mathcal{A}}$ is just the restriction of $\imath^*$ to M. We prefer to work with M rather than with $\operatorname{Spec} \mathcal{O}(M)$ since that is the space we originally started from; however we like to keep $\operatorname{Spec} \mathcal{A}$, even when $\mathcal{A}$ is finitely generated so that its underlying variety is an affine variety, because we will also be interested in the fibers of $\imath^*$ over points which are not closed. Notice that the irreducible components of each fiber of $\pi_{\mathcal{A}}$ are affine varieties and a complete set of equations (from which we may choose a *finite* generating set) for the fiber which contains $m \in M$ is given by

$$\forall f \in \mathcal{A} : f(x) = \chi_m(f). \tag{2.21}$$

Often our statements will be about the general fibers of $\pi_{\mathcal{A}}$: when saying that some property holds for a *general fiber* we mean that it holds for the fiber over a *general point*, i.e., for all closed points which do not belong to a certain divisor. We denote the *Krull dimension* of $\mathcal{A}$ by $\dim \mathcal{A}$; if $\mathcal{A}$ is finitely generated then it is a basic result that $\dim \mathcal{A} = \dim \operatorname{Spec} \mathcal{A}$. The following proposition relates the dimension of $\mathcal{A}$ to the dimension of the fibers of $\pi_{\mathcal{A}}$ (see [Sha] Ch. 1 ¶6).

Proposition 2.37 *All (non-empty) fibers of $\pi_{\mathcal{A}} : M \to \operatorname{Spec} \mathcal{A}$ have co-dimension at most $\dim \mathcal{A}$ and the general fiber has co-dimension precisely $\dim \mathcal{A}$.*

Let us apply this to the case where $\mathcal{A}$ is the algebra of Casimirs of an affine Poisson variety $(M, \{\cdot, \cdot\})$, which we denoted by $\operatorname{Cas}(M)$. As an application of Proposition 2.18, the following proposition shows that the fibers of $M \to \operatorname{Spec} \operatorname{Cas}(M)$ inherit a Poisson structure, thereby giving each irreducible component the structure of an affine Poisson variety.

Proposition 2.38 *Every (non-empty) fiber $\mathcal{F}$ of $\pi_{\mathrm{Cas}(M)}$ inherits a Poisson bracket $\{\cdot,\cdot\}_{\mathcal{F}}$ from $\{\cdot,\cdot\}$ and all Hamiltonian vector fields X_f, $f \in \mathcal{O}(M)$ are tangent to these fibers. Moreover $\mathrm{Rk}\{\cdot,\cdot\}_{\mathcal{F}} \leq \mathrm{Rk}\{\cdot,\cdot\}$ with equality for a general fiber $\mathcal{F}$.*

Proof

It suffices to prove the property for a (non-empty) fiber over a closed point. Let $\mathcal{F}$ be any such fiber and $m \in \mathcal{F}$. In view of (2.21) the ideal of $\mathcal{F}$ is generated by

$$\{f - \chi_m(f) \mid f \in \mathrm{Cas}(M)\}$$

so it is a Poisson ideal of $\mathcal{O}(M)$ and Proposition 2.18 applies. In order to determine the rank of the restricted Poisson bracket we use item *(ii)* of Definition 2.6. It shows that the rank of both structures is the same at every point. This leads at once to the inequality; in order to obtain the equality for a general fiber, just recall that the rank of $\{\cdot,\cdot\}$ is maximal on a Zariski open subset of M. ∎

Referring to Example 2.24 we see that the Poisson structure on M can be reconstructed from the Poisson structure on the fibers of $M \to \mathrm{Spec\,Cas}\,M$, where $\mathrm{Spec\,Cas}\,M$ plays the role of the space of parameters. Therefore we will call $\mathrm{Spec\,Cas}\,M$ the *parameter space* and the map $M \to \mathrm{Spec\,Cas}\,M$ the *parameter map* of the affine Poisson variety $(M,\{\cdot,\cdot\})$. The fibers of the parameter map will also be called *level sets of the Casimirs* because picking a fiber (resp. over a closed point) corresponds to fixing some (resp. all) Casimirs. The decomposition of M into the fibers over closed points is called the *Casimir decomposition*.

Remarks 2.39

1. A Poisson morphism does not necessarily induce a map of the corresponding parameter spaces (for a counterexample, see Example 3.14 below).

2. Since $\mathrm{Cas}(M)$ is a Poisson subalgebra of $\mathcal{O}(M)$ (actually a Lie ideal), $\pi_{\mathrm{Cas}(M)}$ is a Poisson morphism, $\mathrm{Spec\,Cas}(M)$ having the trivial Poisson structure.

3. For special fibers $\mathcal{F}$ of the parameter map the rank of the Poisson structure may be strictly smaller, see Example 2.54 below.

Proposition 2.40 *Let $(M,\{\cdot,\cdot\})$ be any affine Poisson variety. Then*

$$\dim \mathrm{Cas}(M) \leq \mathrm{CoRk}\{\cdot,\cdot\}.$$

Proof

Let $\mathcal{F}$ be a general fiber of the parameter map and let $\{\cdot,\cdot\}_{\mathcal{F}}$ be the induced Poisson structure. Then $\dim M - \dim \mathcal{F} = \dim \mathrm{Cas}(M)$ and $\mathrm{Rk}\{\cdot,\cdot\}_{\mathcal{F}} = \mathrm{Rk}\{\cdot,\cdot\}$. Since $\dim \mathcal{F}$ (resp. $\mathrm{Rk}\{\cdot,\cdot\}_{\mathcal{F}}$) equals the number of independent derivations (resp. Hamiltonian derivations) of $\mathcal{O}(\mathcal{F})$ at a general point of $\mathcal{F}$ we find that

$$\dim \mathrm{Cas}(M) = \dim M - \dim \mathcal{F} \leq \dim M - \mathrm{Rk}\{\cdot,\cdot\} = \mathrm{CoRk}\{\cdot,\cdot\}.$$

∎

When studying integrable Hamiltonian systems we will exclusively be interested in affine Poisson varieties for which the above inequality is an equality, because only in that case, the varieties or manifolds which are traced out by the flows of the integrable vector fields, can be affine (sub-) varieties (of the phase space). It motivates the following definition.

Definition 2.41 Let $(M, \{\cdot, \cdot\})$ be an affine Poisson variety. We say that its algebra of Casimirs is *maximal* if

$$\dim \mathrm{Cas}(M) = \mathrm{CoRk}\{\cdot, \cdot\},$$

i.e., if the general level set of the Casimirs has dimension equal to the rank of the Poisson structure.

Following the above comment, we will want to know if maximality of the algebra of Casimirs is preserved by the different constructions we made (restriction, product, quotient and taking away a divisor). We show that maximality of the algebra of Casimirs is indeed preserved by restriction of the Poisson structure to a general level of (a subalgebra of) the algebra of Casimirs. Similar propositions for the other constructions which we have discussed are easier to obtain and are not made explicit here.

Proposition 2.42 Let $(M, \{\cdot, \cdot\})$ be an affine Poisson variety whose algebra of Casimirs is maximal. Then for any subalgebra $\mathcal{A}$ of $\mathrm{Cas}(M)$ the induced Poisson structure on the general fiber of $M \to \mathrm{Spec}\,\mathcal{A}$ is also maximal.

Proof

Let $\mathcal{A}$ be any subalgebra of $\mathrm{Cas}(M)$ and let $\mathcal{F}$ denote a general fiber of the natural map $M \to \mathrm{Spec}\,\mathcal{A}$ so that $\dim \mathcal{F} = \dim M - \dim \mathcal{A}$. Obviously, if $f \in \mathrm{Cas}(M)$ then $f_{|\mathcal{F}} \in \mathrm{Cas}(\mathcal{F})$ and

$$\dim \mathrm{Cas}\,\mathcal{F} \geq \dim \mathrm{Cas}(M)_{|\mathcal{F}} = \dim \mathrm{Cas}(M) - \dim \mathcal{A}.$$

Since $\mathrm{Cas}(M)$ is maximal and $\mathcal{F}$ is general it follows that

$$\dim \mathcal{F} - \dim \mathrm{Cas}(\mathcal{F}) \leq \dim M - \dim \mathrm{Cas}(M) = \mathrm{Rk}\{\cdot, \cdot\} = \mathrm{Rk}\{\cdot, \cdot\}_{\mathcal{F}},$$

showing that for a general fiber $\mathcal{F}$ the algebra of Casimirs of $\{\cdot, \cdot\}_{\mathcal{F}}$ is maximal. ∎

Example 2.43 The algebra of Casimirs needs not be maximal. The simplest counterexample is given by the following Poisson matrix, defining a Lie-Poisson structure on $\mathbf{C}^3$,

$$\begin{pmatrix} 0 & 0 & \alpha x \\ 0 & 0 & -y \\ -\alpha x & y & 0 \end{pmatrix},$$

coming from a solvable Lie algebra. $F(x, y, z)$ will be a Casimir if and only if

$$\frac{\partial F}{\partial z} = \alpha x \frac{\partial F}{\partial x} - y \frac{\partial F}{\partial y} = 0.$$

These equations are easily solved giving $F = cxy^\alpha + d$ where c and d are integration constants. If $\alpha \notin \mathbf{N}$ then F is not a regular function, if $\alpha \notin \mathbf{Q}$ then the level sets of F are not even algebraic varieties. It should however be remarked that for any $\alpha = p/q \in \mathbf{Q}$ we can restrict the Poisson structure to a general level, which is an affine algebraic surface given by $x^q y^p = C$, where $C \in \mathbf{C}$.

Example 2.44 Taking up Example 2.11 again we see that every polynomial $F(x, y, z)$ appears as a Casimir for some Poisson structure on $\mathbf{C}^3$. Namely, consider the Poisson structure on $\mathbf{C}^3$ defined by the following Poisson matrix (which corresponds to $\chi = 1$ and $\phi = F$):

$$\begin{pmatrix} 0 & \frac{\partial F}{\partial z} & -\frac{\partial F}{\partial y} \\ -\frac{\partial F}{\partial z} & 0 & \frac{\partial F}{\partial x} \\ \frac{\partial F}{\partial y} & -\frac{\partial F}{\partial z} & 0 \end{pmatrix}.$$

Then F is a Casimir of this Poisson structure. All the fibers of the parameter map are two-dimensional and if F is non-constant then they all have rank two, the singularities of these fibers being precisely the points where the rank drops to zero.

Example 2.45 We also give an example to show that the fibers (of the parameter map) lying over closed points may have higher dimension. Consider on $\mathbf{C}^4$ the following Poisson matrix:

$$\begin{pmatrix} 0 & 0 & 0 & x \\ 0 & 0 & 0 & -y \\ 0 & 0 & 0 & -z \\ -x & y & z & 0 \end{pmatrix}.$$

It is easy to verify that the algebra of Casimirs is given by $\mathrm{Cas}(\mathbf{C}^4) = \mathbf{C}[xy, xz]$. Thus the parameter space $\mathrm{Spec}\,\mathrm{Cas}(\mathbf{C}^4)$ is isomorphic to $\mathbf{C}^2$ and the fibers of the parameter map are given by $xy = a$, $xz = b$ for $a, b \in \mathbf{C}$. For $(a, b) \neq (0, 0)$ the fiber is two-dimensional, however $xy = xz = 0$ has $x = 0$ as a component, hence is three-dimensional. Notice that on this special fiber the algebra of Casimirs of the induced Poisson structure is not maximal anymore.

We prove one more proposition about the algebra of Casimirs which may be useful for its computation; the type of argument used in the proof will be used several times in the next section.

Proposition 2.46 *The algebra of Casimirs of an affine Poisson variety $(M, \{\cdot, \cdot\})$ is integrally closed in $\mathcal{O}(M)$.*

Proof

Suppose $F \in \mathcal{O}(M)$ and $\sum_{i=0}^{n} \alpha_i F^i = 0$, with $\alpha_i \in \mathrm{Cas}(M)$, $\alpha_n = 1$ and n minimal. We need to show that $F \in \mathrm{Cas}(M)$. Taking the bracket with any $g \in \mathcal{O}(M)$ we find

$$0 = \left\{ g, \sum_{i=0}^{n} \alpha_i F^i \right\} = \left(\sum_{i=1}^{n} (i-1)\alpha_i F^{i-1} \right) \{g, F\},$$

for all $g \in \mathcal{O}(M)$. Since n was supposed minimal, it follows that $\{g, F\} = 0$ for all $g \in \mathcal{O}(M)$, i.e., $F \in \mathrm{Cas}(M)$. ∎

- The rank decomposition

Another decomposition relates to the rank of the affine Poisson variety at each point. Given $(M, \{\cdot, \cdot\})$ define for $0 \leq i \leq \frac{1}{2}\operatorname{Rk}\{\cdot, \cdot\}$

$$M_i = \{p \in M \mid \operatorname{Rk}_p\{\cdot, \cdot\} \leq 2i\}.$$

Then the components of each M_i are affine varieties: let g be the Poisson matrix of $\{\cdot, \cdot\}$ with respect to an arbitrary system of generators, then M_i can be described as a so-called *determinantal variety* (see [ACGH] Ch. 2)

$$M_i = \{p \in M \mid \text{all determinants of order } 2i + 1 \text{ of } g \text{ vanish at } p\},$$

a description which also gives the equations defining these algebraic sets[7]. Obviously $M_i \subset M_{i+1}$; also for $r = \operatorname{Rk}\{\cdot, \cdot\}/2$ one has $M = M_r$ and the Poisson structure is regular if and only if $M_{r-1} = \emptyset$; finally $M_0 = M$ if and only if the bracket is trivial. We call the (singular) stratification by the M_i the *rank decomposition* of M.

The algebraic sets M_i may be reducible and their components may be of varying dimension. Therefore we define for $0 \leq i \leq \frac{1}{2}\operatorname{Rk}\{\cdot, \cdot\}$ and $j \in \mathbf{N}$

$$M_{ij} = \{p \in M_i \mid \exists \mathcal{D} \text{ irred. comp. of } M_i, p \in \mathcal{D} \text{ and } \dim \mathcal{D} = j\}.$$

Thus M_{ij} is the (finite) union of the j-dimensional irreducible components of M_i. This leads at once to the following polynomial invariant for an affine Poisson variety.

Definition 2.47 Let $(M, \{\cdot, \cdot\})$ be an affine Poisson variety. Its *invariant polynomial* $\rho(M) = \rho(M, \{\cdot, \cdot\}) \in \mathbf{Z}[R, S]$ is defined by

$$\rho(M) = \sum \rho_{ij} R^i S^j, \qquad \rho_{ij} = \#\text{irred. comp. of } M_{ij}.$$

The polynomial can also be represented by a matrix (of minimal size), called the *invariant matrix* of the Poisson structure by taking as (i, j)-th entry the integer ρ_{ij} (labeling of matrix entries starts here from zero).

Proposition 2.48 Let $(M_1, \{\cdot, \cdot\}_1)$ and $(M_2, \{\cdot, \cdot\}_2)$ be two affine Poisson varieties. If there exists a biregular map $\phi : M_1 \to M_2$ which is Poisson, then $\rho(M_1) = \rho(M_2)$.

Proof

We noticed that the inverse of a biregular Poisson morphism is also Poisson. From Proposition 2.17 it follows that for all points x, the rank at x equals the rank at $\phi(x)$. Thus the restriction of ϕ to any of the subvarieties M_r or M_{rs} is a biregular map onto the subvarieties N_r and N_{rs}. It follows that $\rho(M) = \rho(N)$. ∎

---

[7] Strictly speaking the M_i are defined here as affine schemes, i.e., the ideal generated by these determinants is in general not reduced. The invariant, defined below, leads to another invariant when passing to the radical of this ideal, but all properties listed below also hold for this (coarser) invariant.

Proposition 2.49   *Let $(M, \{\cdot, \cdot\})$ be an affine Poisson variety of rank $2r$ and dimension d and let $\rho = \rho(R, S)$ its invariant.*

(1) $\rho(M) = R^r S^d (1 + \mathcal{O}(R^{-1}, S^{-1}))$;

(2) *If M is non-singular the coefficients ρ_{ij} of $\rho = \sum \rho_{ij} R^i S^j$, satisfy $\rho_{ij} = 0$ for $2i > j$.*

(3) *For any $r \le d/2$, there exists an affine Poisson variety whose invariant is $R^r S^d$.*

Proof

Since M is irreducible and $M_{rd} = M$ we have $\rho_{rd} = 1$, all other M_{ij} have by definition lower rank and being given as the intersection of hypersurfaces in M they also have lower dimension. This shows (1).

As for (2) we need to rely on the symplectic foliation, described in Section 4 below; an algebraic proof which would allow to remove the assumption about M being non-singular is still missing (in view of Proposition 2.18 it would suffice to show that the irreducible components of the M_i are affine Poisson subvarieties of M). Through every point of M passes a leaf which inherits a symplectic structure from the Poisson structure, so on the one hand all Hamiltonian vector fields at this point (which span a subspace of dimension equal to the rank $2r$ of the Poisson structure at this point) are tangent to such a leaf, on the other hand such a leaf is entirely contained in the subset M_{2r}; thus every irreducible component of M_{2r} has dimension at least $2r$ showing (2).

For (3) take the canonical Poisson structure of rank $2r$ on $\mathbf{C}^{2d}$ (Example 2.7).   ∎

Before we give a refinement of the invariant, let us consider some first examples.

Example 2.50   An affine Poisson variety is regular if and only if its invariant polynomial is a monomial, i.e., is of the form $R^r S^s$, where $2r$ is the rank and s the dimension of the variety. In particular the invariant polynomial of the trivial structure on an affine Poisson variety of dimension s is S^s.

Example 2.51   For the Poisson structures on $\mathbf{C}^2$, which are defined by a single polynomial $\varphi(x, y) = \{x, y\}$, with $\varphi \ne 0$ we have $\rho = RS^2 + kS$, where k is the number of components of the plane curve defined by $\varphi(x, y) = 0$. Its invariant matrix is thus given by

$$\begin{pmatrix} 0 & k & 0 \\ 0 & 0 & 1 \end{pmatrix}.$$

It follows in particular that the polynomial invariant is not a complete invariant: all non-constant irreducible polynomials $\varphi(x, y)$ lead to a Poisson structure on $\mathbf{C}^2$ with invariant $\rho = RS^2 + S$.

Example 2.52   The Sklyanin brackets and their generalizations (see Example 2.9) lead for the various values of the parameters to a lot of different invariant polynomials, giving an easy proof that many of these Poisson structures are different. We give the different polynomials — which are easily computed — in the following table (the integers i, j, k are taken different and range from 1 to 3; a dash means that the values of the parameters are incompatible with the relation $a_1 b_1 - a_2 b_2 + a_3 b_3 = 0$).

ρ	all $a = 0$	$a_i = a_j = 0$	$a_k = 0$	all $a \neq 0$
all $b = 0$	S^4	$RS^4 + 2S^3$	$RS^4 + S^3 + S^2$	$RS^4 + S^3 + S$
$b_i = b_j = 0$	$RS^4 + 2S^3$	$-$	$RS^4 + 3S^2$	$-$
$b_i = b_k = 0$	$RS^4 + 2S^3$	$RS^4 + S^3 + S^2$	$-$	$-$
$b_i = 0$	$RS^4 + S^3 + S^2$	$-$	$-$	$RS^4 + S^2 + 2S$
$b_k = 0$	$RS^4 + S^3 + S^2$	$RS^4 + S^3 + S$	$RS^4 + 2S^2$	$RS^4 + S^2 + 2S$
all $b \neq 0$	$RS^4 + 3S^2$	$-$	$RS^4 + S^2 + 2S$	$RS^4 + 4S$

Table 1

A more precise description of an affine Poisson variety can be given by combining the above polynomial invariant with Proposition 2.38. We know from that proposition that there corresponds to each point of the affine variety $\operatorname{Spec}\operatorname{Cas}(M)$ a fiber whose irreducible components are affine Poisson varieties. Then we may define a polynomial invariant $\rho_c(M)$ for each $c \in \operatorname{Spec}\operatorname{Cas}(M)$ by

$$\rho_c(M) = \rho\left(\pi^{-1}_{\operatorname{Cas}(M)}(c)\right),$$

under the assumption that the fiber over c is irreducible; if not then the right hand side in this definition is just replaced by the sum over all irreducible components. Thus we label each point of $\operatorname{Spec}\operatorname{Cas}(M)$ by the invariant polynomial of the corresponding fiber over it and obtain in this way a more sensitive invariant for affine Poisson varieties. In the examples which follow we will only consider the fibers over closed points c.

Example 2.53 The simplest non-trivial example is given by the Lie-Poisson structure on the dual of a three-dimensional semi-simple Lie algebra (see Example 2.8). A basis $\{x, y, z\}$ of this space can be chosen such that the corresponding Poisson matrix takes the form

$$\begin{pmatrix} 0 & -z & y \\ z & 0 & x \\ -y & -x & 0 \end{pmatrix}. \tag{2.22}$$

The algebra of Casimirs is clearly given by $\mathbf{C}[x^2 - y^2 - z^2]$, hence $\operatorname{Spec}\operatorname{Cas}(M)$ can be identified with $\mathbf{C}$ by evaluation on the element $x^2 - y^2 - z^2$; we denote the corresponding coordinate by u. Since (2.22) has only rank zero at the origin, which lies in the fiber over zero, we conclude that $\rho = RS^3 + 1$ and

$$\rho_c = \begin{cases} RS^2 & \text{if } u(c) \neq 0, \\ RS^2 + 1 & \text{if } u(c) = 0. \end{cases}$$

It may also be depictured as follows.

Example 2.54 For the Heisenberg algebra the Lie-Poisson structure can be written as $\{x, y\} = \{x, z\} = 0$, $\{y, z\} = x$. As above one finds that the algebra of Casimirs is given by $\mathbf{C}[x]$, and again its spectrum can be identified with $\mathbf{C}$ (with coordinate u) by evaluation on the Casimir x. The Poisson structure has now rank zero on the plane $x = 0$ which is an entire level of the Casimirs (showing that equality in Proposition 2.38 needs not hold for all level sets). It follows that $\rho = RS^3 + S^2$ and

$$\rho_c = \begin{cases} RS^2 & \text{if } u(c) \neq 0, \\ S^2 & \text{if } u(c) = 0. \end{cases}$$

This case is depictured as follows.

Example 2.55 An interesting example is found by taking the Lie-Poisson structure on $\mathfrak{g}^* = \mathfrak{gl}(2)^*$. Consider the following basis

$$X = \begin{pmatrix} 1 & 0 \\ 0 & 0 \end{pmatrix}, \quad Y = \begin{pmatrix} 0 & 1 \\ 0 & 0 \end{pmatrix}, \quad Z = \begin{pmatrix} 0 & 0 \\ 1 & 0 \end{pmatrix}, \quad T = \begin{pmatrix} 0 & 0 \\ 0 & 1 \end{pmatrix},$$

for $\mathfrak{g}$ and let $x, \ldots, t$ be the generators of $\mathcal{O}(\mathfrak{g})$, $\hat{x} = X, \ldots, \hat{t} = T$. The corresponding Poisson matrix is given by

$$\begin{pmatrix} 0 & y & -z & 0 \\ -y & 0 & x-t & y \\ z & t-x & 0 & -z \\ 0 & -y & z & 0 \end{pmatrix}$$

and we have $\mathrm{Cas}(\mathfrak{g}^*) = \mathbf{C}[x+t, xt-yz]$. It follows that $\mathrm{Spec}\,\mathrm{Cas}(\mathfrak{g}^*)$ is in this case isomorphic to $\mathbf{C}^2$; we pick the isomorphism such that the standard coordinates u and v on $\mathbf{C}^2$ correspond to $x+t$ and $xt-yz$ (in that order). Since the rank of the Poisson structure is two except for the points on the line $y = z = 0$, $x = t$, we find that in this case $\rho = RS^4 + S$ and

$$\rho_c = \begin{cases} RS^2 & \text{if } u^2(c) \neq 4v(c), \\ RS^2 + 1 & \text{if } u^2(c) = 4v(c). \end{cases}$$

Example 2.56 The following example will come up later when studying the Toda lattice (Section VII.7). In terms of coordinates $\{t_1, \ldots, t_6\}$ for $\mathbf{C}^6$ we consider the Lie-Poisson structure determined by the Poisson matrix

$$\begin{pmatrix} 0 & {}^tT \\ -T & 0 \end{pmatrix} \text{ with } T = \begin{pmatrix} 0 & -t_2 & t_3 \\ t_1 & 0 & -t_3 \\ -t_1 & t_2 & 0 \end{pmatrix}. \tag{2.23}$$

We will show later (in Paragraph VII.7.1) that $\mathrm{Cas}(\mathbf{C}^6) = \mathbf{C}[t_1 t_2 t_3, t_4 + t_5 + t_6]$, so that $\mathrm{Spec}\,\mathrm{Cas}(\mathbf{C}^6)$ can be identified with $\mathbf{C}^2$, with coordinates u and v, corresponding to $t_1 t_2 t_3$ and $t_4 + t_5 + t_6$ (in that order). By computing a few determinants one sees that the rank is zero on the three-plane $t_1 = t_2 = t_3 = 0$, two on the three four-planes $t_i = t_j = 0$ ($1 \le i < j \le 3$) and four elsewhere. From it one easily obtains the following invariant polynomials:

$$\rho = R^2 S^6 + 3RS^4 + S^3,$$

$$\rho_c = \begin{cases} R^2 S^4 & \text{if } u(c) \ne 0, \\ 3R^2 S^4 + 3RS^3 + S^2 & \text{if } u(c) = 0. \end{cases}$$

It is represented by the following diagram.

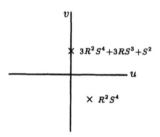

Proposition 2.57 *Let $(M, \{\cdot, \cdot\}_M)$ and $(N, \{\cdot, \cdot\}_N)$ be two affine Poisson varieties and let their product $M \times N$ be equipped with the product bracket. Then*

$$\rho(M \times N) = \rho(M)\rho(N).$$

In particular, if the invariant polynomial of an affine Poisson variety is irreducible then this Poisson variety is not a product (with the product bracket).

Proof

We use as above M_i, N_i and $(M \times N)_i$ as notation for the determinantal varieties associated to M, N and $M \times N$ respectively. The coefficients of the invariant polynomials $\rho(M)$, $\rho(N)$ and $\rho(M \times N)$ are written as ρ_{ij}^1, ρ_{ij}^2 and $\rho_{ij}^\times$. By Proposition 2.21, we have

$$(M \times N)_i = \bigcup_{k+l=i} M_k \times N_l.$$

Using the fact that the irreducible components of a product are precisely the products of irreducible components, we find

$$
\begin{aligned}
\rho^{\times}_{ij} &= \#j\text{-dim. irred. comp. of } (M \times N)_i \\
&= \sum_{k+l=i} \#j\text{-dim. irred. comp. of } M_k \times N_l \\
&= \sum_{k+l=i} \sum_{m+n=j} (\#m\text{-dim. irred. comp. of } M_k)\,(\#n\text{-dim. irred. comp. of } N_l) \\
&= \sum_{k+l=i} \sum_{m+n=j} \rho^1_{km}\rho^2_{ln}.
\end{aligned}
$$

This shows that $\rho(M \times N) = \rho(M)\rho(N)$. ∎

Remark 2.58 It would be interesting to determine the invariant(s) of the Lie-Poisson structure of an arbitrary semi-simple Lie algebra and to relate it to the theory of (co-) adjoint orbits.

3. Integrable Hamiltonian systems and their morphisms

In the study of semi-simple Lie algebras the notion of a Cartan subalgebra plays a dominant role. The corresponding object for affine Poisson spaces is an integrable algebra: a maximal commutative (in this context called involutive) subalgebra. An affine Poisson variety with a fixed choice of integrable algebra is what we call an integrable Hamiltonian system. The study of integrable Hamiltonian systems can be seen as a chapter in Poisson geometry; for example we will see that all propositions which we proved for affine Poisson varieties have their equivalents for integrable Hamiltonian systems.

Our definition is an adaption of the classical definition of an integrable system on a symplectic manifold (see e.g., [AM1]) to the case of an affine Poisson variety. Notice that we do not ask that the rank of the Poisson variety be maximal (or constant). Another difference is that the classical definition demands for having the right number of independent functions in involution, while we ask for having a complete algebra (of the right dimension) of functions in involution, completeness meaning here that this algebra contains every function which is in involution with all the elements of this algebra. On the one hand this adaption is very natural, it is even inevitable if one wants to discuss morphisms and isomorphisms of integrable Hamiltonian systems. On the other hand it is not easy to verify completeness of an involutive algebra, e.g., the (polynomial) algebra generated by a maximal number of functions in involution needs not be complete. Accordingly we will also prove some propositions in this section which will be useful for describing and determining explicitly the integrable algebra in the case of concrete examples.

3.1. Integrable Hamiltonian systems on affine Poisson varieties

Definition 3.1 Let $(M, \{\cdot, \cdot\})$ be an affine Poisson variety. A subalgebra $\mathcal{A}$ of $\mathcal{O}(M)$ is called *involutive* if $\{\mathcal{A}, \mathcal{A}\} = 0$; we say that it is *complete* if moreover for any $f \in \mathcal{O}(M)$ one has $\{f, \mathcal{A}\} = 0 \Leftrightarrow f \in \mathcal{A}$. The triple $(M, \{\cdot, \cdot\}, \mathcal{A})$ is called a *(complete) involutive Hamiltonian system*.

Lemma 3.2 Let $(M, \{\cdot, \cdot\}, \mathcal{A})$ be an involutive Hamiltonian system.
(1) If $\mathcal{A}$ is complete then $\mathcal{A}$ is integrally closed in $\mathcal{O}(M)$;
(2) The integral closure of $\mathcal{A}$ in $\mathcal{O}(M)$ is also involutive and is finitely generated when $\mathcal{A}$ is finitely generated.

Proof

The proof of (1) goes in exactly the same way as the proof of Proposition 2.46, replacing $\mathrm{Cas}(M)$ by $\mathcal{A}$ and $g \in \mathcal{O}(M)$ by $g \in \mathcal{A}$. It is well-known that if $\mathcal{A}$ is finitely generated then its *integral closure* in $\mathcal{O}(M)$ (defined as the set of all elements ϕ of $\mathcal{O}(M)$ for which there exists a monic polynomial with coefficients in $\mathcal{A}$, which has ϕ as a root) is also a finitely generated algebra (see e.g., [AD] Ch. 5). To check that it is involutive, we first check that every element of the integral closure of $\mathcal{A}$ is in involution with all elements of $\mathcal{A}$. Thus, let ϕ be an element of $\mathcal{O}(M)$ for which there exists a polynomial

$$P(X) = X^n + \alpha_1 X^{n-1} + \cdots + \alpha_n$$

for which $P(\phi) = 0$ and with all α_i belonging to $\mathcal{A}$; we assume that the polynomial is of minimal degree. For any $f \in \mathcal{A}$ the equality $\{P(\phi), f\} = 0$ implies as in the proof of Proposition 2.46 that $\{\phi, f\} = 0$, upon using the minimality of P. Using this, it can now be checked by a similar argument that any two functions in the integral closure are in involution.∎

47

Every involutive algebra is contained in an involutive algebra which is complete, but the latter is in general not unique. This is contained in the following lemma.

Lemma 3.3 *Let $(M, \{\cdot, \cdot\}, A)$ be an involutive Hamiltonian system and denote by $\bar{A}$ the integral closure of the field of fractions of A.*

(1) *The subalgebra $\bar{A} \cap \mathcal{O}(M)$ of $\mathcal{O}(M)$ is also involutive;*

(2) *If A is complete then $\bar{A} \cap \mathcal{O}(M) = A$;*

(3) *A is contained in an involutive subalgebra B of $\mathcal{O}(M)$ which is complete; it is unique if $\dim B = \dim A$.*

Proof

Recall (e.g., from [AD] Ch. 5) that $\bar{A} \cap \mathcal{O}(M)$ can be identified as the set of elements ϕ of $\mathcal{O}(M)$ for which there exists a polynomial (which is not necessarily monic) with coefficients in A, which has ϕ as a root. If $\phi \in \bar{A} \cap \mathcal{O}(M)$ and

$$P(X) = \alpha_0 X^n + \alpha_1 X^{n-1} + \cdots + \alpha_n$$

is a polynomial of minimal degree (with coefficients α_i in A) for which $P(\phi) = 0$, then $\{P(\phi), A\} = 0$ implies $\{\phi, A\} = 0$, upon using the minimality of P (again as in the proof of Proposition 2.46). In turn this implies that if ϕ' is another element of $\bar{A} \cap \mathcal{O}(M)$ the equality $\{P(\phi), \phi'\} = 0$ leads to $\{\phi, \phi'\} = 0$. Thus $\bar{A} \cap \mathcal{O}(M)$ is involutive, showing (1); from it (2) follows at once.

If A is involutive but not complete we pass to $A_0 = \bar{A} \cap \mathcal{O}(M)$; if the latter is complete it is the unique involutive subalgebra of $\mathcal{O}(M)$ which contains A and is complete. If not, we add an element $f \in \mathcal{O}(M) \setminus A_0$ for which $\{f, A_0\} = 0$ and repeat the above construction to obtain A_1. Since $\dim A_1 = \dim A_0 + 1$ we are done after a finite number of steps; because of the choice of f the algebra which is obtained is not unique in general (interesting examples of this are given below). ∎

In this text we will only be interested in involutive algebras of the maximal possible dimension, given by the next proposition. We know from Lemma 3.3 that such an algebra A has a unique *completion*, which we will denote by $\mathrm{Compl}(A)$ (or by $\mathrm{Compl}\{f_1, \ldots, f_n\}$ if A is generated by $\{f_1, \ldots, f_n\}$).

Proposition 3.4 *Let $(M, \{\cdot, \cdot\}, A)$ be an involutive Hamiltonian system. Then*

$$\dim A \leq \dim M - \frac{1}{2} \mathrm{Rk}\{\cdot, \cdot\}. \tag{3.1}$$

Proof

Consider a general fiber $\mathcal{F}$ of the map $M \to \mathrm{Spec}\, A$ which is induced by the inclusion map $A \subset \mathcal{O}(M)$. By Proposition 2.37,

$$\dim \mathcal{F} = \dim M - \dim A. \tag{3.2}$$

$\dim \mathcal{F}$ also equals the number of independent derivations of $\mathcal{O}(\mathcal{F})$ at a general point of $\mathcal{F}$ and involutivity of A implies that such derivations can be constructed using functions from A.

To see the latter, recall that the ideal of $\mathcal{F}$ is generated by the functions $f - \chi_m(f)$ where $m \in \mathcal{F}$ is arbitrary but fixed and f ranges over $\mathcal{A}$. For any $g \in \mathcal{A}$ we have

$$X_g(f - \chi_m(f)) = \{f, g\} = 0,$$

hence X_g is tangent to the locus defined by the ideal of $\mathcal{F}$, i.e., to $\mathcal{F}$ and we can construct derivations of $\mathcal{O}(\mathcal{F})$ using elements of $\mathcal{A}$. Next we show that the elements of $\mathcal{A}$ lead to $\dim \mathcal{A} - \dim \operatorname{Cas}(M)$ independent derivations, giving a lower bound for $\dim \mathcal{F}$. Consider a nested sequence of subalgebras

$$\operatorname{Cas} = \mathcal{A}_0 \subset \mathcal{A}_1 \subset \mathcal{A}_2 \subset \cdots \subset \mathcal{A}_r = \mathcal{O}(M),$$

where $\dim \mathcal{A}_{i+1} = \dim \mathcal{A}_i + 1$, in particular $r = \operatorname{Rk}\{\cdot, \cdot\}$. If n_i denotes the number of independent vector fields on M coming from $\mathcal{A}_i$ (i.e., having independent vectors at a general point) then obviously $n_i \leq n_{i+1} \leq n_i + 1$, $n_0 = 0$ and $n_r = r$. It follows that $n_i = i$ for all i. It gives the following lower bound

$$\dim \mathcal{F} \geq \dim \mathcal{A} - \dim \operatorname{Cas}(M). \tag{3.3}$$

Combining (2.40), (3.2) and (3.3) we find

$$\dim \mathcal{A} \leq \frac{1}{2}(\dim M + \dim \operatorname{Cas}(M)) \leq \dim M - \frac{1}{2}\operatorname{Rk}\{\cdot, \cdot\}. \tag{3.4}$$

∎

We finally get to the definition of an integrable Hamiltonian system (on an affine Poisson variety).

Definition 3.5 If $(M\{\cdot, \cdot\})$ is an affine Poisson variety whose algebra of Casimirs is maximal and $\mathcal{A}$ is a complete involutive subalgebra of $\mathcal{O}(M)$ then $\mathcal{A}$ is called *integrable* if

$$\dim \mathcal{A} = \dim M - \frac{1}{2}\operatorname{Rk}\{\cdot, \cdot\}. \tag{3.5}$$

The triple $(M, \{\cdot, \cdot\}, \mathcal{A})$ is then called an *integrable Hamiltonian system* and each non-zero vector field in

$$\operatorname{Ham}(\mathcal{A}) = \{X_f \mid f \in \mathcal{A}\}$$

is called an *integrable vector field*. The dimension of $\mathcal{A}$ is called the *dimension* or the *degrees of freedom* of the integrable Hamiltonian system. M is called its *phase space* and $\operatorname{Spec} \mathcal{A}$ its *base space*.

If $\mathcal{A}_1$ and $\mathcal{A}_2$ are two different subalgebras of $\mathcal{O}(M)$ which make $\mathcal{O}(M)$ into an integrable Hamiltonian system then every non-zero vector field in the intersection $\operatorname{Ham}(\mathcal{A}_1) \cap \operatorname{Ham}(\mathcal{A}_2)$ is called a *super-integrable vector field*.

Remarks 3.6

1. What we call an integrable vector field is in the literature often called an integrable system; the distinction we make is motivated by the fact that the datum of one integrable vector field X_f (or its corresponding Hamiltonian f) does not suffice in general to determine $\mathcal{A}$ (see Examples 3.10 and 3.11 below).

2. In view of (3.4) the condition that the algebra of Casimirs is maximal *follows* from (3.5); it was added in the hypotheses to stress that it is a condition on the Poisson structure — in our approach affine Poisson varieties whose algebra of Casimirs is not maximal do *not* admit integrable Hamiltonian systems.

3. Completeness of the integrable algebra $\mathcal{A}$ implies that $\mathrm{Cas}(M) \subset \mathcal{A}$ and $\mathcal{A}$ can be seen as an intermediate involutive object between $\mathrm{Cas}(M)$ and $\mathcal{O}(M)$; for example, it follows from (3.4) and (3.5) that

$$\dim \mathcal{A} = \frac{1}{2}(\dim M + \dim \mathrm{Cas}(M)),$$

which supports this assertion.

The commutative triangle of inclusions

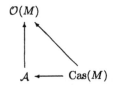

induces, as explained in Paragraph 2.4, the following commutative triangle of dominant (Poisson) morphisms.

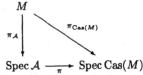

Thus the parameter map $\pi_{\mathrm{Cas}\,M}$, which maps the phase space to the parameter space, can be factorized via the map $\pi_{\mathcal{A}} : M \to \mathrm{Spec}\,\mathcal{A}$ from the phase space to the base space; we call the latter map the *momentum map*. The irreducible components of the fibers of the momentum map are affine varieties which will play a dominant role in this text. We call them the *level sets of the integrable Hamiltonian system* or the *level sets of $\mathcal{A}$* for short.

We now come to the technicallity alluded to at the beginning of this section. We know from Lemma 3.3 abstractly how to complete an involutive algebra $\mathcal{A}$ (say of the maximal possible dimension), but it does not lead to an explicit description of the completion when studying concrete examples. The following proposition gives sufficient and checkable conditions for such an algebra $\mathcal{A}$ to be complete; it will be used several times when we get to the examples.

Proposition 3.7   *Let $(M, \{\cdot, \cdot\})$ be an affine Poisson variety and let $\mathcal{A}$ be an involutive subalgebra of $\mathcal{O}(M)$ of dimension*

$$\dim \mathcal{A} = \dim M - \frac{1}{2} \operatorname{Rk}\{\cdot, \cdot\}.$$

Then $\mathcal{A}$ is complete, hence integrable, if the fibers of $\pi_{\mathcal{A}} : M \to \operatorname{Spec}(\mathcal{A})$ have the following two properties

(1) *the general fiber is irreducible;*

(2) *the fibers over all closed points have the same dimension.*

Proof

Let us suppose that $\mathcal{A}$ is not complete, i.e., $f \notin \mathcal{A}$ and $\{f, \mathcal{A}\} = 0$ for some $f \in \mathcal{O}(M)$. We denote by $\mathcal{A}'$ the algebra generated by f and the elements of $\mathcal{A}$, which has by Proposition 3.4 the same dimension as $\mathcal{A}$. By Lemma 3.3 f belongs to the integral closure of the quotient field of $\mathcal{A}$. Thus $f \in \mathcal{O}(M)$ is a root of a polynomial $Q(t) \in \mathcal{A}[t]$. Consider the following commutative diagram which is induced by the inclusion $\mathcal{A} \subset \mathcal{A}'$.

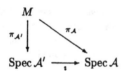

If $Q(t)$ has degree at least two then $\imath$ is a ramified covering map of degree at least two, hence the fiber of $\pi_{\mathcal{A}}$ over a general point P has at least two components, which are the fibers of $\pi_{\mathcal{A}'}$ over the antecedents $\imath^{-1}(P)$. This is in conflict with assumption (1), hence $Q(t)$ is of degree one, $Q(t) = p_1 t + p_2$. Since $f \in \mathcal{O}(M) \setminus \mathcal{A}$ neither p_1 nor p_2 are constant. Therefore there is a closed point P in $\operatorname{Spec} \mathcal{A}$ which corresponds to an algebra homomorphism onto $\mathbf{C}$ which sends both p_1 and p_2 to 0. This closed point is the image under $\imath$ of a point which is not closed, namely the corresponding algebra homomorphism can take any value on f. Then the fibers of $\pi_{\mathcal{A}'}$ over these points have dimension one less than the dimension of the fiber $\pi_{\mathcal{A}}^{-1}(P)$ which is $\dim M - \dim \mathcal{A}$ by assumption (2). Since $\mathcal{A}'$ has the same dimension as $\mathcal{A}$, all fibers of $\pi_{\mathcal{A}'}$ have dimension at least $\dim M - \dim \mathcal{A}$, a contradiction. It follows that $\mathcal{A}$ is complete.   ∎

We have seen in Proposition 2.38 that all Hamiltonian vector fields X_f, $f \in \mathcal{O}(M)$ are tangent to all fibers of the parameter map. Similarly we show now that all *integrable* vector fields X_f, $f \in \mathcal{A}$ are tangent to all fibers of the momentum map; in addition they have the special property to pairwise commute.

Proposition 3.8   *Let $(M, \{\cdot, \cdot\}, \mathcal{A})$ be an integrable Hamiltonian system. Then all Hamiltonian vector fields in $\operatorname{Ham}(\mathcal{A})$ are tangent to all fibers of the momentum map $\pi_{\mathcal{A}} : M \to \operatorname{Spec} \mathcal{A}$ and they all commute; the irreducible components of these fibers are affine varieties and the dimension of the general fiber is $\frac{1}{2} \operatorname{Rk}\{\cdot, \cdot\}$, which coincides with the number of independent vector fields in $\operatorname{Ham}(\mathcal{A})$.*

If $(M, \{\cdot, \cdot\}, \mathcal{A}_1)$ is another integrable Hamiltonian system, then super-integrable vector fields in $\mathrm{Ham}(\mathcal{A}) \cap \mathrm{Ham}(\mathcal{A}_1)$ are tangent to the (strictly smaller) intersection of the fibers of the corresponding maps $\pi_{\mathcal{A}}$ and $\pi_{\mathcal{A}_1}$.

Proof

Let $g \in \mathrm{Ham}(\mathcal{A})$. Then for any $f \in \mathrm{Ham}(\mathcal{A})$ we have $X_g f = \{f, g\} = 0$, hence X_g is tangent to all fibers of $\pi_{\mathcal{A}}$. Clearly these fibers are affine varieties and commutativity of the vector fields in $\mathrm{Ham}(\mathcal{A})$ follows from item (3) in Proposition 2.3. The dimension of a general fiber is $\dim M - \dim \mathcal{A} = \frac{1}{2} \mathrm{Rk}\{\cdot, \cdot\}$ in view of Proposition 2.37. Our claim about super-integrable vector fields follows at once from the first part of the proposition. ∎

We now get to some first examples of integrable Hamiltonian systems, in particular we will give two examples of a super-integrable vector field.

Example 3.9 If $(M, \{\cdot, \cdot\})$ is an affine Poisson variety of rank two whose algebra of Casimirs is maximal, then any function F which does not belong to $\mathrm{Cas}(M)$ leads to an integrable Hamiltonian system. Namely $\mathcal{A} = \mathrm{Compl}\{\mathrm{Cas}(M), F\}$ is obviously involutive and $\dim \mathcal{A} = \dim \mathrm{Cas}(M) + 1 = \dim M - 1$, hence $\mathcal{A}$ is integrable; clearly its level sets are just algebraic curves.

This well-known fact is often expressed by saying that in one degree of freedom all Hamiltonian systems are integrable (although the condition that the algebra of Casimirs should be maximal is never stated explicitly; when assuming implicitly that M has dimension two this condition is of course automatically satisfied).

Example 3.10 Another trivial class of integrable Hamiltonian systems is defined on $\mathbf{C}^n$, with a regular Poisson bracket, by considering linear functions; the example shows that the integrable algebra is not always determined by just one of its (non-trivial) elements. For simplicity let us take the case $n = 4$ with a constant Poisson structure of rank 4. As we know from Example 2.7 linear coordinates q_1, p_1, q_2, p_2 on $\mathbf{C}^4$ may be picked such that $\{q_i, p_j\} = \delta_{ij}$ and $\{q_1, q_2\} = \{p_1, p_2\} = 0$. Take $F = aq_1 + bq_2 + cp_1 + dp_2$ with e.g. $a \neq 0$ and look for a linear function $G = a'q_1 + b'q_2 + c'p_1 + d'p_2$ which is in involution with F. Replacing G by $G - Fa'/a$ if necessary we may assume that $a' = 0$ and we find

$$G = b'q_2 + (db' - bd')p_1 + d'p_2$$

as the most general solution (up to adding multiples of F). Here $b', d' \in \mathbf{C}$ are arbitrary, so that we have essentially a one-parameter family of possibilities for G (parametrized by d'/b'), all leading to an integrable subalgebra $\mathcal{A}$ of $\mathcal{O}(\mathbf{C}^4)$. The Poisson bracket of two of these possibilities for G, is given by

$$\{b'q_2 + (db' - bd')p_1 + d'p_2, b''q_2 + (db'' - bd'')p_1 + d''p_2\} = b'd'' - d'b''$$

which is non-zero if they are different (i.e., non-proportional), in agreement with Proposition 3.4. The general fiber of $\mathcal{A}$ is in this case just a plane and all integrable vector fields $\mathrm{Ham}(\mathcal{A})$ are constant when restricted to each plane. Clearly, all these vector fields are super-integrable and their flow evolves on a (straight) line.

3. Integrable Hamiltonian systems and their morphisms

The above examples are the most trivial classes of examples of integrable Hamiltonian systems — apart from the really trivial class where affine Poisson spaces of rank zero are considered. To increase complexity one may consider Poisson structures which are of higher rank and not constant (in particular they are never regular), also polynomials of higher degree may be considered and an ambient affine variety of higher dimension. It turns out that in these cases it is a non-trivial matter to find integrable Hamiltonian systems. There are of course some trivial ways to obtain new systems from old ones, one may for example take the product of two integrable Hamiltonian systems or rewrite a simple system in a complicated way by changing variables (see Section 3.3), but these results are in reality often only interesting in the other sense, namely for reducing large or complicated integrable Hamiltonian systems to smaller or simpler ones. A general scheme for either constructing integrable Hamiltonian systems or for deciding whether a given Hamiltonian vector field is integrable is not known. We will come back to this in Chapters III and VI.

Example 3.11 Let Σ be a compact oriented topological surface of genus $g \geq 1$ with fundamental group $\pi_1(\Sigma)$ and let G be a reductive algebraic group. Then $\mathrm{Hom}(\pi_1(\Sigma), G)$ is an affine variety on which G acts by conjugation, more precisely if $\rho : \pi_1(\Sigma) \to G$ and $g \in G$ then $g \cdot \rho$ is the homomorphism $\pi_1(\Sigma) \to G$ defined by

$$g \cdot \rho(C) = g(\rho(C))g^{-1},$$

for $C \in \pi_1(\Sigma)$. It turns out (see [Gol]) that the quotient

$$M = \mathrm{Hom}(\pi_1(\Sigma), G)/G$$

(which is an affine variety since G is reductive) has a natural Poisson structure which can very explicitly be described for the classical groups. For simplicity let us consider the case $G = SL(n)$ in the standard representation. For a curve $C \in \pi_1(\Sigma)$ the function

$$f_C : M \to G : \rho \mapsto \mathrm{Trace}(\rho(C))$$

is a well-defined regular function on M and it can be shown that these functions generate $\mathcal{O}(M)$. It was shown by Goldman (see [Gol]) that on such functions a Poisson bracket of maximal rank is given by

$$\{f_C, f_{C'}\} = \sum_{p \in C \# C'} \epsilon(p; C, C') \left(f_{C_p C_p'} - \frac{1}{n} f_C f_{C'} \right). \tag{3.6}$$

The sum runs over the intersection points of C and C' (one may suppose that the curves intersect transversally) and $\epsilon(p; C, C')$ is a sign which is determined by the way the (oriented) curves C and C' intersect at p, upon using the orientation of Σ. Finally, $C_p C_p'$ is the curve on Σ, based at p which is obtained by first following C and then following C'.

A large involutive algebra for this bracket is obtained as follows. Σ can be decomposed (in several ways) into so-called *trinions*; a trinion, also called a *pair of pants*, is just a three-holed sphere and such a decomposition will consist of $2g - 2$ trinions (in the case of genus two there exist precisely two such decompostions) Each trinion being bounded by three curves (which are identified two by two) one gets $3g - 3$ curves on Σ and what is important here is that they are non-intersecting. Calling these curves $C_1, \ldots, C_{3g-3}$ we find from Goldman's formula (3.6) that the functions $f_{C_1}, \ldots, f_{C_{3g-3}}$ are in involution; thus one obtains an involutive algebra

$\mathcal{A} = \text{Compl}\{f_{C_1}, \dots, f_{C_{3g-3}}\}$ and its dimension is computed to be $3g - 3$. Since the rank of the Poisson bracket is maximal, $\mathcal{A}$ will be integrable if and only if

$$3g - 3 = \dim M - \frac{1}{2} \text{Rk}\{\cdot, \cdot\} = \frac{1}{2} \dim M,$$

i.e., for $\dim M = 6g - 6$. Since $\pi_1(\Sigma)$ has a system of $2g$ generators, which are bound by one relation, $\dim \text{Hom}(\pi_1(\Sigma), G)$ has dimension $(2g - 1) \dim G$, hence M has dimension $(2g - 2) \dim G$ and $\mathcal{A}$ is integrable if and only if

$$6g - 6 = (2g - 2) \dim G,$$

i.e., for $\dim G = 3$. Since we restricted ourselves to $G = SL(n)$ we find that $\mathcal{A}$ is only integrable for $G = SL(2)$; it is clear from the above pictures that the Hamiltonian vector fields corresponding to all functions f_{C_i} are actually super-integrable.

3.2. Morphisms of integrable Hamiltonian systems

In parallel with our discussion of morphisms of affine Poisson varieties we now turn to morphisms of integrable Hamiltonian systems.

Definition 3.12 Let $(M_1, \{\cdot, \cdot\}_1, \mathcal{A}_1)$ and $(M_2, \{\cdot, \cdot\}_2, \mathcal{A}_2)$ be two integrable Hamiltonian systems, then a *morphism* $\phi : (M_1, \{\cdot, \cdot\}_1, \mathcal{A}_1) \to (M_2, \{\cdot, \cdot\}_2, \mathcal{A}_2)$ is a morphism $\phi : M_1 \to M_2$ with the following properties:

(1) ϕ is a Poisson morphism;
(2) $\phi^* \text{Cas}(M_2) \subset \text{Cas}(M_1)$;
(3) $\phi^* \mathcal{A}_2 \subset \mathcal{A}_1$.

Schematically, regularity of the map and (2) and (3) can be represented as follows:

$$
\begin{array}{ccccc}
\text{Cas}(M_2) & \longrightarrow & \mathcal{A}_2 & \longrightarrow & \mathcal{O}(M_2) \\
\phi^* \downarrow & & \phi^* \downarrow & & \phi^* \downarrow \\
\text{Cas}(M_1) & \longrightarrow & \mathcal{A}_1 & \longrightarrow & \mathcal{O}(M_1)
\end{array}
\qquad (3.7)
$$

A morphism $\phi : (M_1, \{\cdot, \cdot\}_1, \mathcal{A}_1) \to (M_2, \{\cdot, \cdot\}_2, \mathcal{A}_2)$ which is biregular has an inverse which is automatically a morphism: we call such a map an *isomorphism* (it forces all inclusion maps in the diagram to be bijective).

From the very definition it is clear that the composition of two morphisms is a morphism (hence we have a category). It is also immediate that for any biregular map $\phi : M_1 \to M_2$ and for any integrable Hamiltonian system $(M_1, \{\cdot, \cdot\}_1, \mathcal{A}_1)$ there exists a unique Poisson bracket $\{\cdot, \cdot\}_2$ on M_2 and a unique integrable algebra $\mathcal{A}_2 \subset \mathcal{O}(M_2)$ such that $\phi : (M_1, \{\cdot, \cdot\}_1, \mathcal{A}_1) \to (M_2, \{\cdot, \cdot\}_2, \mathcal{A}_2)$ is an isomorphism; explicitly $\mathcal{A}_2 = \phi^* \mathcal{A}_1$ and

$$\{f, g\}_2 = \left(\phi^{-1}\right)^* \{\phi^* f, \phi^* g\}_1 \qquad \forall f, g \in \mathcal{O}(M_2).$$

Conditions (1) and (2) are conditions at the level of the Poisson structures, rather than on the level of the integrable algebras. Condition (2) resp. (3) implies that ϕ induces a morphism of the corresponding parameter spaces resp. base spaces, as is shown in the following proposition.

Proposition 3.13 *Let* $\phi : (M_1, \{\cdot,\cdot\}_1, \mathcal{A}_1) \to (M_2, \{\cdot,\cdot\}_2, \mathcal{A}_2)$ *be a morphism of integrable Hamiltonian systems. Then ϕ induces a morphism*

$$\tilde{\phi} : \operatorname{Spec} \operatorname{Cas}(M_1) \to \operatorname{Spec} \operatorname{Cas}(M_2)$$

which makes the following diagram commutative,

$$
\begin{array}{ccc}
M_1 & \xrightarrow{\ \phi\ } & M_2 \\
{\scriptstyle \pi_{\operatorname{Cas}(M_1)}}\downarrow & & \downarrow{\scriptstyle \pi_{\operatorname{Cas}(M_2)}} \\
\operatorname{Spec} \operatorname{Cas}(M_1) & \xrightarrow[\tilde{\phi}]{} & \operatorname{Spec} \operatorname{Cas}(M_2)
\end{array}
$$

as well as a morphism

$$\hat{\phi} : \operatorname{Spec} \mathcal{A}_1 \to \operatorname{Spec} \mathcal{A}_2$$

which makes the following diagram commutative.

$$
\begin{array}{ccc}
M_1 & \xrightarrow{\ \phi\ } & M_2 \\
{\scriptstyle \pi_{\mathcal{A}_1}}\downarrow & & \downarrow{\scriptstyle \pi_{\mathcal{A}_2}} \\
\operatorname{Spec} \mathcal{A}_1 & \xrightarrow[\hat{\phi}]{} & \operatorname{Spec} \mathcal{A}_2
\end{array}
$$

If $\phi^* \operatorname{Cas}(M_2) = \operatorname{Cas}(M_1)$ *(resp.* $\phi^* \mathcal{A}_2 = \mathcal{A}_1$*) then* $\tilde{\phi}$ *(resp.* $\hat{\phi}$*) is injective.*

Proof

The first assertions are immediate from diagram (3.7) by taking spectra; also surjectivity of ϕ^* implies injectivity at the level of the corresponding spectra. ∎

Said differently, condition (3) in Definition 3.12 implies that each level set of $\mathcal{A}_1$ is mapped into a level set of $\mathcal{A}_2$ and if $\phi^* \mathcal{A}_2 = \mathcal{A}_1$ then different level sets of $\mathcal{A}_1$ are mapped into different level sets of $\mathcal{A}_2$; condition (2) can be given a similar interpretation. We further illustrate the meaning and relations between the three conditions in Definition 3.12 in the following examples and propositions.

Example 3.14 Let us show that in Definition 3.12 neither (2) nor (3) follow from (1). Consider $\mathbf{C}^4$ (with coordinates q_1, q_2, p_1, p_2) with the canonical Poisson structure $\{q_i, p_j\} = \delta_{ij}$, $\{q_i, q_j\} = \{p_i, p_j\} = 0$, and $\mathbf{C}^3$ (with coordinates q_1, q_2, p_1) with $\{q_1, p_1\} = 1$ and q_2 as Casimir. We look at this $\mathbf{C}^3$ as the $q_1 q_2 p_1$-plane in $\mathbf{C}^4$ and denote by ϕ the projection map along p_2. Then ϕ is a Poisson morphism, however $\phi^* q_2$ is not a Casimir of $\mathbf{C}^4$, showing that (1) does not imply (2). Notice that in this case ϕ does not induce a map $\tilde{\phi}$ as in Proposition 3.13. Taking two different functions on $\mathbf{C}^2$ (i.e., the algebras generated by them) shows that (1) does not imply (3).

There is however a large class of morphisms for which condition (2) in Definition 3.12 follows from (1), namely that of universally closed morphisms; these include the proper morphisms and, in particular, the finite morphisms (see [Har] pp. 95–105). We prove this in the following proposition, however we restrict ourselves to the case of finite morphisms, since we will only use the result in this case (the proof however generalizes verbatim to the case of universally closed morphisms).

Proposition 3.15 *Let $(M_1, \{\cdot, \cdot\}_1)$ and $(M_2, \{\cdot, \cdot\}_2)$ be two affine Poisson varieties and suppose that $\phi : M_1 \to M_2$ is a finite morphism (for example a (possibly ramified) covering map). If ϕ is a Poisson morphism then $\phi^* \operatorname{Cas}(M_2) \subset \operatorname{Cas}(M_1)$; if ϕ is moreover dominant then $\operatorname{Cas}(M_1)$ is the integral closure of $\phi^* \operatorname{Cas}(M_2)$ in $\mathcal{O}(M_1)$.*

Proof
Let us show that if ϕ is finite then for any $f \in \operatorname{Cas}(M_2)$, $\phi^* f$ is in involution with all elements of $\mathcal{O}(M_1)$. The main property which is used about finite (or universally closed) morphisms is that if $\phi : M_1 \to M_2$ is such a morphism then $\mathcal{O}(M_1)$ is integral over $\phi^* \mathcal{O}(M_2)$. Thus any element $g \in \mathcal{O}(M_1)$ is a root of a monic polynomial P (of minimal degree) with coefficients in $\phi^* \mathcal{O}(M_2)$. As in the proof of Proposition 2.46 we find

$$0 = \{\phi^* f, P(g)\} = P'(g)\{\phi^* f, g\}$$

where P' denotes the derivative of the polynomial P. By minimality of P we find $\{\phi^* f, g\} = 0$ as desired. We have shown that $\phi^* \operatorname{Cas}(M_2) \subset \operatorname{Cas}(M_1)$.

Next we take an element $g \in \operatorname{Cas}(M_1)$ and call P its polynomial as above, with coefficients in $\phi^* \mathcal{O}(M_2)$. We show that P has actually its coefficients in $\phi^* \operatorname{Cas}(M_2)$, thereby proving that $\operatorname{Cas}(M_1)$ is the integral closure of $\phi^* \operatorname{Cas}(M_2)$. To do this, let $\phi^* f \in \phi^* \mathcal{O}(M_2)$ be arbitrary, then

$$
\begin{aligned}
0 &= \{\phi^* f, P(g)\}_1 \\
&= \{\phi^* f, g^n + \phi^* \alpha_1 g^{n-1} + \cdots + \phi^* \alpha_n\}_1 \\
&= \{\phi^* f, \phi^* \alpha_1\}_1 g^{n-1} + \cdots + \{\phi^* f, \phi^* \alpha_n\}_1 \\
&= \phi^* \{f, \alpha_1\}_2 g^{n-1} + \cdots + \phi^* \{f, \alpha_n\}_2.
\end{aligned}
$$

Since this polynomial has its coefficients in $\phi^* \mathcal{O}(M_2)$ and since P was supposed of minimal degree, we find that $\phi^* \{f, \alpha_i\} = 0$ for all i. Since ϕ is dominant it follows that $\{f, \alpha_i\} = 0$ for all $f \in \mathcal{O}(M_2)$, so that $\alpha_i \in \operatorname{Cas}(M_2)$ for $i = 1, \ldots, n$. ∎

It can be seen in a similar way that if $\phi : (M_1, \{\cdot, \cdot\}_1, \mathcal{A}_1) \rightarrow (M_2, \{\cdot, \cdot\}_2, \mathcal{A}_2)$ is a morphism of integrable Hamiltonian systems which is finite and dominant then $\mathcal{A}_1$ is the integral closure of $\phi^* \mathcal{A}_2$ in $\mathcal{O}(M_1)$ (for a proof, use completeness of $\mathcal{A}_1$). It leads to the following corollary.

Corollary 3.16 *Let $(M_1, \{\cdot, \cdot\}_1, \mathcal{A}_1) \rightarrow (M_2, \{\cdot, \cdot\}_2, \mathcal{A}_2)$ be a morphism which is finite and whose image is an affine subvariety of M_2. Then ϕ is the composition of an injective and a surjective morphism.*

Proof

We know from Proposition 2.16 that, as a Poisson morphism, ϕ can be decomposed via $(\phi(M_1), \{\cdot, \cdot\})$ say $\phi = \imath \circ \bar{\phi}$. Define

$$\mathcal{A} = \left\{ f \in \mathcal{O}(\phi(M_1)) \mid \bar{\phi}^* f \in \mathcal{A}_1 \right\}.$$

For $f, g \in \mathcal{A}$ we have $\bar{\phi}^* \{f, g\} = \{\bar{\phi}^* f, \bar{\phi}^* g\} = 0$; by injectivity of $\bar{\phi}^*$ we see that $\mathcal{A}$ is involutive. If $\{f, \mathcal{A}\} = 0$ then $\{\bar{\phi}^* f, \mathcal{A}_1\} = 0$ since $\mathcal{A}_1$ is the integral closure of $\phi^* \mathcal{A}$ in $\mathcal{O}(M_1)$. Then $\bar{\phi}^* f \in \mathcal{A}_1$ by completeness of $\mathcal{A}_1$ and $\mathcal{A}$ is also complete. Finally the dimension count for $\phi(M_1)$ is the same as the one for M_1 since ϕ is finite. It follows that $(\phi(M_1), \{\cdot, \cdot\}, \mathcal{A})$ is an integrable Hamiltonian system. Clearly $\imath$ and $\bar{\phi}$ are morphisms of integrable Hamiltonian systems. ∎

Example 3.17 If a Poisson morphism $\phi : (M_1, \{\cdot, \cdot\}_1) \rightarrow (M_2, \{\cdot, \cdot\}_2)$ is finite but not dominant then $\mathrm{Cas}(M_1)$ may be larger than the integral closure of $\phi^* \mathrm{Cas}(M_2)$ in $\mathcal{O}(M_1)$. Take for example for $(M_2, \{\cdot, \cdot\}_2)$ the Lie-Poisson structure for the Heisenberg algebra (Example 2.54), for M_1 the plane $x = 0$ with the trivial Poisson structure and for ϕ the inclusion map. Then $\mathrm{Cas}(M_2) = \mathbf{C}[x]$ hence $\phi^* \mathrm{Cas}(M_2) = \mathbf{C}$, while $\mathrm{Cas}(M_1) = \mathcal{O}(M_1)$.

Example 3.18 Even if a Poisson morphism $\phi : (M_1, \{\cdot, \cdot\}_1) \rightarrow (M_2, \{\cdot, \cdot\}_2)$ is finite and dominant then $\mathrm{Cas}(M_1)$ may be different from $\phi^* \mathrm{Cas}(M_2)$. Take for example on $\mathbf{C}^3$ the Poisson structure from Example 3.14 and consider the finite covering map $\phi : \mathbf{C}^3 \rightarrow \mathbf{C}^3$ given by $\phi(q_1, p_1, q_2) = (q_1, p_1, q_2^2)$. Obviously this is a Poisson morphism; however the Casimir q_2 is not of the form $\phi^* F$ for any function $F \in \mathcal{O}(\mathbf{C}^3)$. Notice that $\bar{\phi} : \mathbf{C} \rightarrow \mathbf{C}$ is in this case not injective, being given by $\bar{\phi}(q_2) = q_2^2$. A similar remark applies to condition (3) in Definition 3.12.

3.3. Constructions of integrable Hamiltonian systems

In Section 2.3 we gave several constructions to build new affine Poisson varieties from old ones. Using these we now give the corresponding constructions for integrable Hamiltonian systems on them. We first show that an integrable Hamiltonian system restricts to a general fiber of the parameter map.

Proposition 3.19 *Let $(M, \{\cdot, \cdot\}, \mathcal{A})$ is an integrable Hamiltonian system and $\mathcal{F}$ an irreducible component of a general level of the Casimirs. Then $(\mathcal{F}, \{\cdot, \cdot\}_{|\mathcal{F}}, \mathcal{A}_{|\mathcal{F}})$ is an integrable Hamiltonian system and the inclusion map is a morphism. The property also holds for the general levels of any subalgebra of the Casimirs.*

Proof

Let $\mathcal{B}$ be any subalgebra of $\mathrm{Cas}(M)$ and let $\mathcal{F}$ be an irreducible component of a general fiber of $M \to \mathrm{Spec}\,\mathcal{B}$. We know already from Proposition 2.38 that $\mathcal{F}$ has an induced Poisson structure and from Proposition 2.42 that the algebra of Casimirs of this structure is maximal. If we restrict $\mathcal{A}$ to $\mathcal{F}$ then we get again an involutive algebra $\mathcal{A}_{\mathcal{F}}$ which is complete since $\mathcal{A}$ is complete and $\mathcal{F}$ is general. Thus it suffices to compute the dimension of $\mathcal{A}_{\mathcal{F}}$,

$$\dim \mathcal{F} - \dim \mathcal{A}_{\mathcal{F}} = \dim M - \dim \mathcal{B} - (\dim \mathcal{A} - \dim \mathcal{B}) = \frac{1}{2}\,\mathrm{Rk}\{\cdot,\cdot\} = \frac{1}{2}\,\mathrm{Rk}\{\cdot,\cdot\}_{\mathcal{F}}.$$

This shows that $\mathcal{A}_{\mathcal{F}}$ is an integrable algebra. Clearly the inclusion map is a morphism. ∎

Definition 3.20 Any integrable Hamiltonian system obtained from $(M,\{\cdot,\cdot\},\mathcal{A})$ by Proposition 3.19 is called a *trivial subsystem*.

One may think of a trivial subsystem as being obtained by fixing the values of some of the Casimirs.

Example 3.21 In the examples one has however to be careful when picking a particular fiber $\mathcal{F}$ (i.e., in the choice of values assigned to (some of) the Casimirs). Namely one has to check that $\mathcal{F}$ is general enough in the sense that both the dimension and rank of $\mathcal{F}$ coincide with those of a general fiber. The dimension of a special fiber $\mathcal{F}$ may be higher and/or its rank may be lower; then

$$\dim \mathcal{F} - \frac{1}{2}\,\mathrm{Rk}\{\cdot,\cdot\}_{\mathcal{F}} > \dim \mathcal{A} \geq \dim \mathcal{A}_{|\mathcal{F}},$$

so $(\mathcal{F},\{\cdot,\cdot\}_{|\mathcal{F}},\mathcal{A}_{|\mathcal{F}})$ is not an integrable Hamiltonian system. Reconsider e.g. Example 2.54: none of the integrable Hamiltonian systems on $\mathbf{C}^3$ for this Poisson structure will lead to an integrable Hamiltonian system on the fiber $x = 0$, since the induced Poisson structure on that fiber is trivial, while $\mathcal{A}_{|\mathcal{F}} \neq \mathcal{O}(\mathcal{F})$

Proposition 3.22 *For* $i \in \{1,2\}$ *let* $(M_i,\{\cdot,\cdot\}_i,\mathcal{A}_i)$ *be an integrable Hamiltonian system and let* π_i *denote the natural projection map* $M_1 \times M_2 \to M_i$ *Then*

$$(M_1 \times M_2, \{\cdot,\cdot\}_{M_1 \times M_2}, \pi_1^*\mathcal{A}_1 \otimes \pi_2^*\mathcal{A}_2) \tag{3.8}$$

is an integrable Hamiltonian system and the projection maps π_i *are morphisms. Each level set of the integrable Hamiltonian system is a product of a level set of* $(M_1,\{\cdot,\cdot\}_1,\mathcal{A}_1)$ *and a level set of* $(M_2,\{\cdot,\cdot\}_2,\mathcal{A}_2)$.

Proof

The Poisson-part of this proposition was already given in Proposition 2.21. As for involutivity,

$$\{\pi_1^*\mathcal{A}_1 \otimes \pi_2^*\mathcal{A}_2, \pi_1^*\mathcal{A}_1 \otimes \pi_2^*\mathcal{A}_2\}_{M_1 \times M_2} = \pi_1^*\{\mathcal{A}_1,\mathcal{A}_1\}_1 + \pi_2^*\{\mathcal{A}_2,\mathcal{A}_2\}_2 = 0.$$

We count dimensions:

$$\dim \pi_1^* \mathcal{A}_1 \otimes \pi_2^* \mathcal{A}_2 = \dim \mathcal{A}_1 + \dim \mathcal{A}_2$$

$$= \dim M_1 - \frac{1}{2} \mathrm{Rk}\{\cdot,\cdot\}_1 + \dim M_2 - \frac{1}{2} \mathrm{Rk}\{\cdot,\cdot\}_2$$

$$= \dim(M_1 \times M_2) - \frac{1}{2} \mathrm{Rk}\left(\{\cdot,\cdot\}_{M_1 \times M_2}\right).$$

Since $\pi_1^* \mathcal{A}_1 \otimes \pi_2^* \mathcal{A}_2$ is complete and involutive with respect to the product bracket, this computation shows that $\pi_1^* \mathcal{A}_1 \otimes \pi_2^* \mathcal{A}_2$ is integrable. Since for each of the projection maps π_i one has $\pi_i^* \mathcal{A}_i \subset \pi_1^* \mathcal{A}_1 \otimes \pi_2^* \mathcal{A}_2$, these projection maps are morphisms. The fibers of the momentum map are given by the fibers of $M_1 \times M_2 \to \mathrm{Spec}(\pi_1^* \mathcal{A}_1 \otimes \pi_2^* \mathcal{A}_2)$, that is, of the product map $M_1 \times M_2 \to \mathrm{Spec}\,\mathcal{A}_1 \times \mathrm{Spec}\,\mathcal{A}_2$ hence all fibers are products of level sets of $\mathcal{A}_1$ and $\mathcal{A}_2$. ∎

It is easy to show in addition that $\mathrm{Ham}(\pi_1^* \mathcal{A}_1 \otimes \pi_2^* \mathcal{A}_2)$ contains a super-integrable vector field if $\mathrm{Ham}(\mathcal{A}_1)$ (or $\mathrm{Ham}(\mathcal{A}_2)$) does.

Definition 3.23 We call (3.8) the *product* of $(M_1, \{\cdot,\cdot\}_1, \mathcal{A}_1)$ and $(M_2, \{\cdot,\cdot\}_2, \mathcal{A}_2)$.

A construction which is related to (but different from) the product construction and which will be used several times in the next chapters, is obtained when dealing with integrable Hamiltonian systems which depend on parameters. By this we mean that we have an affine Poisson variety $(M, \{\cdot,\cdot\})$ and for all possible values c of a set of parameters we have an integrable algebra $\mathcal{A}_c$ on it. This set of parameters is assumed here to be the points on an affine variety N and we assume that $\mathcal{A}_c$ (i.e., its elements) depends regularly on c. Then we can build a big affine Poisson variety which contains all the integrable Hamiltonian systems $(M, \{\cdot,\cdot\}, \mathcal{A}_c)$ as trivial subsystems. This is given by the following proposition.[8]

Proposition 3.24 *If N is an affine variety and for each $c \in N$ an integrable Hamiltonian system $(M, \{\cdot,\cdot\}_M, \mathcal{A}_c)$, depending regularly on c is given on an affine Poisson variety $(M, \{\cdot,\cdot\})$ then $M \times N$ has a structure of an affine Poisson variety $(M \times N, \{\cdot,\cdot\})$ and $\mathcal{O}(M \times N)$ contains an integrable subalgebra $\mathcal{A}$ such that each $(M, \{\cdot,\cdot\}_M, \mathcal{A}_c)$ is isomorphic to a trivial subsystem of $(M \times N, \{\cdot,\cdot\}, \mathcal{A})$ via the inclusion maps*

$$\phi_c : M \to M \times N : m \mapsto (m, c).$$

Proof

For N one takes the trivial structure so that $\mathrm{Cas}(N) = \mathcal{O}(N)$ which makes $M \times N$ into a Poisson manifold. The algebra of Casimirs on this product is maximal since the one on M is maximal and $\mathrm{Cas}(M \times N) = \mathrm{Cas}(M) \otimes \mathcal{O}(N)$. The fact that $\mathcal{A}_c$ depends regularly on c means that there exists a subalgebra $\mathcal{A}$ of $\mathcal{O}(M \times N)$ which restricts to $\mathcal{A}_c$ on the fiber over c of the projection $p : M \times N \to N$. Clearly its dimension is given by $\dim \mathcal{A} = \dim \mathcal{A}_c + \dim N$

---

[8] The proposition generalizes to the situation considered in Example 2.24, namely when $\pi : P \to N$ is a dominant morphism, for each $n \in N$, $\{\cdot,\cdot\}_n$ is a Poisson bracket on the fiber $\pi^{(-1)}(n)$ and $\mathcal{A}_n$ is an involutive subalgebra of $\mathcal{O}\left(\pi^{(-1)}(n)\right)$ which is integrable for general n; both $\{\cdot,\cdot\}_n$ and $\mathcal{A}_n$ are supposed to depend regularly on $n \in N$. Proposition 3.24 corresponds to the special case $P = M \times N$ considered at the end of Example 2.24.

so that $\dim \mathcal{A} = \dim(M \times N) - \frac{1}{2} \text{Rk}\{\cdot, \cdot\}$; since $\mathcal{A}$ is complete and involutive it is integrable. Since $\mathcal{O}(N)$ is a subalgebra of $\text{Cas}(M \times N)$ the fiber over p is a level set of the Casimirs and the restriction of the Poisson structure corresponds to the one on M via the morphism ϕ_c which is an isomorphism when restricted to such a fiber. ∎

The next construction we discuss is that of taking a quotient. This is of interest, because many of the classical integrable Hamiltonian systems possess discrete or continuous symmetry groups. The algebraic setup which we use here has the virtue to allow to pass easily to the quotient (one does not need to worry about the action being free, picking regular values and so on).

Proposition 3.25 *Let G be a finite or reductive group and consider a Poisson action $\chi : G \times M \to M$, where $(M, \{\cdot, \cdot\})$ is an affine Poisson variety. If $\mathcal{A}$ is an involutive algebra such that for each $g \in G$ the biregular map $\chi_g : M \to M$ defined by $\chi_g(m) = \chi(g, m)$ leaves $\mathcal{A}$ invariant, i.e., $\chi_g^* \mathcal{A} \subset \mathcal{A}$, then $(M/G, \{\cdot, \cdot\}_0, \mathcal{A}^G)$ is an involutive Hamiltonian system and the quotient map π is a morphism. Here $\{\cdot, \cdot\}_0$ is the quotient bracket on M/G given by Proposition 2.25. If G is finite then $(M/G, \{\cdot, \cdot\}_0, \mathcal{A}^G)$ is integrable.*

Proof
   Involutivity of $\mathcal{A}^G$ is immediate from Proposition 2.25. Suppose now that G is finite. Then completeness of $\mathcal{A}$ implies completeness of $\mathcal{A} \cap \mathcal{O}(M)^G$. As for dimensions, since G is a finite group we have

$$\dim \mathcal{A} \cap \mathcal{O}(M)^G = \dim \mathcal{A}$$
$$= \dim M - \frac{1}{2} \text{Rk}\{\cdot, \cdot\}$$
$$= \dim M/G - \frac{1}{2} \text{Rk}\{\cdot, \cdot\}_0,$$

where we used in the first equality that $\dim \mathcal{O}(M)^G = \dim \mathcal{O}(M)$ and $\mathcal{A} \subset \mathcal{O}(M)$. Similarly one shows that the algebra of Casimirs is maximal, being given by $\text{Cas}(M) \cap \mathcal{O}(M)^G$. Thus $\mathcal{A} \cap \mathcal{O}(M)^G$ is integrable; obviously $\pi^*(\mathcal{A} \cap \mathcal{O}(M)^G) \subset \mathcal{A}$, hence the quotient map is a morphism. ∎

We will encounter a lot of examples later. Here are some first observations.

Example 3.26 A special case occurs when $\mathcal{A} \subset \mathcal{O}(M)^G$ (which implies $\text{Cas}(M/G) \subset \mathcal{O}(M)^G$). Namely, in this case each level set of $(M, \{\cdot, \cdot\}, \mathcal{A})$ is stable for the action of G and the level sets of $(M/G, \{\cdot, \cdot\}_0, \mathcal{A})$ are precisely the quotients of the level sets of $(M, \{\cdot, \cdot\}, \mathcal{A})$. A similar result applies for the level sets of the Casimirs in case $\text{Cas}(M/G) \subset \mathcal{O}(M)^G$.

Example 3.27 The quotient construction leads to a lot of new integrable Hamiltonian systems which look interesting. One may e.g. start with an integrable Hamiltonian system $(M, \{\cdot, \cdot\}, \mathcal{A})$ and consider its square $(M \times M, \{\cdot, \cdot\}_{M \times M}, \mathcal{A} \otimes \mathcal{A})$. The group $\mathbf{Z}_2$ acts on $M \times M$ by interchanging the factors in the product. Obviously this is a Poisson action and the action leaves $\mathcal{A} \otimes \mathcal{A}$ invariant, thereby leading to a quotient. The level sets which correspond to the diagonal are symmetric products of the original level sets.

Notice that the group G in Proposition 3.25 can be seen as a subgroup of the automorphism group of M. For future use we introduce also the slightly more general notion of a quasi-automorphism.

Definition 3.28 Let $(M, \{\cdot, \cdot\}, \mathcal{A})$ be an integrable Hamiltonian system. An *automorphism* is an isomorphism $(M, \{\cdot, \cdot\}, \mathcal{A}) \to (M, \{\cdot, \cdot\}, \mathcal{A})$. More generally, if $\{\cdot, \cdot\}_1$ and $\{\cdot, \cdot\}_2$ are two Poisson brackets on M then an isomorphism $\phi : (M, \{\cdot, \cdot\}_1, \mathcal{A}) \to (M, \{\cdot, \cdot\}_2, \mathcal{A})$ is called a *quasi-automorphism*.

The final construction is to remove a divisor from phase space.

Proposition 3.29 *Let $(M, \{\cdot, \cdot\}, \mathcal{A})$ be an integrable Hamiltonian system and let $f \in \mathcal{O}(M)$ be a function which is not constant. Then there exists an integrable Hamiltonian system $(N, \{\cdot, \cdot\}_N, \mathcal{A}_N)$ and a morphism $(N, \{\cdot, \cdot\}_N, \mathcal{A}_N) \to (M, \{\cdot, \cdot\}, \mathcal{A})$ which is dominant, having the complement (in M) of the zero locus of f as image.*

Proof
Most of the proof (the Poisson part) was given in Proposition 2.35 and we use the notation of that proposition. We start with the case $f \in \mathcal{A}$. If we define $\mathcal{A}_N = \pi^* \mathcal{A}[t]$ then $\mathcal{A}_N$ is involutive since π is a Poisson morphism and it has the right dimension in order to be integrable. We need to verify completeness. Let $\sum_{i=0}^n f_i t^i \in \mathcal{O}(N)$ then

$$\left\{ \sum_{i=0}^n f_i t^i, \mathcal{A}_N \right\}_N = 0 \iff \sum_{i=0}^n \{f_i, \pi^* \mathcal{A}[t]\}_N t^i = 0$$

$$\implies \sum_{i=0}^n \{f_i, \pi^* \mathcal{A}\}_N f^{n-i} = 0$$

$$\iff \sum_{i=0}^n \{f_i, \mathcal{A}\} f^{n-i} = 0$$

$$\iff \left\{ \sum_{i=0}^n f_i f^{n-i}, \mathcal{A} \right\} = 0$$

$$\iff \sum_{i=0}^n f_i f^{n-i} \in \mathcal{A}$$

$$\iff \sum_{i=0}^n f_i f^{n-i} t^n \in \mathcal{A}_N$$

$$\iff \sum_{i=0}^n f_i t^i \in \mathcal{A}_N.$$

Since $\mathcal{A}_N$ is involutive the last line also implies the first line, so we have established the desired equivalence.

If $f \notin \mathcal{A}$ then an explicit description of $\mathcal{A}_N$ is still available if $(M, \{\cdot, \cdot\}, \mathcal{A})$ satisfies the conditions of Proposition 3.7. In that case the fibers of $N \to \operatorname{Spec} \pi^* \mathcal{A}$ also satisfy the conditions of Proposition 3.7 hence $\pi^* \mathcal{A}$ is complete and $\mathcal{A}_N = \pi^* \mathcal{A}$. In general one has $\mathcal{A}_N = \operatorname{Compl}(\pi^* \mathcal{A})$ and a more explicit description is not available. ∎

3.4. Compatible and multi-Hamiltonian integrable systems

We now introduce a few concepts which relate to compatible integrable Hamiltonian systems.

Definition 3.30 Let $\{\cdot,\cdot\}_i$, $i = 1,\ldots,n$ be n (linearly independent) compatible Poisson brackets on an affine variety M. If $(M,\{\cdot,\cdot\}_i, \mathcal{A})$ is an integrable Hamiltonian system for each $i = 1,\ldots,n$ then these systems are called *compatible integrable Hamiltonian systems*. Any non-zero vector field Y on M which is integrable (in particular Hamiltonian) with respect to all Poisson structures $\{\cdot,\cdot\}_i$, i.e., for which there exist $f_1,\ldots,f_n \in \mathcal{A}$ such that

$$Y = \{\cdot, f_1\}_1 = \ldots = \{\cdot, f_n\}_n,$$

is called a *multi-Hamiltonian* (*bi-Hamiltonian* if $n = 2$) vector field, since it is Hamiltonian in many different ways; any of the integrable Hamiltonian systems $(M,\{\cdot,\cdot\}_i, \mathcal{A})$ is then called an *integrable multi-Hamiltonian system* (*bi-Hamiltonian* when $n = 2$).

Remark 3.31 We do not demand in the definition of an integrable multi-Hamiltonian system that all the integrable vector fields be multi-Hamiltonian. Although this condition is satisfied in Examples 3.33 and 3.34 it is far too restrictive in general.

All propositions and basic constructions given above are easily adapted to the case of compatible or multi-Hamiltonian structures, but this will not be made explicit here. Just one example: an action of a reductive group which is a Poisson action with respect to both Poisson structures of two compatible integrable Hamiltonian systems yields on the quotient two compatible integrable Hamiltonian systems. Here are some properties which are specific to compatible integrable Hamiltonian systems.

Proposition 3.32
(1) *Compatible integrable Hamiltonian systems have the same level sets;*
(2) *The Poisson brackets of compatible integrable Hamiltonian systems have the same rank, which also equals the rank of a general linear combination of these Poisson structures;*
(3) *If $(M,\{\cdot,\cdot\}_i, \mathcal{A})$ are compatible integrable Hamiltonian system then for a general linear combination $\{\cdot,\cdot\}_\lambda$ of the Poisson structures the system $(M,\{\cdot,\cdot\}_\lambda, \mathcal{A})$ is an integrable Hamiltonian system.*

Proof
The proof of (1) is obvious since the level sets are determined by $\mathcal{A}$ only. Since $\mathrm{Rk}\{\cdot,\cdot\}_i = 2\dim M - 2\dim \mathcal{A}$ we find that the rank of all structures $\{\cdot,\cdot\}_i$ is equal. To determine the rank of a linear combination of these structures one looks at the corresponding Poisson matrix (with respect to a system of generators of $\mathcal{O}(M)$) which is given by the same linear combination of the Poisson matrices of the structures $\{\cdot,\cdot\}_i$. Now a general linear combination of invertible matrices is invertible, which applied to a non-singular minor of size $\mathrm{Rk}\{\cdot,\cdot\}_i$ leads to (2). For a linear combination $\{\cdot,\cdot\}_\lambda$ of (maximal) rank $\mathrm{Rk}\{\cdot,\cdot\}_i$ one has that $\{\mathcal{A},\mathcal{A}\}_\lambda = 0$ and $\dim \mathcal{A} = \dim M - \frac{1}{2}\mathrm{Rk}\{\cdot,\cdot\}_i$, hence $(M,\{\cdot,\cdot\}_\lambda, \mathcal{A})$ is an integrable Hamiltonian system, showing (3). ∎

3. Integrable Hamiltonian systems and their morphisms

We will encounter in this text many (non-trivial) examples of compatible integrable Hamiltonian systems and of integrable multi-Hamiltonian systems. Here are two simple examples of integrable bi-Hamiltonian systems.

Example 3.33 Consider the Poisson structures $\{\cdot,\cdot\}_1$ and $\{\cdot,\cdot\}_2$ on $\mathbf{C}^4$ (with coordinates q_1, q_2, p_1 and p_2) defined by the Poisson matrices

$$\begin{pmatrix} 0 & 0 & 1 & 0 \\ 0 & 0 & 0 & 1 \\ -1 & 0 & 0 & 0 \\ 0 & -1 & 0 & 0 \end{pmatrix} \quad \text{and} \quad \begin{pmatrix} 0 & 0 & 0 & 1 \\ 0 & 0 & 1 & 0 \\ 0 & -1 & 0 & 0 \\ -1 & 0 & 0 & 0 \end{pmatrix}.$$

For $\mathcal{A} \subset \mathcal{O}(\mathbf{C}^4)$ take those functions which are independent of q_1 and q_2. Then both Poisson structures are compatible and since their integrable vector fields are of the form

$$\left\{ f\frac{\partial}{\partial q_1} + g\frac{\partial}{\partial q_2} \;\middle|\; f, g \in \mathcal{A} \right\},$$

they are all bi-Hamiltonian.

Example 3.34 Recall from Example 2.11 that the matrix

$$u \begin{pmatrix} 0 & \frac{\partial F}{\partial z} & -\frac{\partial F}{\partial y} \\ -\frac{\partial F}{\partial z} & 0 & \frac{\partial F}{\partial x} \\ \frac{\partial F}{\partial y} & -\frac{\partial F}{\partial x} & 0 \end{pmatrix}$$

defines for any u and F in $\mathcal{O}(\mathbf{C}^3)$ a Poisson structure on $\mathbf{C}^3$; F is assumed non-constant here in order to obtain a non-trivial Poisson structure. Let us denote this Poisson structure by $\{\cdot,\cdot\}_{u,F}$. If G is any other non-constant element of $\mathcal{O}(\mathbf{C}^3)$ then $\{\cdot,\cdot\}_{u,F} + \{\cdot,\cdot\}_{u,G} = \{\cdot,\cdot\}_{u,F+G}$ hence $\{\cdot,\cdot\}_{u,F}$ and $\{\cdot,\cdot\}_{u,G}$ are compatible and, assuming that F and G are independent, $\mathcal{A} = \mathrm{Compl}\{F, G\}$ defines an integrable Hamiltonian system on $(\mathbf{C}^3, \{\cdot,\cdot\}_{u,F})$. However, by interchanging the roles of F and G we find that $\mathcal{A}$ also defines an integrable Hamiltonian system on $(\mathbf{C}^3, \{\cdot,\cdot\}_{u,G})$, hence leading to a pair of compatible integrable Hamiltonian systems. Since moreover the Hamiltonian vector fields with respect to *both* Poisson structures are given by

$$\{ u\psi \nabla F \times \nabla G \mid \psi \in \mathcal{A} \}$$

we conclude that $\mathcal{A}$ defines an integrable bi-Hamiltonian system on $\mathbf{C}^3$.

Closely related to the concept of an integrable multi-Hamiltonian system is that of a multi-Hamiltonian hierarchy. Let us define this in the case of a bi-Hamiltonian hierarchy and explain its use. Let $\{\cdot,\cdot\}_1$ and $\{\cdot,\cdot\}_2$ be two compatible Poisson brackets on M. Then a sequence of functions $\{f_i \mid i \in \mathbf{Z}\}$ is called a *bi-Hamiltonian hierarchy* if

$$\{\cdot, f_i\}_2 = \{\cdot, f_{i+1}\}_1, \qquad (i \in \mathbf{Z}).$$

The following property is essentially due to Lenard and Magri.

Proposition 3.35 *All functions f_i of a bi-Hamiltonian hierarchy $\{f_i \mid i \in \mathbf{Z}\}$ are in involution with respect to both Poisson brackets (hence with respect to any linear combination). If one of these functions is a Casimir (for either of the structures) then all these f_i are also in involution with the elements of any other bi-Hamiltonian hierarchy.*

Proof

If $\{f_i \mid i \in \mathbf{Z}\}$ forms a hierarchy, then for any $i < j \in \mathbf{Z}$

$$
\begin{aligned}
\{f_i, f_j\}_1 &= \{f_i, f_{j-1}\}_2 \\
&= \{f_{i+1}, f_{j-1}\}_1 \\
&= \quad \ldots \\
&= \{f_j, f_i\}_1,
\end{aligned}
$$

so $\{f_i, f_j\}_1 = 0$ by skew-symmetry. They are also in involution with respect to the second bracket since $\{f_i, f_j\}_2 = \{f_i, f_{j+1}\}_1$. In the same way, if $\{g_j \mid j \in \mathbf{Z}\}$ is another bi-Hamiltonian hierarchy and f_k is a Casimir, say of $\{\cdot, \cdot\}_1$ then for any $i, j \in \mathbf{Z}$

$$
\{f_i, g_j\}_1 = \{f_k, g_{i+j-k}\}_1 = 0.
$$

∎

The above proposition leads to many interesting integrable Hamiltonian systems; said differently it can be used to give an elegant proof of the involutivity of many integrable Hamiltonian systems.

4. Integrable Hamiltonian systems on other spaces

In this section we wish to consider briefly integrable Hamiltonian systems on spaces other than affine algebraic varieties. One possible generalization is to consider spaces which are not necessarily algebraic, but have a differential structure (real or complex analytic), at least on a dense open subset. Examples include smooth manifolds, analytic varieties and orbifolds. Note however that extra generality comes also from the fact that one can often choose which algebra of functions on these spaces to consider, for example one may consider an affine variety with its algebra of rational functions; however these algebras should be reasonably big in order to lead to integrable Hamiltonian system, as is clear from the example of a projective algebraic variety with its regular functions (which are only the constant functions). Another possible generalization, closely related to the problem raised by the latter example is to consider (reasonable) ringed spaces or schemes. We will only consider the first generalization here.

4.1. Poisson spaces

At first we define a general class of spaces, which includes both affine algebraic varieties and smooth manifolds, on which it is possible to define the notion of an integrable Hamiltonian system.

Definition 4.1 Let M be a topological space which has at least on a dense open subset a smooth (or holomorphic) structure. Also let $\mathcal{R}$ be an algebra of functions on M which is big enough to distinguish (smooth) points in M, and whose elements are smooth (resp. holomorphic) on a dense open subset of M. A *Poisson bracket* on $(M, \mathcal{R})$ is as in the case of affine varieties a Lie bracket $\{\cdot, \cdot\} : \mathcal{R} \times \mathcal{R} \to \mathcal{R} : (f, g) \mapsto \{f, g\}$, which satisfies the Leibniz rule in each of its arguments. We call $(M, \mathcal{R}, \{\cdot, \cdot\})$ (or $(M, \{\cdot, \cdot\})$ for short) a *Poisson space;* in the special case that M is a manifold and $\mathcal{R} = C^\infty(M)$ (resp. $\mathcal{R} = C^\omega(M)$) $(M, \{\cdot, \cdot\})$ is called a *Poisson manifold* (resp. *analytic Poisson manifold*).

The algebra of *Hamiltonian vector fields* and the algebra of *Casimirs* are defined as in case of affine Poisson varieties. The Hamiltonian vector fields are of course only (real or holomorphic) vector fields on the non-singular part of the space. On this non-singular part a Poisson tensor representing the Poisson bracket can be defined and also there is a notion of rank at a non-singular point. Notice that all this was in the case of affine Poisson spaces even defined at the singular points.

Example 4.2 The example which originated the theory of Poisson brackets and Poisson manifolds is that of symplectic manifolds. A *symplectic manifold* (M, ω) is a manifold equipped with a closed two-form ω (a *symplectic two-form*) which is non-degenerate (as a bilinear form on each tangent space). A vector field X_F is associated to any function $f \in C^\infty(M)$ by

$$\omega(X_f, \cdot) = df(\cdot)$$

and a skew-symmetric bracket is defined on smooth functions by

$$\{f, g\} = \omega(X_f, X_g).$$

Notice that this new definition of X_f is consistent with the definition $X_f = \{\cdot, f\}$ which we gave in the case of affine Poisson varieties.

Clearly $\{\cdot,\cdot\}$ is a derivation in each of its arguments and the Jacobi identity for this bracket is equivalent to the fact that ω is closed. Thus a symplectic manifold is a Poisson manifold in a natural way. Such a Poisson manifold is regular and its dimension equals its rank (in particular it is even). Conversely every regular Poisson manifold of maximal rank is a symplectic manifold in a natural way. In turn, the main examples of symplectic manifolds are provided by the cotangent bundle to any manifold and by Kähler manifolds. The literature on symplectic manifolds is immense. See e.g. [AL], [AM1] and [LM3].

A fundamental property of symplectic manifolds is that they admit locally so-called canonical coordinates (the Darboux Theorem). The following theorem provides the proper generalization of this property to Poisson manifolds. This theorem is due to A. Weinstein; for a proof we refer to [CW].

Theorem 4.3 *Let* $(M,\{\cdot,\cdot\})$ *be a Poisson manifold and let* $p \in M$ *be arbitrary. There exists a coordinate neighborhood* V *of* p *with coordinates* $(q_1,\ldots,q_r,p_1,\ldots,p_r,y_1\ldots,y_s)$ *centered at* p, *such that*

$$\{\cdot,\cdot\}_V = \sum_{i=1}^r \frac{\partial}{\partial q_i} \wedge \frac{\partial}{\partial p_i} + \frac{1}{2}\sum_{k,l=1}^s \phi_{kl}(y)\frac{\partial}{\partial y_k} \wedge \frac{\partial}{\partial y_l},$$

where the functions ϕ_{kl} *are smooth functions which vanish at* p.

The rank of $\{\cdot,\cdot\}$ is $2r$ but is not necessarily constant on a neighborhood of p. When the rank is constant on a neighborhood of p the neighborhood V can be chosen such that, on V, the functions ϕ_{kl} vanish, yielding the following *canonical brackets* for the above coordinates:

$$\{q_i,q_j\} = \{p_i,p_j\} = \{q_i,y_k\} = \{p_i,y_k\} = \{y_k,y_l\} = 0, \qquad \{q_i,p_j\} = \delta_{ij}, \qquad (4.1)$$

where $1 \leq i,j \leq r$ and $1 \leq k,l \leq s$. In this form Weinstein's Theorem is usually referred to as the *Darboux Theorem* and the above local coordinates are called *Darboux coordinates* or *canonical coordinates*. The Darboux Theorem may be refrased by saying that the rank of the Poisson manifold at a point where it is locally constant is the only local invariant of a Poisson manifold. A stronger version of the Darboux Theorem says that a collection of independent functions (around the point) which satisfy canonical commutation relations can be extended to a complete set of canonical coordinates. In this stronger form the Darboux Theorem is false for affine Poisson variety, consider for example on $\mathbf{C}^2$ the Poisson bracket $\{x,y\} = x$ at a point not on the Y-axis and let the incomplete collection consist just of $\{y\}$. The only way to complete it with f such that $\{f,y\} = 1$, is to take $f = \ln(x)$ which is not a regular function on any Zariski open subset of $\mathbf{C}^2$. Canonical coordinates (which are regular on a Zariski open subset) exist however for this bracket, for example one has $\{\frac{1}{x},-yx\} = 1$, (clearly canonical coordinates which are regular on $\mathbf{C}^2$ do not exist). It is unlikely that a set of independent regular (on an open subset) functions, satisfying commutation relations as in the Darboux Theorem, can be found for any affine Poisson variety, but a counterexample (if any) is missing.

Although there is a notion of rank at each point of a Poisson manifold, it is not true as in the case of affine Poisson varieties that the rank is constant on an open dense subset of the Poisson manifold which may result in some nasty behavior of the algebra of Casimirs. Consider the following example.

Example 4.4 We first construct a bump Poisson structure on the plane $\mathbf{R}^2$. Let φ be a non-zero function on $\mathbf{R}^2$ whose support $\text{Supp}(\varphi)$ is compact and connected. Clearly $\{x, y\} = \varphi(x, y)$ defines a Poisson bracket on $\mathbf{R}^2$ and there is an open subset where the rank is two but also an open subset where the rank is zero. Moreover its algebra of Casimirs is non-trivial since it contains all functions whose support is disjoint from $\text{Supp}(\varphi)$. Thus $\text{Supp}(\varphi)$ is a level set as well as every point in $M \setminus \text{Supp}(\varphi)$. The former level set is never a manifold (in the best case it might be a manifold with boundary, but it is in general singular as well). Of course all this is typical for the smooth case; when analytic brackets are considered then the rank *is* constant on an open dense subset, the fibers of a (real or complex) analytic map will be analytic varieties and so on.

We have discussed in Paragraph 2.4 two decompositions of affine Poisson varieties, the Casimir decomposition and the rank decomposition. From what we said it is clear that the rank decomposition does not have its counterparts in a smooth setting. There is however in the case of Poisson manifolds another decomposition (singular foliation) the *symplectic decomposition* or *symplectic foliation* which is very useful. Its name stems from the fact that the Poisson structure restricts to a regular structure of maximal rank on each leaf, hence the Poisson structure permits to define a symplectic structure on each leaf. On an affine Poisson variety the leaves of the symplectic foliation need not be algebraic (as e.g. in the Example 2.43) and they (i.e., equations for them) are difficult to determine explicitly in general (for example it is a well-known result that in the Lie-Poisson case (see Example 2.8) the symplectic leaves coincide with the co-adjoint orbits, i.e., the orbits of the corresponding group G acting on $\mathfrak{g}^*$ via the co-adjoint action; even in low dimensions these orbits may be very hard to compute).

The easiest way to obtain the symplectic foliation is by using Weinstein's Theorem. Indeed, a subvariety of M around p is obtained by taking $y_1 = \ldots = y_l = 0$ and along this subvariety $\{\cdot, \cdot\}$ restricts to a symplectic structure and this (local) subvariety is the only one containing p on which $\{\cdot, \cdot\}$ restricts to a Poisson bracket of maximal rank. Hence we may globalize this construction to find a unique symplectic leaf passing through each point. Notice that these leaves are immersed submanifolds and not closed submanifolds in general; each leaf may even be dense in M, as is shown in the following example (the example also shows that, even in the case of Poisson manifolds, the algebra of Casimirs needs not be maximal).

Example 4.5 Take on $\mathbf{R}^3$ an orthogonal basis e_1, e_2, e_3 with $e_3 = (1, \alpha, \beta)$ where $1, \alpha$ and β are linearly independent over $\mathbf{Q}$. The bivector $e_1 \wedge e_2$ determines by parallel translation a Poisson structure on $\mathbf{R}^3$ which descends to a Poisson structure $\{\cdot, \cdot\}$ on the torus $\mathbf{R}^3/\mathbf{Z}^3$. All symplectic leaves are two-dimensional, but they are dense, hence none of them can be a level set of the Casimirs, such level sets being always closed.

As a final remark about the symplectic foliation, we wish to point out that Weinstein's proof is easily seen to be valid also in the holomorphic case, yielding a holomorphic symplectic foliation on any holomorphic Poisson manifold. For affine Poisson varieties this leads to a holomorphic symplectic foliation on its smooth part (which is a complex manifold).

In the following definition we generalize Definition 2.15 to the case of general Poisson spaces.

Definition 4.6 Let $(M_1, \mathcal{R}_1, \{\cdot, \cdot\}_1)$ and $(M_2, \mathcal{R}_2, \{\cdot, \cdot\}_2)$ be two Poisson spaces, then a map $\phi : M_1 \to M_2$ is called a a *Poisson morphism* if

(1) $\phi^* \mathcal{R}_2 \subset \mathcal{R}_1$,
(2) $\phi^* \{f, g\}_2 = \{\phi^* f, \phi^* g\}_1$, for all $f, g \in \mathcal{R}_2$.

A Poisson morphism which has an inverse is called a *Poisson isomorphism*.

In terms of integral curves the relevance of Poisson morphisms for (integrable) Hamiltonian systems (as defined below) is formulated by the following proposition.

Proposition 4.7 *Let $(M_1, \{\cdot, \cdot\}_1)$ and $(M_2, \{\cdot, \cdot\}_2)$ be two Poisson manifolds and suppose that $\phi : M_1 \to M_2$ is a Poisson morphism. Then the integral curves of a Hamiltonian vector field X_H, $H \in C^\infty(M_2)$ which intersect $\phi(M_1)$ are entirely contained in $\phi(M_1)$ and are the projections under ϕ of the integral curves of $X_{\phi^* H}$.*

Proof
If γ is an integral curve of $\phi^* H$ then $\phi \circ \gamma$ is an integral curve of H. Indeed, let g_i be any local coordinates, then

$$(g_i \circ \phi \circ \gamma)^{\cdot} = \{g_i \circ \phi, H \circ \phi\} \circ \gamma = \{g_i, H\} \circ \phi \circ \gamma.$$

If $P \in \phi(M_1) \subset M_2$, let $Q \in M_1$ be lying over P, then the above computation shows that the integral curve of $H \circ \phi$ through Q projects (via ϕ) onto the (unique) integral curve of H passing through P — in particular this integral curve cannot leave $\phi(M_1)$. ∎

We wish to point out that a similar proposition, stating that *all* integral curves of X_H are projections of integral curves of $X_{\phi^* H}$, is given in [Wei2] (Lemma 1.2 p. 528), but this cannot be true: it would imply surjectivity of the map ϕ (at least onto the non-singular part).

Even when dealing with integrable Hamiltonian systems on symplectic manifolds one should by the above proposition consider Poisson morphisms rather than symplectic maps. It is seen from the following simple example that the two concepts do not agree in general and that the above proposition needs not hold for symplectic maps.

Example 4.8 Take $M_2 = \mathbf{R}^4$ (with coordinates x_1, y_1, x_2, y_2) and $M_1 \subset \mathbf{R}^4$ the plane given by $x_2 = y_2 = 0$. On both M_1 and M_2 we put the standard symplectic structure: $\omega_1 = dx_1 \wedge dy_1$ and $\omega_2 = dx_1 \wedge dy_1 + dx_2 \wedge dy_2$. Then there are obvious projection and inclusion maps

$$\pi : \mathbf{R}^4 \to \mathbf{R}^2 \qquad \text{and} \qquad \imath : \mathbf{R}^2 \to \mathbf{R}^4,$$

and it is easy to check that π is Poisson but not symplectic and $\imath$ is symplectic but not Poisson.

Example 4.9 Let us show by a simple modification of the previous example that Proposition 4.7 needs not be true for symplectic maps. Instead of the obvious inclusion map we consider now the symplectic map

$$\phi : \mathbf{R}^2 \to \mathbf{R}^4 : (x_1, y_1) \mapsto (x_1, y_1, x_1, 0).$$

The function x_2 on $\mathbf{R}^4$ has all integral curves parallel to the Y_2 axis, hence none of them is included in the image of ϕ.

The polynomial invariant which we associated to affine Poisson varieties does not generalize to general Poisson spaces since the rank decomposition may not lead to (a finite number of) reasonable spaces, so it may not be clear how to count "components". For analytic brackets our construction goes however over verbatim. A lot of attention has been given over the last few years to global invariants for symplectic manifolds, a good introduction and more references are given in [AL].

4.2. Integrable Hamiltonian systems on Poisson spaces

As for integrable Hamiltonian systems on general Poisson spaces we would like to copy Definition 3.5, but a few modifications are needed.

At first, we wish the rank of the Poisson space to be constant on some open dense subset, otherwise we may run into complications such as in Example 4.4 in which at some open subset the level sets of the integrable Hamiltonian system are given by the levels of the Casimirs and in some other open subset they are given by the level sets of the integrable algebra. In such case the Poisson space can be split in two, so it is a mild assumption that the rank is constant on an open dense subset; this constant is then called the *rank* of the Poisson space.

Second, the notions of spectrum and dimension for an algebra $\mathcal{A} \subset \mathcal{R}$ need to be modified. our algebras $\mathcal{A}$ have no spectrum nor a dimension; the dimension is naturally replaced by the number of independent functions (we say that a collection of functions is *independent* if their differentials are independent at every point of some open dense subset). As for the spectrum, which we needed in order to define the momentum map, we could take $\mathrm{Hom}(\mathcal{A}, \mathbf{R})$ (resp. $\mathrm{Hom}(\mathcal{A}, \mathbf{C})$) or the real spectrum (in the case of manifolds) but this may be a very complicated (and ugly) object; in particular we will not have a smooth or holomorphic projection map $M \to \mathrm{Hom}(\mathcal{A}, \mathbf{R})$; however for any system of generators $f_1, \ldots, f_n$ as above, we will have a smooth (resp. holomorphic) map $M \to \mathbf{R}^n$ (resp. $M \to \mathbf{C}^n$).

Third, it is not clear at all how to show for general Poisson spaces that some algebra is complete (in the sense of Definition 3.1). Recall that we insisted in having completeness in order not to call two systems non-isomorphic while their algebras have the same completion. A solution to this is not to insist on completeness in the definition of an integrable Hamiltonian system but to call two systems isomorphic when some involutive extension of their integrable algebras coincide.

These remarks lead to the following definition.

Definition 4.10 Let $(M, \mathcal{R}, \{\cdot, \cdot\})$ be a Poisson space which is of constant rank on an open dense subset of M and whose algebra of Casimirs is maximal, i.e., it contains $\dim M - \mathrm{CoRk}\{\cdot, \cdot\}$ independent functions. An involutive subalgebra $\mathcal{A}$ of $\mathcal{R}$ is called *integrable* if it contains $\dim M - \frac{1}{2}\mathrm{Rk}\{\cdot, \cdot\}$ independent functions. The quadruple $(M, \mathcal{R}, \{\cdot, \cdot\}, \mathcal{A})$ is then called an *integrable Hamiltonian system* and each non-zero vector field in

$$\mathrm{Ham}(\mathcal{A}) = \{X_f \mid f \in \mathcal{A}\}$$

is called an *integrable vector field*.

Example 4.11   In its original form the three body Toda lattice is given on $\mathbf{R}^6$ with the standard symplectic structure $\sum dq_i \wedge dp_i$ by the algebra generated by the following two smooth functions:

$$H = \frac{1}{2}\sum_{k=1}^{3} p_k^2 + \sum_{k=1}^{3} e^{q_{k+1}-q_{k-1}},$$

$$I = p_1 p_2 p_3 - \sum_{k=1}^{3} p_k e^{q_{k+1}-q_{k-1}}.$$

Since the translations

$$(q_1, q_2, q_3, p_1, p_2, p_3) \mapsto (q_1 + a, q_2 + a, q_3 + a, p_1, p_2, p_3)$$

define a Poisson action, the quotient of $\mathbf{R}^6$ by these translations, which is $\mathbf{R}^5$ inherits a Poisson structure. It leads on every hyperplane $p_1 + p_2 + p_3 = c$ ($c \in \mathbf{R}$ any fixed constant) to a symplectic structure. Since the group action leaves the functions H and I invariant they descend to this quotient and since they are in involution they are also in involution on the quotient. Clearly they are also independent, hence the algebra generated by H and I is integrable.

In view of the exponentials this is not what we called an integrable Hamiltonian system on an affine Poisson space; it is however closely related to one, see Section VII.7.

Example 4.12   A second example is given by the elliptic Calogero-Moser system, studied in detail (especially from the point of view of algebraic geometry) by Treibich and Verdier (see [TV]). The setup is the same as for the Toda lattice above but the exponentials are replaced by the Weierstrass $\wp$ function. In the simplest case of three "particles" the involutive algebra is generated by the following two meromorphic functions ($\wp$ is the Weierstrass function associated to a fixed elliptic curve)

$$H = \frac{1}{2}\sum_{k=1}^{3} p_k^2 - 2\sum_{k=1}^{3} \wp(q_{k+1} - q_{k-1}),$$

$$K = \frac{1}{3}\sum_{k=1}^{3} p_k^3 - 2\sum_{k=1}^{3} (p_{k+1} - p_{k-1})\wp(q_{k+1} - q_{k-1}).$$

As in the Toda case there are many different versions of the Calogero system (rational, trigoniometric, relativistic, ...) and as in that case they are all closely related to integrable Hamiltonian systems on affine Poisson varieties.

Finally, here is the definition of a morphism of integrable Hamiltonian system on Poisson spaces. Notice that in property (3) below we do not ask that $\phi^* \mathcal{A}_2 \subset \mathcal{A}_1$, in accordance with the third remark, just before Definition 4.10.

Definition 4.13   Let $(M_1, \mathcal{R}_1, \{\cdot, \cdot\}_1, \mathcal{A}_1)$ and $(M_2, \mathcal{R}_2, \{\cdot, \cdot\}_2, \mathcal{A}_2)$ be integrable Hamiltonian systems, then a map $\phi : M_1 \to M_2$ is a *morphism* if it has the following properties.

(1) ϕ is a Poisson morphism,
(2) $\phi^* \operatorname{Cas}(M_2) \subset \operatorname{Cas}(M_1)$;
(3) $\phi^* \mathcal{A}_2 \subset \mathcal{A}_3$, where $\mathcal{A}_3 \subset \mathcal{R}_1$ is an involutive algebra which contains $\mathcal{A}_1$.

Chapter III

Integrable Hamiltonian systems and symmetric products of curves

1. Introduction

This chapter is devoted to the construction and a geometric study of a big family of integrable Hamiltonian systems. The phase space is $\mathbf{C}^{2d}$, equipped with an infinite dimensional vector space of Poisson structures: for each non-zero $\varphi \in \mathbf{C}[x,y]$ we construct (in Paragraph 2.2) a Poisson bracket $\{\cdot,\cdot\}_d^\varphi$ which makes $(\mathbf{C}^{2d}, \{\cdot,\cdot\}_d^\varphi)$ into an affine Poisson variety. Each of these brackets has maximal rank $2d$ (in particular the algebra of Casimirs is trivial) and they are all compatible. An explicit formula for all these brackets is given; they grow in complexity (i.e., degree) with φ so that only the first members are (modified) Lie-Poisson structures.

What is surprising is that all these structures (for fixed d) have many integrable algebras in common; moreover a system of generators of these algebras are given by a very compact and simple formula. Namely there is one integrable algebra corresponding to each polynomial $F(x,y)$ in two variables (it is assumed here that the polynomial depends on y). The magical formula is given by

$$H(\lambda) = F(\lambda, v(\lambda)) \bmod u(\lambda);$$

in this formula $u(\lambda)$ is a monic polynomial of degree d and $v(\lambda)$ is a polynomial of degree less than d and the $2d$ coefficients of these two polynomials are the coordinates on $\mathbf{C}^{2d}$. The integrable algebra is obtained from this formula by taking

$$\mathcal{A}_{F,d} = \mathbf{C}[H_0, \ldots, H_{d-1}],$$

where H_i is the coefficient of λ^i in $H(\lambda)$. It leads to many integrable Hamiltonian systems and for fixed $F(x,y)$ they are all compatible; the integrable vector fields which correspond to them are however different so that these do not give integrable multi-Hamiltonian systems. Their

71

integrability is shown in Paragraph 2.3. We will look at the special case for which $F(x, y) = y^2 - f(x)$ in Paragraph 2.4; in this case we are able to write down Lax equations for the vector fields.

A closer study of the fibers of the momentum map reveals the meaning of the polynomial $F(x, y)$. We describe the fiber $\mathcal{F}_{F,d}$ over $(H_0, \ldots, H_{d-1}) = (0, \ldots, 0)$ and obtain a description of the other fibers by a slight change in F. If the algebraic curve Γ_F (in $\mathbf{C}^2$) defined by $F(x, y) = 0$ is non-singular then $\mathcal{F}_{F,d}$ is non-singular and we show that in this case the fiber $\mathcal{F}_{F,d}$ is isomorphic to an affine part of the d-fold symmetric product of the algebraic curve Γ_F (we also give an explicit description of the divisor which is missing). This shows that basically all our systems (for different F) are different and that the d-fold symmetric products of any curve (smoothly embedded in $\mathbf{C}^2$) appears as a level set of some integrable Hamiltonian system. We deduce from the description of the general fibers of the momentum map a description of their real parts. For $d = 2$ (when surfaces are obtained as level sets) the description is easily visualized and shows at once that a large family of topological types is present. The level sets are described in Paragraph 3.2 and their real parts in Paragraph 3.3.

The effect of changing the Poisson structure (keeping $F(x, y)$ and d fixed) manifests itself only at the level of the integrable vector fields (the Poisson structure is not seen from the fibers of the momentum map since these depend on $F(x, y)$ and d only). These vector fields are all tangent to the same fibers and span the tangent space at each (non-singular) point, hence these vector fields must be related; they are in the present example even related in a very simple way, however these vector fields are different for all choices of φ so that changing φ also leads to different (i.e., non-isomorphic) systems. The effect of varying the Poisson structure is given in Paragraph 3.5.

Later in the text we will refer on several occasions to the systems described in this chapter. For a futher generalization of these systems, in which $F(x, y)$ is replaced by a family of algebraic curves, we refer to [Van5]. For a more abstract, but less explicit, construction of these systems, where $\mathbf{C}^2$ is replaced by any Poisson surface, see [Bot].

2. The systems and their integrability

In this section we show how there is associated to every polynomial $F(x, y)$ an algebra of functions which is integrable with respect to a family of compatible Poisson structures on $\mathbf{C}^{2d}$, which is parametrized by the set of all polynomials $\varphi(x, y)$ in two variables.

2.1. Notation

$\mathbf{C}^{2d}$ is viewed throughout this chapter as the space of pairs of polynomials $(u(\lambda), v(\lambda))$, with $u(\lambda)$ monic of degree d and $v(\lambda)$ of degree less than d, via

$$\begin{aligned}
u(\lambda) &= \lambda^d + u_{d-1}\lambda^{d-1} + \cdots + u_1\lambda + u_0, \\
v(\lambda) &= \phantom{\lambda^d + {}} v_{d-1}\lambda^{d-1} + \cdots + v_1\lambda + v_0,
\end{aligned} \tag{2.1}$$

so the coefficients u_i and v_i serve as coordinates on $\mathbf{C}^{2d}$. Some formulas below are simplified by denoting $u_d = 1$.

For any rational function $r(\lambda)$, we denote by $[r(\lambda)]_+$ its polynomial part and we let $[r(\lambda)]_- = r(\lambda) - [r(\lambda)]_+$. If $f(\lambda)$ is any polynomial and $g(\lambda)$ is a monic polynomial, then $f(\lambda) \bmod g(\lambda)$ denotes the polynomial of degree less than $\deg g(\lambda)$, defined by

$$f(\lambda) \bmod g(\lambda) = g(\lambda) \left[\frac{f(\lambda)}{g(\lambda)} \right]_-,$$

so $f(\lambda) = f(\lambda) \bmod g(\lambda) + h(\lambda)g(\lambda)$ for a unique polynomial $h(\lambda)$, and $f(\lambda) \bmod u(\lambda)$ is easily computed as the rest obtained by the Euclidean division algorithm.

2.2. The compatible Poisson structures $\{\cdot, \cdot\}_d^\varphi$

Any polynomial $\varphi(x, y)$ specifies a Poisson bracket on $\mathbf{C}^2$ by $\{y, x\} = \varphi(x, y)$, hence also on the cartesian product $(\mathbf{C}^2)^d = \mathbf{C}^2 \times \cdots \times \mathbf{C}^2$ (by taking the product bracket). Explicitly

$$\{y_i, x_j\} = \delta_{ij}\varphi(x_j, y_i), \qquad \{x_i, x_j\} = \{y_i, y_j\} = 0, \tag{2.2}$$

where (x_i, y_i) are the coordinates on the i-th factor, coming from the chosen coordinates on $\mathbf{C}^2$. Let Δ denote the closed subset of $(\mathbf{C}^2)^d$ defined by

$$\Delta = \{((x_1, y_1), (x_2, y_2), \ldots, (x_d, y_d)) \mid x_i = x_j \text{ for some } i \neq j\},$$

and consider the map $S : (\mathbf{C}^2)^d \setminus \Delta \to \mathbf{C}^{2d}$, given by

$$((x_1, y_1), (x_2, y_2), \ldots, (x_d, y_d)) \mapsto (u(\lambda), v(\lambda)) = \left(\prod_{i=1}^d (\lambda - x_i), \sum_{i=1}^d y_i \prod_{j \neq i} \frac{\lambda - x_j}{x_i - x_j} \right). \tag{2.3}$$

This map can be interpreted as a morphism of affine Poisson varieties upon using Proposition II.2.35. This is done as follows. Define

$$M_1 = \{(x_0, (x_1, y_1), \ldots, (x_d, y_d)) \mid x_0 \prod_{i<j}(x_i - x_j)^2 = 1\} \subset \mathbf{C} \times (\mathbf{C}^2)^d,$$

and

$$M_2 = \{(t, u(\lambda), v(\lambda)) \mid t \cdot \mathrm{disc}(u(\lambda)) = 1\} \subset \mathbf{C} \times \mathbf{C}^{2d}.$$

Then we have a commutative diagram

$$
\begin{array}{ccc}
M_1 & \xrightarrow{\;\bar{S}\;} & M_2 \\
\Big\downarrow{\scriptstyle p_1} & & \Big\downarrow{\scriptstyle p_2} \\
(\mathbf{C}^2)^d \setminus \Delta & \xrightarrow{\;S\;} & \mathbf{C}^{2d}
\end{array}
$$

with $\bar{S}$ a morphism between the affine varieties M_1 and M_2. By Proposition II.2.35 the Poisson bracket $\{\cdot, \cdot\}$ on $(\mathbf{C}^2)^d$ leads to a Poisson bracket on M_1, also denoted by $\{\cdot, \cdot\}$, upon using the relation $x_0 \prod_{i<j}(x_i - x_j)^2 = 1$; namely one adds the brackets

$$\{x_0, x_i\} = 0, \qquad \text{and} \qquad \{x_0, y_k\} = -x_0^2 \left\{ \prod_{i<j}(x_i - x_j)^2, y_k \right\},$$

the latter being computed from

$$\{x_0, y_k\} \prod_{i<j}(x_i - x_j)^2 + \left\{ \prod_{i<j}(x_i - x_j)^2, y_k \right\} x_0 = 0.$$

The natural action of the permutation group S_d on $(\mathbf{C}^2)^d$ lifts to a free action of S_d on M_1 and since it is a Poisson action on $(\mathbf{C}^2)^d$ it is also a Poisson action on $(M_1, \{\cdot, \cdot\})$. Now $\bar{S}$ is a $d! : 1$ (unramified) covering morphism onto the affine variety M_2 and $\bar{S}$ is invariant for the action of S_d on M_1, hence M_2 may be identified with the quotient M_1/S_d. By Proposition II.2.25 M_2 has a unique Poisson structure such that $\bar{S}$ is a Poisson morphism. It will be denoted by $\{\cdot, \cdot\}_d^\varphi$.

We would like to transport this Poisson structure on M_2 to $\mathbf{C}^{2d}$ by the morphism $p_2 : M_2 \to \mathbf{C}^{2d}$. Of course this is in general impossible, however in the present case it turns out that there *does* exist a (unique) Poisson structure on $\mathbf{C}^{2d}$, also denoted by $\{\cdot, \cdot\}_d^\varphi$, such that p_2 is a Poisson morphism. It is given in the following proposition.

Proposition 2.1 *There exists a (unique) Poisson structure $\{\cdot, \cdot\}_d^\varphi$ on $\mathbf{C}^{2d}$ such that $p_2 : M_2 \to \mathbf{C}^{2d}$ is a Poisson morphism. In terms of the coordinates u_i, v_i for $\mathcal{O}(\mathbf{C}^{2d})$ the Poisson bracket is given by*

$$
\begin{aligned}
&\{u(\lambda), u_j\}_d^\varphi = \{v(\lambda), v_j\}_d^\varphi = 0, \\
&\{u(\lambda), v_j\}_d^\varphi = \{u_j, v(\lambda)\}_d^\varphi = \varphi(\lambda, v(\lambda)) \left[\frac{u(\lambda)}{\lambda^{j+1}}\right]_+ \bmod u(\lambda), \qquad 0 \le j \le d-1.
\end{aligned}
\tag{2.4}
$$

Except for the zero bracket $\{\cdot,\cdot\}_d^0$, all Poisson brackets $\{\cdot,\cdot\}_d^\varphi$ are of rank 2d and they are all compatible.

As a special and most important case, if y and x are canonical variables, i.e., $\varphi(x,y) = 1$, then the Poisson structure $\{\cdot,\cdot\}_d^\varphi$, also denoted by $\{\cdot,\cdot\}_d$, is regular; the nonzero part of the Poisson bracket (2.4) reduces in this case to

$$\{u(\lambda), v_j\}_d = \{u_j, v(\lambda)\}_d = \left[\frac{u(\lambda)}{\lambda^{j+1}}\right]_+ \tag{2.5}$$

and its matrix of Poisson brackets with respect to the coordinate functions u_i and v_j, takes the form

$$P = \begin{pmatrix} 0 & U \\ -U & 0 \end{pmatrix} \quad \text{where} \quad U = \begin{pmatrix} 0 & 0 & \cdots & 0 & 1 \\ 0 & 0 & \cdots & 1 & u_{d-1} \\ \vdots & \vdots & & \vdots & \vdots \\ 0 & 1 & \cdots & u_3 & u_2 \\ 1 & u_{d-1} & \cdots & u_2 & u_1 \end{pmatrix}.$$

In terms of $\{\cdot,\cdot\}_d$, the Poisson structure $\{\cdot,\cdot\}_d^\varphi$ is given by

$$\begin{aligned}
\{u(\lambda), f\}_d^\varphi &= \varphi(\lambda, v(\lambda)) \{u(\lambda), f\}_d \bmod u(\lambda), \\
\{v(\lambda), f\}_d^\varphi &= \varphi(\lambda, v(\lambda)) \{v(\lambda), f\}_d \bmod u(\lambda),
\end{aligned} \tag{2.6}$$

where f is any element of $\mathcal{O}(\mathbf{C}^{2d})$.

Proof

We compute explicitly on M_2 the Poisson brackets of $u_0, \ldots, u_{d-1}, v_0, \ldots, v_{d-1}$ (without worrying about their brackets with t) and observe that they are independent of t; by Proposition II.2.4 this leads to a Poisson bracket $\{\cdot,\cdot\}_d^\varphi$ on $\mathbf{C}^{2d}$ which makes p_2 into a Poisson morphism (the unicity of this bracket is immediate). Clearly $\{u(\lambda), u(\mu)\}_d^\varphi = 0$. If $1 \leq j \leq d$, then

$$\begin{aligned}
\{u_{d-j}, v(\lambda)\}_d^\varphi &= (-1)^j \left\{ \sum_{i_1 < i_2 < \cdots < i_j} x_{i_1} x_{i_2} \cdots x_{i_j}, \sum_{l=1}^{d} y_l \prod_{k \neq l} \frac{\lambda - x_k}{x_l - x_k} \right\}_d^\varphi \\
&= (-1)^j \sum_{i_1 < i_2 < \cdots < i_j} \sum_{l=1}^{d} \{x_{i_1} x_{i_2} \cdots x_{i_j}, y_l\}_d^\varphi \prod_{k \neq l} \frac{\lambda - x_k}{x_l - x_k} \\
&= (-1)^{j-1} \sum_{i_1 < i_2 < \cdots < i_j} \sum_{t=1}^{j} x_{i_1} x_{i_2} \cdots \widehat{x_{i_t}} \cdots x_{i_j} \varphi(x_{i_t}, y_{i_t}) \prod_{k \neq i_t} \frac{\lambda - x_k}{x_{i_t} - x_k} \\
&= (-1)^{j-1} \sum_{l \notin \{i_1 < i_2 < \cdots < i_{j-1}\}} x_{i_1} x_{i_2} \cdots x_{i_{j-1}} \varphi(x_l, y_l) \prod_{k \neq l} \frac{\lambda - x_k}{x_l - x_k} \\
&= (-1)^{j-1} \sum_{l=1}^{d} \varphi(x_l, y_l) \prod_{k \neq l} \frac{\lambda - x_k}{x_l - x_k} (-1)^{j-1} \sum_{m=0}^{j-1} x_l^m u_{d-j+m+1} \\
&= \sum_{l=1}^{d} \sum_{m=0}^{j-1} x_l^m u_{d-j+m+1} \varphi(x_l, y_l) \prod_{k \neq l} \frac{\lambda - x_k}{x_l - x_k}.
\end{aligned}$$

Substituting $\lambda = x_l$ in the right hand side one sees that $\{u_{d-j}, v(\lambda)\}_d^\varphi$ is the (unique) polynomial in λ of degree less than d which takes at $\lambda = x_l$ the value $\sum_{m=0}^{j-1} x_l^m u_{d-j+m+1} \varphi(x_l, v(x_l))$, for $l = 1, \ldots, d$. As the x_l are the zeros of $u(\lambda)$ and since $y_l = v(x_l)$ the same is true for $\sum_{m=0}^{j-1} \lambda^m u_{d-j+m+1} \varphi(\lambda, v(\lambda)) \bmod u(\lambda)$, and we find

$$
\begin{aligned}
\{u_{d-j}, v(\lambda)\}_d^\varphi &= \sum_{m=0}^{j-1} \lambda^m u_{d-j+m+1} \varphi(\lambda, v(\lambda)) \bmod u(\lambda) \\
&= \varphi(\lambda, v(\lambda)) \left[\frac{u(\lambda)}{\lambda^{d-j+1}} \right]_+ \bmod u(\lambda),
\end{aligned}
$$

which proves the second equality in (2.4). For the first equality in (2.4), notice that

$$
\begin{aligned}
\{u(\lambda), v(\mu)\}_d^\varphi &= \sum_{l=1}^d \left\{ \prod_{i=1}^d (\lambda - x_i), y_l \right\}_d^\varphi \prod_{j \neq l} \frac{\mu - x_j}{x_l - x_j} \\
&= \sum_{l=1}^d \varphi(x_l, y_l) \prod_{j \neq l} \frac{(\lambda - x_j)(\mu - x_j)}{x_l - x_j},
\end{aligned}
$$

is symmetric in λ and μ, which leads at once to $\{u(\lambda), v_i\}_d^\varphi = \{u_i, v(\lambda)\}_d^\varphi$.

In order to show that $\{v_i, v_j\}_d^\varphi = 0$, let us simplify the formulas by chosing $\varphi = 1$; for general φ the result then follows from (2.6). By construction

$$
\{v(\lambda), v(\mu)\}_d = \sum_{i,k=1}^d y_k \left(\prod_{j \neq i} \frac{\lambda - x_j}{x_i - x_j} \right) \left\{ y_i, \prod_{l \neq k} \frac{\mu - x_l}{x_k - x_l} \right\} - (i \leftrightarrow k, \ \mu \leftrightarrow \lambda), \tag{2.7}
$$

where $(i \leftrightarrow k, \ \mu \leftrightarrow \lambda)$ denotes a term similar to the first one, obtained by exchanging i with k as well as μ with λ. As for the first term, its terms corresponding to $k = i$ are given by

$$
-\sum_{i=1}^d y_i \left(\prod_{j \neq i} \frac{\lambda - x_j}{x_i - x_j} \right) \left(\prod_{l \neq i} \frac{\mu - x_l}{x_i - x_l} \right) \sum_{l \neq i} \frac{1}{x_i - x_l},
$$

which is symmetric in λ and μ so that these cancel when substracting the symmetric term in (2.7). The remaining terms, which correspond to $i \neq k$ can be rewritten as

$$
\sum_{k \neq i} y_k \left(\prod_{j \neq i} \frac{\lambda - x_j}{x_i - x_j} \right) \left(\prod_{l \neq k, i} \frac{\mu - x_l}{x_k - x_l} \right) \frac{\mu - x_k}{(x_i - x_k)^2},
$$

a polynomial which evaluates to 0 for all $(\lambda, \mu) = (x_s, x_t)$ with $s \neq t$. It follows that both terms in (2.7) are of degree less than d which agree on the d^2 couples $(\lambda, \mu) = (x_s, x_t)$ and we may conclude that $\{v(\lambda), v(\mu)\}_d = 0$.

Compatibility of the brackets derives from the formula

$$
\{\cdot, \cdot\}_d^\varphi + \{\cdot, \cdot\}_d^\psi = \{\cdot, \cdot\}_d^{\varphi + \psi}
$$

which is an easy consequence of (2.4).

2. The systems and their integrability

Notice that it is also seen from formula (2.4) that $\{\cdot,\cdot\}_d^\varphi$ really is a map from $\mathcal{O}(\mathbf{C}^{2d}) \times \mathcal{O}(\mathbf{C}^{2d}) \to \mathcal{O}(\mathbf{C}^{2d})$: it suffices to use that if a polynomial (in several variables) is reduced by using a *monic* polynomial then the result is also a polynomial (in all these variables).

For $\varphi = 1$ one obtains (2.5), because the degree of $\left[\frac{u(\lambda)}{\lambda^{j+1}}\right]_+$ is less than d for any $j = 0, \ldots, d-1$, which also leads at once to the matrix representation of $\{\cdot,\cdot\}_d$ — since the determinant of this matrix equals 1, it is regular of rank $2d$. Note also that if $d > 1$ then $\{\cdot,\cdot\}_d$ is not compatible with the standard Poisson structure on $\mathbf{C}^{2d}$.

To see where the rank of the Poisson structure $\{\cdot,\cdot\}_d^\varphi$ fails to be maximal, we need to investigate the determinant of the matrix of Poisson brackets $\{u_i, v_j\}_d^\varphi$. Using elementary properties of determinants one finds that for any values $x_1, \ldots, x_d$,

$$\det\left(\{u_i, v(x_{j+1})\}_d^\varphi\right)_{0\le i,j\le d-1} = \det\left(\{u_i, v_j\}_d^\varphi\right)_{0\le i,j\le d-1} \prod_{k<l}(x_k - x_l). \tag{2.8}$$

Choosing $x_1, \ldots, x_d$ to be the roots of $u(\lambda)$, we get from (2.4)

$$\det\left(\{u_i, v(x_{j+1})\}_d^\varphi\right)_{0\le i,j\le d-1} = \det\left(\varphi(x_{j+1}, v(x_{j+1}))\left[\frac{u(\lambda)}{\lambda^{i+1}}\right]_{+|\lambda=x_{j+1}}\right)_{0\le i,j\le d-1}$$

$$= \det\left(\left[\frac{u(\lambda)}{\lambda^{i+1}}\right]_{+|\lambda=x_{j+1}}\right)_{0\le i,j\le d-1} \prod_{m=1}^d \varphi(x_m, v(x_m))$$

$$= \det\left(\{u_i, v(x_{j+1})\}_d\right)_{0\le i,j\le d-1} \prod_{m=1}^d \varphi(x_m, v(x_m))$$

$$\overset{(i)}{=} (-1)^{[d/2]}\prod_{k<l}(x_k - x_l) \prod_{m=1}^d \varphi(x_m, v(x_m)),$$

where in (i) we used (2.8) for $\varphi = 1$. It follows that (even if $u(\lambda)$ has multiple roots)

$$\det\left(\{u_i, v_j\}_d^\varphi\right)_{0\le i,j\le d-1} = (-1)^{[d/2]}\prod_{m=1}^d \varphi(x_m, v(x_m)),$$

on all of $\mathbf{C}^{2d}$, hence the Poisson structure is of lower rank on the locus $\prod_{j=1}^d \varphi(x_j, v(x_j)) = 0$, which for given φ and d is easy written as the equation of an algebraic hypersurface in $\mathbf{C}^{2d}$.

Finally, (2.6) follows immediately from the Leibniz property of Poisson brackets. ∎

If φ depends only on x and has degree at most d, then $\{\cdot,\cdot\}_d^\varphi$ is a modified Lie-Poisson structure (see Example II.2.14). Explicitly, for $\varphi = x^n$, $0 \le n < d$ the Poisson matrix P_n with respect to the above coordinates, taken in the order $u_{d-1}, \ldots, u_0, v_{d-1}, \ldots, v_0$, is given by

$$P_n = \begin{pmatrix} 0 & 0 & U_n & 0 \\ 0 & 0 & 0 & -U_n' \\ -U_n & 0 & 0 & 0 \\ 0 & U_n' & 0 & 0 \end{pmatrix}$$

where

$$
U_n = \begin{pmatrix} 0 & 0 & \cdots & 0 & 1 \\ 0 & 0 & \cdots & 1 & u_{d-1} \\ \vdots & \vdots & & \vdots & \vdots \\ 0 & 1 & \cdots & u_{n+3} & u_{n+2} \\ 1 & u_{d-1} & \cdots & u_{n+2} & u_{n+1} \end{pmatrix} \quad \text{and} \quad U_n' = \begin{pmatrix} u_{n-1} & u_{n-2} & \cdots & u_1 & u_0 \\ u_{n-2} & u_{n-3} & \cdots & u_0 & 0 \\ \vdots & \vdots & & \vdots & \vdots \\ u_1 & u_0 & \cdots & 0 & 0 \\ u_0 & 0 & \cdots & 0 & 0 \end{pmatrix}.
$$

In particular, if $0 < n < d$ then the bracket is a product bracket (of a regular Poisson structure of rank $2(d-n)$ on $\mathbf{C}^{2(d-n)}$ and a non-regular Poisson structure of rank $2n$ on $\mathbf{C}^{2n}$). For $n = d$ one finds a Lie-Poisson bracket which is given by

$$
P_d = \begin{pmatrix} 0 & -U_d' \\ U_d' & 0 \end{pmatrix}, \quad \text{where} \quad U_d' = \begin{pmatrix} u_{d-1} & u_{d-2} & \cdots & u_1 & u_0 \\ u_{d-2} & u_{d-3} & \cdots & u_0 & 0 \\ \vdots & \vdots & & \vdots & \vdots \\ u_1 & u_0 & \cdots & 0 & 0 \\ u_0 & 0 & \cdots & 0 & 0 \end{pmatrix}.
$$

For $\varphi = \sum_{i=0}^{d} c_i x^i$ the corresponding Poisson matrix is given by $\sum_{i=0}^{d} c_i P_i$. Notice that these are the only $\varphi(x,y)$ for which $\{\cdot,\cdot\}_d^\varphi$ is a modified Lie-Poisson structure and that $\varphi(x,y) = cx^d$, $(c \in \mathbf{C})$ is the only one which gives a Lie-Poisson structure.

For $\varphi(x,y) = x^n$, $0 \le n \le d$ it is easy to compute the invariant polynomial (defined in Definition II.2.47) of the Poisson structure, which we will denote by $\rho_{d,n}$. Namely, since U_n is non-singular it suffices to look at the matrix U_n' whose determinant is zero if and only if $u_0 = 0$. By induction U_n' has co-rank at least k if and only if $u_0 = u_1 = \cdots = u_{k-1} = 0$. In conclusion the rank is at most $2d$ on all of $\mathbf{C}^{2d}$, it is of rank at most $2d - 2$ on the hyperplane $u_0 = 0, \ldots$, it is of rank at most $2d - 2n$ on the $(2d - n)$-dimensional space $u_0 = u_1 = \cdots = u_{n-1} = 0$ and on the latter space it is regular. Thus we have established the following formula for the polynomial associated to the Poisson structure $\{\cdot,\cdot\}_d^{x^n}$ $(0 \le n \le d)$:

$$
\rho_{d,n} = R^d S^{2d} + R^{d-1} S^{2d-1} + \cdots + R^{d-n} S^{2d-n}
$$
$$
= R^{d-n} S^{2d-n} (1 + RS + R^2 S^2 + \cdots + R^n S^n).
$$

Notice that the fact that $\rho_{d,n}$ is reducible reflects the fact that the Poisson structure is a product. The polynomial has the simple $(d+1) \times (2d+1)$ matrix representation

$$
\begin{pmatrix} 0 & 0 \\ 0 & I_{n+1} \end{pmatrix}
$$

where I is the identity matrix of size $n + 1$.

2.3. Polynomials in involution for $\{\cdot,\cdot\}_d^\varphi$

We will now show how an arbitrary polynomial $F(x,y)$ leads to a natural set of d polynomials on $\mathbf{C}^{2d}$ which have the remarkable property to be in involution for all the Poisson structures $\{\cdot,\cdot\}_d^\varphi$. These polynomials generate a d-dimensional algebra (under the assumption that $F(x,y)$ is not independent of y), hence they define an integrable Hamiltonian system on $\mathbf{C}^{2d}$ for any structure $\{\cdot,\cdot\}_d^\varphi$. Since all the brackets are compatible this means that we have for each $F(x, y)$ (which is not independent of y) a large class of compatible integrable Hamiltonian systems. They are however not multi-Hamiltonian as we will see in Section 3.5 below (see however also Paragraph VI.3 and [Van5]).

Let $F(x, y) \in \mathbf{C}[x, y] \setminus \mathbf{C}[x]$ and let us view $\mathbf{C}^d$ as the space of polynomials (say in λ) of degree less than d. Then there is a natural map $\hat{H}_{F,d}$ from $(\mathbf{C}^2)^d \setminus \Delta$ to $\mathbf{C}^d$, which assigns to a d-tuple $((x_1, y_1), \dots, (x_d, y_d))$ the unique polynomial in $\mathbf{C}[\lambda]$ of degree less than d, which takes for $\lambda = x_i$ the value $F(x_i, y_i)$ (for $i = 1, \dots, d$). We thereby arrive at the following commutative diagram

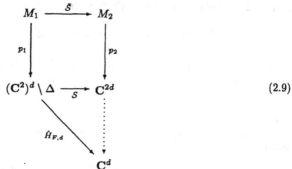

$$(2.9)$$

in which the existence of the dotted arrow is guaranteed by the following lemma.

Lemma 2.2 *There exists a (unique) morphism $H_{F,d} : \mathbf{C}^{2d} \to \mathbf{C}^d$ such that the triangle in (2.9) is commutative. $H_{F,d}$ is explicitly given by*

$$H_{F,d}(u(\lambda), v(\lambda)) = F(\lambda, v(\lambda)) \bmod u(\lambda). \tag{2.10}$$

Proof

$p_1^* \hat{H}_{F,d} : M_1 \to \mathbf{C}^d$ is a morphism which is invariant for the action of S_d hence it can be factorized via the quotient M_1/S_d which we identified with M_2. This means that we have a morphism $p_3 : M_2 \to \mathbf{C}^d$. It associates to $(t, u(\lambda), v(\lambda)) \in M_2$ the unique polynomial (in λ) of degree at most $d - 1$ whose value for $\lambda = x_i$, x_i any root of $u(\lambda)$, is given by $F(x_i, y_i)$. A compact formula for p_3 can be given:

$$p_3(t, u(\lambda), v(\lambda)) = F(\lambda, v(\lambda)) \bmod u(\lambda).$$

To check this formula, note that the right hand side is clearly a polynomial (in λ) of degree at most $d - 1$ and for any x_i which is a root of $u(\lambda)$ it evaluates to $F(x_i, v(x_i)) = F(x_i, y_i)$. Since the map p_3 does not depend on t it can be factorized in turn via p_2, i.e., $p_3 = p_2^* H_{F,d}$ and $H_{F,d}$ is explicitly given by (2.10). ∎

The d components of the map $H_{F,d}$ define d regular functions (polynomials) on $\mathbf{C}^{2d}$, which will be simply denoted by $H_{d-1}, \dots, H_0$ (omitting the dependence on F and d in the notation), i.e., $H_{F,d}(u(\lambda), v(\lambda)) = H_{d-1}\lambda^{d-1} + H_{d-2}\lambda^{d-2} + \dots + H_0$.

The main result of this section is the following.

Proposition 2.3 *For any polynomial $F(x,y) \in \mathbf{C}[x,y] \setminus \mathbf{C}[x]$, let $\mathcal{A}_{F,d} = \mathbf{C}[H_0, \ldots, H_{d-1}]$, where $H_0, \ldots, H_{d-1}$ are the coefficients (in λ) of $H(\lambda) = F(\lambda, v(\lambda)) \bmod u(\lambda)$. The triple $(\mathbf{C}^{2d}, \{\cdot, \cdot\}_d^\varphi, \mathcal{A}_{F,d})$ defines for any non-zero $\varphi(x,y) \in \mathbf{C}[x,y]$ an integrable Hamiltonian system and these systems are all compatible.*

Before proving this proposition we prove a key lemma and write down explicit equations for the Hamiltonian vector fields $X_{H_i}^\varphi = \{\cdot, H_i\}_d^\varphi$.

Lemma 2.4 *Let $p(\lambda)$, $q(\lambda)$ and $r(\lambda)$ be polynomials, with $\deg q(\lambda) \geq \deg r(\lambda)$ and let $i \in \mathbf{N}$.*

$$
\begin{aligned}
&(1) \quad r(\lambda)\left[\lambda^{-i}q(\lambda)\right]_+ \bmod q(\lambda) = r(\lambda)\left[\lambda^{-i}q(\lambda)\right]_+ - q(\lambda)\left[\lambda^{-i}r(\lambda)\right]_+, \\
&(2) \quad \sum_{l=1}^{\deg q} \mu^{l-1}p(\lambda)\left[\lambda^{-l}q(\lambda)\right]_+ \bmod q(\lambda) = \sum_{l=1}^{\deg q} \lambda^{l-1}p(\mu)\left[\mu^{-l}q(\mu)\right]_+ \bmod q(\mu).
\end{aligned}
\tag{2.11}
$$

Proof

For the proof of (1) note that if $\deg r(\lambda) \leq \deg q(\lambda)$ then the right hand side of the identity

$$
r(\lambda)\left[\lambda^{-i}q(\lambda)\right]_+ - q(\lambda)\left[\lambda^{-i}r(\lambda)\right]_+ = -r(\lambda)\left[\lambda^{-i}q(\lambda)\right]_- + q(\lambda)\left[\lambda^{-i}r(\lambda)\right]_-
$$

is a polynomial of degree less than $\deg q(\lambda)$, hence also the left hand side. To show (2) we may assume that $\deg p(\lambda) < \deg q(\lambda)$ because the equality depends only on $p(\lambda) \bmod q(\lambda)$. Then

$$
\begin{aligned}
\sum_{l=1}^{\deg q} \lambda^{l-1}p(\mu)\left[\mu^{-l}q(\mu)\right]_+ \bmod q(\mu) &\overset{(i)}{=} \sum_{l=1}^{\deg q} \lambda^{l-1}\left(p(\mu)\left[\mu^{-l}q(\mu)\right]_+ - q(\mu)\left[\mu^{-l}p(\mu)\right]_+\right) \\
&\overset{(ii)}{=} \sum_{l=1}^{\deg q} \mu^{l-1}\left(p(\lambda)\left[\lambda^{-l}q(\lambda)\right]_+ - q(\lambda)\left[\lambda^{-l}p(\lambda)\right]_+\right) \\
&= \sum_{l=1}^{\deg q} \mu^{l-1}p(\lambda)\left[\lambda^{-l}q(\lambda)\right]_+ \bmod q(\lambda).
\end{aligned}
$$

In (i) we applied part (1) of this lemma; the exchange property in (ii) is proven at once by expanding the polynomials or by induction on $\deg q(\lambda)$. ∎

Proposition 2.5 *The coefficients H_i of $F(\lambda, v(\lambda)) \bmod u(\lambda)$ determine d polynomial vector fields $X_{H_i}^\varphi$ on $\mathbf{C}^{2d}$, which are explicitly given by*

$$
\begin{aligned}
X_{H_i}^\varphi u(\lambda) &= \varphi(\lambda, v(\lambda)) \frac{\partial F}{\partial y}(\lambda, v(\lambda)) \left[\frac{u(\lambda)}{\lambda^{i+1}}\right]_+ \bmod u(\lambda), \\
X_{H_i}^\varphi v(\lambda) &= \varphi(\lambda, v(\lambda)) \left[\frac{F(\lambda, v(\lambda))}{u(\lambda)}\right]_+ \left[\frac{u(\lambda)}{\lambda^{i+1}}\right]_+ \bmod u(\lambda).
\end{aligned}
\tag{2.12}
$$

Moreover, the following remarkable identities hold for all $0 \leq i, j \leq d - 1$:

$$\{u_i, H_j\}_d^\varphi = \{u_j, H_i\}_d^\varphi \quad and \quad \{v_i, H_j\}_d^\varphi = \{v_j, H_i\}_d^\varphi. \tag{2.13}$$

Proof

Writing X_{H_i} as a shorthand for $X_{H_i}^1$, we first compute $X_{H_i} u(\lambda) = \{u(\lambda), H_i\}_d$, which we obtain as the coefficient of μ^{d-i} in $\{u(\lambda), H_{F,d}(u(\mu), v(\mu))\}_d$.

$$
\begin{aligned}
\{u(\lambda), H_{F,d}(u(\mu), v(\mu))\}_d &= \sum_{j=0}^{d-1} \{u(\lambda), v_j\}_d \frac{\partial H_{F,d}}{\partial v_j}(u(\mu), v(\mu)) \\
&= \sum_{j=0}^{d-1} \left[\frac{u(\lambda)}{\lambda^{j+1}} \right]_+ \frac{\partial H_{F,d}}{\partial v_j}(u(\mu), v(\mu)) \\
&= \sum_{j=0}^{d-1} \sum_{k=0}^{d-j-1} u_{d-k} \lambda^{d-j-k-1} \frac{\partial F}{\partial y}(\mu, v(\mu)) \mu^j \bmod u(\mu) \\
&= \sum_{l=1}^{d-1} \sum_{j=0}^{d-l} u_{j+l} \lambda^{l-1} \frac{\partial F}{\partial y}(\mu, v(\mu)) \mu^j \bmod u(\mu) \\
&= \sum_{l=1}^{d} \lambda^{l-1} \frac{\partial F}{\partial y}(\mu, v(\mu)) \left[\frac{u(\mu)}{\mu^l} \right]_+ \bmod u(\mu) \\
&= \sum_{l=1}^{d} \mu^{l-1} \frac{\partial F}{\partial y}(\lambda, v(\lambda)) \left[\frac{u(\lambda)}{\lambda^l} \right]_+ \bmod u(\lambda)
\end{aligned}
$$

where we used the exchange property (2.11) in the last step. Since H_i is the coefficient of λ^i in $H(\lambda)$ this leads to equation (2.12) for $X_{H_i} u(\lambda)$ in case $\varphi(x, y) = 1$. In a similar way $X_{H_i} v(\lambda)$ is found, the computation of $\frac{\partial}{\partial u_j} H_{F,d}(u(\mu), v(\mu))$ is however more involved: let $0 \leq j \leq d-1$ then

$$
\begin{aligned}
\frac{\partial}{\partial u_j} (F(\mu, v(\mu)) \bmod u(\mu)) &= \frac{\partial}{\partial u_j} \left(u(\mu) \left[\frac{F(\mu, v(\mu))}{u(\mu)} \right]_- \right) \\
&= -\frac{\partial}{\partial u_j} \left(u(\mu) \left[\frac{F(\mu, v(\mu))}{u(\mu)} \right]_+ \right) \\
&= -u(\mu) \left(\frac{\mu^j}{u(\mu)} \left[\frac{F(\mu, v(\mu))}{u(\mu)} \right]_+ - \left[\frac{\mu^j}{u(\mu)} \frac{F(\mu, v(\mu))}{u(\mu)} \right]_+ \right) \\
&\overset{(i)}{=} -u(\mu) \left[\frac{\mu^j}{u(\mu)} \left[\frac{F(\mu, v(\mu))}{u(\mu)} \right]_+ \right]_- \\
&= -\mu^j \left[\frac{F(\mu, v(\mu))}{u(\mu)} \right]_+ \bmod u(\mu).
\end{aligned}
$$

In (i) we used that if $R = R(\mu)$ and $P = P(\mu)$ are rational functions, with $[R]_+ = 0$, then

$$R[P]_+ - [RP]_+ = R[P]_+ - [R[P]_+]_+ = [R[P]_+]_- \ .$$

Granted this, we obtain as above

$$\{v(\lambda), H_{F,d}(u(\mu), v(\mu))\}_d = \sum_{l=0}^{d-1} \mu^l \left[\frac{u(\lambda)}{\lambda^{l+1}}\right]_+ \left[\frac{F(\lambda, v(\lambda))}{u(\lambda)}\right]_+ \mod u(\lambda),$$

which leads at once to the expression (2.12) for $X_{H_i} v(\lambda)$ in case $\varphi(x, y) = 1$. Having obtained the formulas (2.12) for $X_{H_i} u(\lambda)$ and $X_{H_i} v(\lambda)$, the formulas for $X_{H_i}^\varphi u(\lambda)$ and $X_{H_i}^\varphi v(\lambda)$, are obtained at once upon using (2.6).

Finally, the exchange property (2.11) implies that λ and μ are everywhere interchangeable in the above computations so we get $\{u(\lambda), H_{F,d}(u(\mu), v(\mu))\}_d^\varphi = \{u(\mu), H_{F,d}(u(\lambda), v(\lambda))\}_d^\varphi$, which is tantamount to the identity $\{u_i, H_j\}_d^\varphi = \{u_j, H_i\}_d^\varphi$. The second formula in (2.13) follows in the same way. ∎

Proof of Proposition 2.3

We first prove that $\{H_i, H_{F,d}(u(\lambda), v(\lambda))\}_d^\varphi = 0$ for $0 \le i \le d-1$, which shows that $\mathcal{A}_{F,d}$ is involutive. To make the proof more transparent, we use the following abbreviations:

$$F_y = \frac{\partial F}{\partial y}(\lambda, v(\lambda)), \quad F_{(u)} = \frac{F(\lambda, v(\lambda))}{u(\lambda)} \quad \text{and} \quad U_i = \frac{\varphi(\lambda, v(\lambda))}{u(\lambda)} \left[\frac{u(\lambda)}{\lambda^{i+1}}\right]_+,$$

so that (2.12) is rewritten as $X_{H_i}^\varphi u(\lambda) = u(\lambda) [U_i F_y]_-$ and $X_{H_i}^\varphi v(\lambda) = u(\lambda) \left[U_i [F_{(u)}]_+\right]_-$. Then

$$\{H_{F,d}(u(\lambda), v(\lambda)), H_i\}_d^\varphi = X_{H_i}^\varphi \left(u(\lambda) \left[\frac{F(\lambda, v(\lambda))}{u(\lambda)}\right]_-\right)$$

$$= X_{H_i}^\varphi u(\lambda) [F_{(u)}]_- + u(\lambda) \left[\frac{X_{H_i}^\varphi F(\lambda, v(\lambda))}{u(\lambda)} - \frac{F_{(u)} X_{H_i}^\varphi u(\lambda)}{u(\lambda)}\right]_-$$

$$= u(\lambda) \left[[U_i F_y]_- [F_{(u)}]_- + F_y \left[U_i [F_{(u)}]_+\right]_- - F_{(u)} [U_i F_y]_-\right]_-$$

$$\overset{(i)}{=} u(\lambda) \left[-[U_i F_y]_- [F_{(u)}]_+ + F_y U_i [F_{(u)}]_+\right]_-$$

$$= u(\lambda) \left[[U_i F_y]_+ [F_{(u)}]_+\right]_-$$

$$= 0.$$

In (i) we used the fact that F_y is a polynomial, i.e., $[F_y]_- = 0$.

We now show that the d coefficients of $H_{F,d}(u(\lambda), v(\lambda)) = F(\lambda, v(\lambda)) \mod u(\lambda)$ are independent, showing that $\dim \mathcal{A}_{F,d} = d = \dim \mathbf{C}^{2d} - \frac{1}{2} \operatorname{Rk} \{\cdot, \cdot\}_d^\varphi$. Clearly the last d coefficients $\tilde{H}_{d-1}, \ldots, \tilde{H}_0$ of $F(\lambda, v(\lambda))$ are independent because v_i appears only in $\tilde{H}_{d-1}, \ldots, \tilde{H}_i$ (it does appear since $F(x, y) \notin \mathbf{C}[x]$). Reducing $F(\lambda, v(\lambda))$ modulo $u(\lambda)$ amounts to substracting from $\tilde{H}_i$ polynomials of lower degree in the variables v_j, so it cannot make these functions dependent and the independence of $\{H_0, \ldots, H_{d-1}\}$ follows.

Finally $\mathcal{A}_{F,d}$ is also complete. Namely we will show (in Proposition 3.3) that the general fiber of $M \to \operatorname{Spec}(\mathcal{A}_{F,d})$ is irreducible and (in Lemma 3.4) that all fibers have the same dimension (d). Thus $\mathcal{A}_{F,d}$ satisfies all conditions of Proposition II.3.7 which implies its completeness. ∎

Amplification 2.6 Suppose that $F(x, y)$ and $F'(x, y)$ differ only by a polynomial $c(x)$ which is independent of y and is of degree less than d in x, say

$$F(x, y) = F'(x, y) + \sum_{i=0}^{d-1} c_i x^i,$$

then

$$F(\lambda, v(\lambda)) \bmod u(\lambda) = F'(\lambda, v(\lambda)) \bmod u(\lambda) + \sum_{i=0}^{d-1} c_i \lambda^i,$$

hence the polynomials in involution which they determine are up to constants the same and we find $\mathcal{A}_{F,d} = \mathcal{A}_{F',d}$, that is both systems are isomorphic. We might reformulate our result by saying that — for $\varphi(x, y)$ fixed — we have associated an integrable Hamiltonian system to a family

$$F_c(x, y) = \left\{ F(x, y) + \sum_{i=0}^{d-1} c_i x^i \mid c_i \in \mathbf{C} \right\}.$$

Suppose now that a bigger family M is given, i.e., $F_c(x, y)$ depends on one or several extra parameters which we suppose to parametrize an affine algebraic variety. One observes that the Hamiltonians $H_1, \ldots, H_d$ depend polynomially on the coefficients of F_c, hence also on the parameters c, so by Proposition II.3.24 there is an integrable Hamiltonian system on $M \times \mathbf{C}^{2d}$ with $\mathcal{O}(M)$ as its algebra of Casimirs and with projection $M \times \mathbf{C}^{2d} \to M$ such that the fiber of this morphism over any closed point $c \in M$ is precisely our original system on $\mathbf{C}^{2d}$ corresponding to the polynomial $F_c(x, y)$ where the parameters have been given the fixed value c. We will often prefer to work on these bigger systems, see Chapters VI and VII. For a further generalization, in which arbitrary families of algebraic curves are considered, see [Van5].

2.4. The hyperelliptic case

We now turn to a case which will be important later: the case that $F(x, y) = y^2 - f(x)$ for some polynomial $f(x)$. We call it the *hyperelliptic case* because $F(x, y) = 0$ now defines a hyperelliptic curve (see Paragraph IV.2.6). In the following proposition we give Lax equations for the hyperelliptic case (Lax equations will be explained in more detail in Section V.5; in this section a Lax equation is no more than a neat way to write down the differential equations describing an integrable vector field).

Proposition 2.7 If $F(x, y)$ has the hyperelliptic form $F(x, y) = y^2 - f(x)$ for some polynomial $f(x)$ then the differential equations describing the vector fields $X_{H_i}^\varphi$ are written in the Lax form (with spectral parameter λ)

$$X_{H_i}^\varphi A(\lambda) = \left[A(\lambda), [B_i(\lambda)]_+ \right], \tag{2.14}$$

where

$$A(\lambda) = \begin{pmatrix} v(\lambda) & u(\lambda) \\ w(\lambda) & -v(\lambda) \end{pmatrix}, \qquad B_i(\lambda) = \frac{\varphi(\lambda, v(\lambda))}{u(\lambda)} \left[\frac{u(\lambda)}{\lambda^{i+1}} \right]_+ A(\lambda)$$

and

$$w(\lambda) = -\left[\frac{F(\lambda, v(\lambda))}{u(\lambda)}\right]_+.$$

The spectral curve $\det(A(\lambda) - \mu\,\mathrm{Id}) = 0$, *which is preserved by the flow of the vector fields* $X^\varphi_{H_i}$, *is given by* $\mu^2 - f(\lambda) = H_{F,d}(u(\lambda), v(\lambda))$.

Proof

If we define the polynomial $w(\lambda)$ as stated above, then equations (2.12) are rewritten as

$$
\begin{aligned}
X^\varphi_{H_i} u(\lambda) &= 2\varphi(\lambda, v(\lambda))v(\lambda)\left[\frac{u(\lambda)}{\lambda^{i+1}}\right]_+ - 2u(\lambda)\left[\varphi(\lambda, v(\lambda))\frac{v(\lambda)}{u(\lambda)}\left[\frac{u(\lambda)}{\lambda^{i+1}}\right]_+\right]_+, \\
X^\varphi_{H_i} v(\lambda) &= -\varphi(\lambda, v(\lambda))w(\lambda)\left[\frac{u(\lambda)}{\lambda^{i+1}}\right]_+ + u(\lambda)\left[\varphi(\lambda, v(\lambda))\frac{w(\lambda)}{u(\lambda)}\left[\frac{u(\lambda)}{\lambda^{i+1}}\right]_+\right]_+,
\end{aligned}
\tag{2.15}
$$

upon using

$$\frac{\partial F}{\partial y}(\lambda, v(\lambda)) = 2v(\lambda).$$

From (2.15) we can compute $X^\varphi_{H_i} w(\lambda)$; observe how in this calculation the explicit dependence on F disappears completely!

$$
\begin{aligned}
X^\varphi_{H_i} w(\lambda) &= -X^\varphi_{H_i}\left[\frac{F(\lambda, v(\lambda))}{u(\lambda)}\right]_+ \\
&= -2\left[\frac{v(\lambda)}{u(\lambda)}X^\varphi_{H_i}v(\lambda)\right]_+ + \left[\frac{F(\lambda, v(\lambda))}{u(\lambda)}\frac{X^\varphi_{H_i}u(\lambda)}{u(\lambda)}\right]_+ \\
&= 2\left[v(\lambda)\left[\frac{w(\lambda)\varphi(\lambda, v(\lambda))}{u(\lambda)}\left[\frac{u(\lambda)}{\lambda^{i+1}}\right]_+\right]_-\right] - w(\lambda)\left[\frac{X^\varphi_{H_i}u(\lambda)}{u(\lambda)}\right]_+ \\
&= 2w(\lambda)\left[\varphi(\lambda, v(\lambda))\frac{v(\lambda)}{u(\lambda)}\left[\frac{u(\lambda)}{\lambda^{i+1}}\right]_+\right]_+ - 2v(\lambda)\left[\varphi(\lambda, v(\lambda))\frac{w(\lambda)}{u(\lambda)}\left[\frac{u(\lambda)}{\lambda^{i+1}}\right]_+\right]_+.
\end{aligned}
$$

This leads at once to the above Lax equations. The associated spectral curve is computed as follows:

$$
\begin{aligned}
\det(A(\lambda) - \mu\,\mathrm{Id}) &= \mu^2 - v^2(\lambda) + u(\lambda)\left[\frac{v^2(\lambda) - f(\lambda)}{u(\lambda)}\right]_+ \\
&= \mu^2 - f(\lambda) - u(\lambda)\left[\frac{v^2(\lambda) - f(\lambda)}{u(\lambda)}\right]_- \\
&= \mu^2 - f(\lambda) - H_{F,d}(u(\lambda), v(\lambda)).
\end{aligned}
$$

∎

For example, if we restrict ourselves to $d = 1$ (i.e., one degree of freedom), then $u(\lambda) = \lambda + u_0$, $v(\lambda) = v_0$ and

$$
\begin{aligned}
H_{F,1}(u_0, v_0) &= \big(v^2(\lambda) - f(\lambda)\big) \bmod u(\lambda) \\
&= \big(v_0^2 - f(\lambda)\big) \bmod (\lambda + u_0) \\
&= v_0^2 - f(-u_0),
\end{aligned}
$$

and $\{\cdot, \cdot\}_1^1$ is the standard bracket on $\mathbf{C}^2$, so we find that for $\varphi = 1$ the hyperelliptic case in one degree of freedom corresponds exactly to the case of polynomial potentials on the line.

84

3. The geometry of the level manifolds

In this section we determine the nature of the (general) level sets of $(\mathbf{C}^{2d}, \{\cdot,\cdot\}_d^\varphi, \mathcal{A}_{F,d})$. These level sets are the fibers of the momentum map $\mathbf{C}^{2d} \to \operatorname{Spec} \mathcal{A}_{F,d}$ where, as we have seen, $\operatorname{Spec} \mathcal{A}_{F,d}$ is isomorphic to $\mathbf{C}^d$. In particular they are independent of the polynomial $\varphi(x,y)$ which dictates the Poisson structure (the impact of $\varphi(x,y)$ presents itself only at the level of the integrable vector fields and is discussed in Section 3.5). It was generally believed that the general level set of an integrable Hamiltonian system with polynomial invariants is an (affine part of) a complex torus (Abelian variety) or an extensions of a complex torus by $\mathbf{C}^{*n}$ (see Chapters IV and V). It will turn out that the level sets encountered in these examples are of a different nature. We will also look at the real parts of the smooth fibers: whereas the real parts of Abelian varieties are quite special (see [Sil]), it will turn out that we find here a very rich class of topological types which appear as real parts of the fibers of the momentum map.

3.1. The real and complex level sets

Since $\operatorname{Spec}(\mathcal{A}_{F,d})$ is isomorphic to $\mathbf{C}^d$ and since the functions $H_0, \ldots, H_{d-1}$ are independent, the fibers over closed points are given by the level sets of $H_0, \ldots, H_{d-1}$ or equivalently by the level sets of $H_{F,d}(u(\lambda), v(\lambda))$. Since $H_{F,d}(u(\lambda), v(\lambda))$ is defined as $F(\lambda, v(\lambda)) \bmod u(\lambda)$, the fiber over an arbitrary polynomial $c(\lambda)$ of degree smaller than d is the same as the fiber over 0 for $H_{F',d}$, where $F'(x,y) = F(x,y) - c(x)$. Therefore it suffices to describe the fiber lying over 0 for all polynomials $F(x,y)$. We denote this fiber by $\mathcal{F}_{F,d}$; thus, by definition, $\mathcal{F}_{F,d}$ is given by

$$\mathcal{F}_{F,d} = \left\{ (u(\lambda), v(\lambda)) \in \mathbf{C}^{2d} \mid \left[\frac{F(\lambda, v(\lambda))}{u(\lambda)} \right]_- = 0 \right\}. \tag{3.1}$$

The *real level sets* are defined as follows: we denote the fixed point set of the complex conjugation map $\tau : \mathbf{C}^{2d} \to \mathbf{C}^{2d} : z \mapsto \bar{z}$ as $\operatorname{Fix}(\tau)$ and we define

$$\mathcal{F}_{F,d}^{\mathbf{R}} = \operatorname{Fix}(\tau) \cap \mathcal{F}_{F,d}. \tag{3.2}$$

$(\mathcal{F}_{F,d}, \tau)$ is a *real* algebraic variety (see [Sil]), whose real part is $\mathcal{F}_{F,d}^{\mathbf{R}}$; in fact, if $F(x,y)$ is a *real* polynomial then the level sets $\mathcal{F}_{F,d}^{\mathbf{R}}$ are nothing but the level sets of the corresponding real integrable Hamiltonian system (obtained by replacing in all definitions $\mathbf{C}$ by $\mathbf{R}$, see Paragraph II.4.2). We determine the non-singular real and complex fibers in the following two propositions.

Proposition 3.1 *If the algebraic curve $\Gamma_F \subset \mathbf{C}^2$ defined by $F(x,y) = 0$ is non-singular, then the fiber $\mathcal{F}_{F,d}^{\mathbf{R}} \subset \mathbf{C}^{2d}$ is also non-singular.*

Proof

$\mathcal{F}_{F,d}^{\mathbf{R}}$ will be smooth if and only if $H_{F,d}$ is submersive at each point of $\mathcal{F}_{F,d}^{\mathbf{R}}$, i.e., if and only if

$$\operatorname{Rk} \left(\frac{\partial H_i}{\partial u_{d-1}}, \ldots, \frac{\partial H_i}{\partial u_0}, \frac{\partial H_i}{\partial v_{d-1}}, \ldots, \frac{\partial H_i}{\partial v_0} \right)_{0 \le i \le d-1} = d, \quad \text{along } \mathcal{F}_{F,d}^{\mathbf{R}}.$$

From the proof of Proposition 2.3 and the definition of $\mathcal{F}^{\mathbf{R}}_{F,d}$, the j-th and $(d+j)$-th columns of this matrix are respectively given by

$$\lambda^j \frac{F(\lambda, v(\lambda))}{u(\lambda)} \bmod u(\lambda) \quad \text{and} \quad \lambda^j \frac{\partial F}{\partial y}(\lambda, v(\lambda)) \bmod u(\lambda).$$

It is therefore sufficient to show that if Γ_F is smooth then the dimension of the linear space

$$\left(R_1(\lambda) \frac{F(\lambda, v(\lambda))}{u(\lambda)} + R_2(\lambda) \frac{\partial F}{\partial y}(\lambda, v(\lambda)) \right) \bmod u(\lambda), \quad \deg R_i(\lambda) < d, \tag{3.3}$$

equals d. Let $\lambda_1, \ldots, \lambda_r$ be the distinct roots of $u(\lambda)$, λ_i having multiplicity s_i. We claim that

$$\frac{F(\lambda_i, v(\lambda_i))}{u(\lambda_i)} = 0 \quad \text{and} \quad \frac{\partial F}{\partial y}(\lambda_i, v(\lambda_i)) = 0 \tag{3.4}$$

cannot hold simultaneously if Γ_F is smooth. For otherwise $(x_i, y_i) = (\lambda_i, v(\lambda_i))$ would be a singular point of Γ_F: if (3.4) holds then clearly $\frac{\partial F}{\partial y}(x_i, y_i) = 0$, but also $F(x_i, y_i) = \frac{\partial F}{\partial x}(x_i, y_i) = 0$ because in this case $F(x, y_i)$ has a double zero at $x = x_i$.

The dimension of (3.3) is now investigated by using the fact that for any polynomial $p(\lambda)$, the value of $p(\lambda) \bmod u(\lambda)$ at λ_i is just $p(\lambda_i)$, and the values of the first $s_i - 1$ derivatives of $p(\lambda) \bmod u(\lambda)$ at λ_i are given by the values of the corresponding derivatives of $p(\lambda)$ at λ_i (s_i is the multiplicity of λ_i in $p(\lambda)$). Let us suppose that the different roots of $u(\lambda)$ are ordered such that $\lambda_1, \ldots, \lambda_t$ are also zeros of $\frac{\partial F}{\partial y}(\lambda, v(\lambda))$, while $\lambda_{t+1}, \ldots, \lambda_r$ are not. As a first restriction, let $R_1(\lambda)$ (resp. $R_2(\lambda)$) be such that its first $s_i - 1$ derivatives vanish at λ_i for $t + 1 \leq i \leq r$ (resp. $1 \leq i \leq t$). As a further restriction it is (by the first restriction and as (3.4) cannot happen) now easy to see that $R_1(\lambda)$ (resp. $R_2(\lambda)$) can be determined such that the polynomial given by (3.3) and the first $s_i - 1$ derivatives of (3.3) take any given values at λ_i for $1 \leq i \leq t$ (resp. $t + 1 \leq i \leq r$). These d conditions are independent, hence the dimension of (3.3) equals d and $\mathcal{F}^{\mathbf{R}}_{F,d}$ is smooth. ∎

Proposition 3.2 *The curve $\Gamma_F \subset \mathbf{C}^2$ is non-singular if and only if the fiber $\mathcal{F}_{F,d} \subset \mathbf{C}^{2d}$ is non-singular.*

Proof

If Γ_F has a singular point $P_1 = (x_1, y_1)$, choose for $i = 2, \ldots, d - 1$ an extra point $P_i = (x_i, y_i)$ on Γ_F and define $(u(\lambda), v(\lambda)) = \mathcal{S}((x_1, y_1), \ldots, (x_d, y_d)) \in \mathcal{F}_{F,d}$. All polynomials given by (3.3) vanish for $\lambda = x_1$, hence they span a linear space of dimension less than d. Thus $H_{F,d}$ is not submersive at $(u(\lambda), v(\lambda))$ and $\mathcal{A}_{F,d}$ is singular at this point. This shows the if part of the proposition; the only if part is proven verbatim as in the real case (Proposition 3.1). ∎

It will be seen that a clear understanding of the structure of the complex level sets $\mathcal{F}_{F,d}$ (for Γ_F smooth), leads also to a precise description of their real parts $\mathcal{F}^{\mathbf{R}}_{F,d}$.

3.2. The structure of the complex level manifolds

We will show that $\mathcal{F}_{F,d}$ is an affine part of the d-fold symmetric product $\mathrm{Sym}^d \Gamma_F$ of $\Gamma_F \subset \mathbf{C}^2$. Recall (e.g. from [Gun]) that $\mathrm{Sym}^d \Gamma_F$ is defined as the orbit space of the natural action of the permutation group S_d on the cartesian product $\Gamma_F^d = \Gamma_F \times \cdots \times \Gamma_F$ (d factors), i.e.,

$$\mathrm{Sym}^d \Gamma_F = \Gamma_F^d / S_d.$$

$\mathrm{Sym}^d \Gamma_F$ inherits its structure as an algebraic variety from the algebraic structure of Γ_F. Moreover if Γ_F is smooth then the same holds true for $\mathrm{Sym}^d \Gamma_F$: namely each point $P = \langle P_1^{m_1}, \dots, P_r^{m_r} \rangle \in \mathrm{Sym}^d \Gamma$ (with all P_i different; m_i is the multiplicity of P_i in P) has a neighborhood which is isomorphic to a neighborhood of $(\langle P_1^{m_1} \rangle, \dots, \langle P_r^{m_r} \rangle)$ in $\mathrm{Sym}^{m_1} \Gamma_F \times \cdots \times \mathrm{Sym}^{m_r} \Gamma_F$, and a point $\langle P_i^{m_i} \rangle$ on the diagonal of $\mathrm{Sym}^{m_i} \Gamma_F$ admits local coordinates given by the m_i elementary symmetric functions of the m_i coordinate functions on $\Gamma_F^{m_i}$.

Proposition 3.3 *If the algebraic curve Γ_F in $\mathbf{C}^2$, defined by $F(x,y) = 0$ is non-singular, then $\mathcal{F}_{F,d}$ is biholomorphic to the (Zariski) open subset of $\mathrm{Sym}^d \Gamma_F$, obtained by removing from it the divisor*

$$\mathcal{D}_{F,d} = \left\{ \langle P_1, \dots, P_d \rangle \mid \exists i,j : 1 \le i < j \le d, \left(\begin{array}{l} x(P_i) = x(P_j) \text{ with } P_i \ne P_j, \text{ or} \\ P_i = P_j \text{ is a ramification point of } x \end{array} \right) \right\}.$$

In particular $\mathcal{F}_{F,d}$ is irreducible.

Proof

- *Construction of the map $\phi_{F,d} : \mathcal{F}_{F,d} \to \mathrm{Sym}^d \Gamma_F \setminus \mathcal{D}_{F,d}$*

Given a point $(u(\lambda), v(\lambda)) \in \mathcal{F}_{F,d}$, a point in $\mathrm{Sym}^d \Gamma_F$ is associated to it as follows: for every root λ_i of $u(\lambda)$ one has $F(\lambda_i, v(\lambda_i)) = 0$, because $\left[\frac{F(\lambda, v(\lambda))}{u(\lambda)} \right]_- = 0$, so each root λ_i of $u(\lambda)$ determines a point $(\lambda_i, v(\lambda_i))$ on Γ_F. Thus there corresponds to $(u(\lambda), v(\lambda)) \in \mathcal{F}_{F,d}$ an unordered set of d points $\langle P_1, \dots, P_d \rangle \in \mathrm{Sym}^d \Gamma_F$, where P_i is defined by $(x(P_i), y(P_i)) = (\lambda_i, v(\lambda_i))$. Clearly, if $x(P_i) = x(P_j)$ then $P_i = P_j$; therefore, to show that $\langle P_1, \dots, P_d \rangle$ stays away from $\mathcal{D}_{F,d}$ we only need to prove that $P_i = P_j$ cannot occur for $i \ne j$ if P_i is a *ramification point* for x, i.e., if $y(P_i)$ is a multiple root of $F(x(P_i), y)$ (as a polynomial in y). As $P_i = P_j$ ($i \ne j$) implies that $u(\lambda)$ has a multiple root $x(P_i)$, in such a case $F(x, y(P_i))$ would have a multiple root $x = x(P_i)$, again because $\left[\frac{F(\lambda, v(\lambda))}{u(\lambda)} \right]_- = 0$. If moreover P_i is a ramification point of x then also $\frac{\partial F}{\partial y}(x(P_i), y(P_i)) = 0$ and it follows that $(x(P_i), y(P_i))$ is a singular point of Γ_F, a contradiction.

- *$\mathcal{D}_{F,d}$ is a divisor on $\mathrm{Sym}^d \Gamma_F$*

This means that $\mathcal{D}_{F,d}$ is given locally as the zero locus of a holomorphic function. If $\langle P_1, \dots, P_g \rangle \in \mathcal{D}_{F,d}$ let the set of indices $\{1, \dots, d\}$ be decomposed as $S_1 \cup \cdots \cup S_n$, such that all points P_i, with i running through one of the subsets S_j, have the same x-coordinate, which is disjoint from the x-coordinates of the points which correspond to the other subsets. For each P_i ($i = 1, \dots, d$) let x_i denote the lifting of x to a small neighborhood of $\langle P_1, \dots, P_d \rangle$ (corresponding to the factor P_i). Then a local defining equation of $\mathcal{D}_{F,d}$ is given by

$$\prod_{i=1}^{n} \prod_{\substack{j,k \in S_i \\ j < k}} (x_j - x_k) = 0.$$

• $\phi_{F,d}$ *is a biholomorphism*

We first construct the inverse of $\phi_{F,d}$, which is closely related to the map $\mathcal{S}$, as given by (2.3). Let $\langle P_1, \ldots, P_d \rangle \in \operatorname{Sym}^d \Gamma_F \setminus \mathcal{D}_{F,d}$. Clearly $u(\lambda)$ is taken as

$$u(\lambda) = \prod_{i=1}^{d} (\lambda - x(P_i)). \tag{3.5}$$

If all $x(P_i)$ are different then $v(\lambda)$ is uniquely determined as the polynomial of degree $d - 1$ whose value at $\lambda = x(P_i)$ is $y(P_i)$, i.e., $v(\lambda)$ is given by

$$v(\lambda) = \sum_{l=1}^{d} y(P_l) \prod_{k \neq l} \frac{\lambda - x(P_k)}{x(P_l) - x(P_k)} \tag{3.6}$$

and is holomorphic there. If two values coincide, say $x(P_1) = x(P_2)$, then $P_1 = P_2$ is not a ramification point (since we stay away from $\mathcal{D}_{F,d}$), hence the equation $F(x, y) = 0$ can be solved uniquely as $y = f(x)$ in a neighborhood of $P_1 = P_2$. For P_1' and P_2' in this neighborhood, substitute

$$f(x(P_i')) = f(x(P_1)) + (x(P_i') - x(P_1)) \frac{df}{dx}(x(P_1)) + \mathcal{O}\left(x(P_1') - x(P_1)\right)^2, \qquad (i = 1, 2)$$

for y_1 and y_2 in (3.6), to obtain that $v(\lambda)$ has no poles as $P_1', P_2' \to P_1$, hence extends to a holomorphic function on the larger subset where at most two points coincide. Since the complement of this larger subset in $\operatorname{Sym}^d \Gamma_F \setminus \mathcal{D}_{F,d}$ is of codimension at least two, $v(\lambda)$ extends to a holomorphic function on $\operatorname{Sym}^d \Gamma_F \setminus \mathcal{D}_{F,d}$. It also follows that this holomorphic function is the inverse of $\phi_{F,d}$ on all of $\operatorname{Sym}^d \Gamma_F \setminus \mathcal{D}_{F,d}$: if the point P_i has multiplicity s_i, then the first $s_i - 1$ derivatives of $v(\lambda)$ at $x(P_i)$ coincide with those of $f(\lambda)$ at $x(P_i)$, hence $F(\lambda, y(P_i))$ has a zero of order s_i at $\lambda = x(P_i)$. Finally, the inverse of a holomorphic bijection between complex manifolds is always holomorphic (see [GH] Ch. 0.2), hence $\phi_{F,d}$ is a biholomorphism. Since the symmetric product of any non-singular curve is irreducible, the same holds true for $\mathcal{F}_{F,d}$. ∎

Having a biholomorphism between affine varieties does not imply that this is an isomorphism. However this is so in the present case, as can be proven by using Newton's interpolation Theorem. This was done by Mumford for a special case (where the curve is hyperelliptic of genus d); since his (quite long) proof applies verbatim to the general case it is not repeated here (see [Mum5] pp. 3.23 – 3.25).

Lemma 3.4 *Even if $\mathcal{F}_{F,d}$ is singular its dimension equals d.*

Proof

We may still associate to each point in $\mathcal{F}_{F,d}$ a divisor consisting of d points on the singular curve Γ_F defined by $F(x, y) = 0$. The d-fold symmetric product of such a curve is singular but is still of dimension d. Therefore it suffices to show that the constructed map is finite on a Zariski open subset. In fact, if we stay away from the singular points in both $\mathcal{F}_{F,d}$ and in $\operatorname{Sym}^d \Gamma_F$ then the proof of Proposition 3.3 applies to show that this map is bijective on these subsets. Thus all fibers (over closed points) have dimension d. ∎

Proposition 3.3 and Lemma 3.4 finish the proof of Proposition 2.3 upon applying Proposition II.3.7.

3.3. The structure of the real level manifolds

Since $\mathcal{F}_{F,d}^{\mathbf{R}}$ is given as $\mathcal{F}_{F,d} \cap \mathrm{Fix}(\tau)$, it consists of those polynomials $(u(\lambda), v(\lambda)) \in \mathcal{F}_{F,d}^{\mathbf{R}}$ whose coefficients are all real. We figure out what this means for the corresponding point in $\mathrm{Sym}^d \Gamma_F$.

Proposition 3.5 *Under the biholomorphism $\phi_{F,d}$, the real level manifolds $\mathcal{F}_{F,d}^{\mathbf{R}}$ correspond to the set of all unordered d-tuples of points $\langle P_1, \ldots, P_d \rangle$ on Γ_F, consisting only of real points $P_i \in \mathbf{R}^2 \cap \Gamma_F$ and complex conjugated pairs $P_i = \bar{P}_j$, each ramification point (of x) occurring at most once, and $x(P_i) = x(P_j)$ only if $P_i = P_j$. Moreover its manifold structure derives from the structure of the d-fold symmetric product of Γ_F.*

Proof

$u(\lambda)$ is real if and only if its roots consist only of real roots and roots which occur in complex conjugate pairs. Obviously, if $v(\lambda)$ is real, $v(\lambda)$ and its derivatives take complex conjugate values when evaluated at complex conjugate points (and real values at real points). Also, if a polynomial of degree smaller than d is specified in d points that are real or occur in complex conjugated pairs then that polynomial is real (it is sufficient to prove this in case in which all points are distinct, which is easily done by using (3.6)). Since $v(x_i) = y_i$, this means that the real polynomials $(u(\lambda), v(\lambda))$ on $\mathcal{F}_{F,d}$ correspond to those points $\langle P_1, \ldots, P_d \rangle$ in $\mathrm{Sym}^d \Gamma_F$ consisting of real points $P_i = (x(P_i), y(P_i)) \in \mathbf{R}^2$ and complex conjugated pairs $P_j = (x(P_j), y(P_j)) = \left(\overline{x(P_k)}, \overline{y(P_k)} \right) = \bar{P}_k$, but not belonging to $\mathcal{D}_{F,d}$, i.e., the multiplicity of each ramification point (of x) is at most one, and $x(P_i) = x(P_j)$ only if $P_i = P_j$. ∎

Proposition 3.5 can be used to obtain a precise description of the topology of the real level manifolds $\mathcal{F}_{F,d}^{\mathbf{R}}$, as we show now for $d = 2$ (for $d = 1$, $\mathcal{F}_{F,d}^{\mathbf{R}}$ is just $\Gamma_F \cap \mathbf{R}^2$, the real part of Γ_F). For a fixed F such that Γ_F is smooth, let the connected components of $\Gamma_F \cap \mathbf{R}^2$ (if any) be denoted by $\Gamma_1, \ldots, \Gamma_s$ and define for $1 \leq i, j, k \leq s$ and $i < j$,

$$\Gamma_{00} = \{ \langle P, \bar{P} \rangle \mid P \in \Gamma_F, \, x(P) \notin \mathbf{R} \},$$
$$\Gamma_{ij} = \{ (P_1, P_2) \in \Gamma_i \times \Gamma_j \mid x(P_1) = x(P_2) \Rightarrow P_1 = P_2 \},$$
$$\Gamma_{kk} = \left\{ \langle P_1, P_2 \rangle \in \frac{\Gamma_k \times \Gamma_k}{S_2} \mid x(P_1) = x(P_2) \Rightarrow (P_1 = P_2 \text{ and is not a ramification point}) \right\}.$$

Then the union of Γ_{00} with all the sets Γ_{ij} and Γ_{kk} is easy identified with $\phi_{F,2}(\mathcal{A}_{F,2})$, the surface to be described. One observes that the only paths in it which are not contained in $\mathbf{R}^2$, are in Γ_{00}, in fact Γ_{00} connects exactly the surfaces Γ_{kk}. Therefore, if $i \neq j$ then Γ_{ij} is not connected to any other Γ_{mn}, Γ_{kk}, nor to Γ_{00}.

Therefore we first concentrate on such a subset Γ_{ij}, say on Γ_{12}. If the intervals $x(\Gamma_1)$ and $x(\Gamma_2)$ are disjoint, then $\Gamma_{12} = \Gamma_1 \times \Gamma_2$, so Γ_{12} is either homeomorphic to a torus, a cylinder or a disc, depending on whether the components Γ_1 and Γ_2 are closed or open. If $x(\Gamma_1)$ and $x(\Gamma_2)$ have a point P in common, then one finds again these surfaces, but with a number of punctures (holes), equal to

$$\prod_{i=1}^{2} \#\{ Q \in \Gamma_i \mid x(Q) = x(P) \}.$$

If $x(\Gamma_1)$ and $x(\Gamma_2)$ have an interval in common, Γ_{12} may even disconnect in different pieces. The structure of these pieces is easily determined from a picture of the real part of the curve. Namely, on a square representing $\Gamma_1 \times \Gamma_2$, the divisor $\{(P_1, P_2) \in \Gamma_i \times \Gamma_j \mid x(P_1) = x(P_2)\}$ is drawn by counting points on the vertical lines $x = $ constant, the only care one needs to take is that if Γ_1 (or Γ_2) is closed, then an origin should be marked on it, and if one passes this origin, one needs to pass over the corresponding edge of the rectangle. The following table shows some examples (all possibilities for which Γ_1 and Γ_2 are closed, and for which x is $2:1$ when restricted to Γ_1 and Γ_2).

Γ_1 and Γ_2	Divisor	Component Γ_{12}	Picture
		torus	
		torus minus point	
		(torus minus disc) + disc	
		two cylinders	
		cylinder + disc	
		two discs	

Table 2

In the same way Γ_{kk} is investigated by drawing the divisor

$$\left\{ (P_1, P_2) \in \Gamma_i \times \Gamma_i \mid x(P_1) = x(P_2) \text{ and } \begin{pmatrix} P_1 \neq P_2 \text{ or} \\ P_1 = P_2 \text{ is a ramification point of } x \end{pmatrix} \right\}.$$

on a rectangle representing $\Gamma_i \times \Gamma_i$. Either triangle cut off from the rectangle by its main diagonal then represents $\frac{\Gamma_i \times \Gamma_i}{S_2}$ and Γ_{ii} is the complement of the divisor in the triangle.

For example, consider a component Γ_1 as in Figure 1.a below. Then Figure 1.b shows a torus with a circle on it (the anti-diagonal of the rectangle), which is the divisor $\mathcal{D}$ to be removed. The resulting piece Γ_{11} is drawn in Figure 1.c and is redrawn in a simpler way in Figure 1.d. For every Γ_i such a piece is found and will be glued to Γ_{00} precisely along

90

the part of its boundary which comes from the diagonal in the rectangle (the solid lines in Figure 1.d).

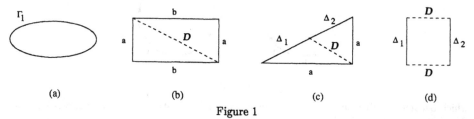

(a)   (b)   (c)   (d)

Figure 1

In order to explain how Γ_{00} is described, we recall the classical picture of a (smooth, complete) algebraic curve $\bar{\Gamma}$. An equation $F(x, y) = 0$ of such a curve defines a $m : 1$ ramified covering map to $\mathbf{P}^1$ by $(x, y) \mapsto x$, when m is the degree of $F(x, y)$ in y. This may be visualized by drawing concentric spheres (called *sheets*), on which there are marked some non-intersecting intervals (called *cuts*, every cut is equally present on all sheets). The topology is such that if you are walking on a sheet i and pass a cut j (from a fixed side) then you move to a sheet $p_j(i)$, each p_j being a permutation of $\{1, \ldots, m\}$. It is clear that the datum of cuts and their corresponding permutations determines the topology of the curve completely. Since each cut connects two ramification points (of x), these cuts may, for a real curve, be taken on the real axis and orthogonal to it.

Γ_{00} is now given as follows. Consider the described picture for the smooth completion $\bar{\Gamma}_F$ of Γ_F. Clearly the conjugation map interchanges the upper and lower hemispheres and is fixed on the equator(s) $\{P \in \bar{\Gamma}_F \mid x(P) \in \mathbf{R} \cup \infty\}$. It follows that the open upper (lower) hemispheres give precisely Γ_{00}. A convenient way to represent them is by drawing a disc for each upper hemisphere and labeling the different parts of the boundary which correspond to the horizontal and vertical cuts. A moment's thought reveals that the different sheets are to be connected along those lines which correspond to the vertical cuts, while the pieces Γ_{kk} are to be connected to the corresponding horizontal cuts. This gives a topological model of $\Gamma_{00} \cup \bigcup_{k=1}^{s} \Gamma_{kk}$ as a disc with holes. The following example may highlight the different steps.

Example 3.6 We consider a hyperelliptic curve $F(x, y) = y^2 - f(x) = 0$, where f has five real zeros and one pair of complex conjugate zeros. The curve has genus three and its graph and related representation as a cover of $\mathbf{P}^1$ are given by

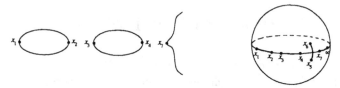

Figure 2

where the imaginary ramification points (of x) are not seen from the graph. For Γ_{00} we get two upper hemispheres

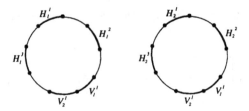

Figure 3

which become one disc after gluing the vertical cut V_1^1, V_2^1. We also get two subsets Γ_{11} and Γ_{22} given as in Figure 1.d by

Figure 4

and one disconnected piece Γ_{33} (since $\infty \in \Gamma_3$).

Figure 5

Now glue Figures 3, 4 and 5 according to their labels H_j^i to find a disc with two holes. Since the other components of $\mathcal{A}_{F,2}$ are direct products of the real components, we find that

$$\mathcal{A}_{F,2} \simeq \text{one torus + two cylinders + one disc with two holes.}$$

It is shown in the same way that, if $F(x, y)$ is of the form

$$F(x, y) = y^2 + \prod_{i=1}^{n}(x - \alpha_i) \prod_{j=1}^{m} \left(x^2 + \beta_j^2\right),$$

with $\alpha_i, \beta_j \in \mathbf{R}$ (all α_i being different, as well as all β_i^2), then

$$\mathcal{A}_{F,2} \simeq \binom{(n-1)/2}{2} \text{ tori } + \frac{n-1}{2} \text{ cylinders + one disc with } g-1 \text{ holes} \qquad \text{if } n \text{ is odd,}$$
$$\mathcal{A}_{F,2} \simeq \binom{n}{2} \quad \text{ tori } + \qquad\qquad\qquad \text{one disc with } g \text{ holes} \qquad \text{if } n \text{ is even,}$$

where g is the genus $\left[\frac{n+1}{2}\right] + m - 1$ of the curve $F(x, y) = 0$.

3.4. Compactification of the complex level manifolds

We now discuss the (smooth) compactification of the manifolds $\mathcal{F}_{F,d}$. There is one obvious and natural compactification, namely the compact manifold $\text{Sym}^d \bar{\Gamma}_F$, defined in a similar way as $\text{Sym}^d \Gamma_F$; as above $\bar{\Gamma}_F$ denotes the smooth compactification of Γ_F. However $\mathcal{F}_{F,d}$ has the disadvantage that none of the vector fields X_{H_i} extends holomorphically to it — a compactification such that at least one of these vector fields extends in a holomorphic way to it, will simply be called *good*. The interest in good compactifications is that it allows one to integrate the corresponding vector fields in terms of theta functions, or degenerations of theta functions. The purpose of this paragraph is to show that even for very simple choices of $F(x,y)$, a good compactification of $\mathcal{F}_{F,d}$ does not exist. We believe that this is true for almost all choices of $F(x,y)$. A special class of examples for which a good compactification *does* exist is considered in Chapter V.

At first we compute, for fixed $F(x,y)$ how the vector fields X_{H_i} behave on the compact manifold $\text{Sym}^d \bar{\Gamma}_F$, which relates to $\mathcal{F}_{F,d}$ (which we identified with $\text{Sym}^d \Gamma_F \setminus \mathcal{D}_{F,d}$) as follows:

$$\text{Sym}^d \bar{\Gamma}_F = \mathcal{F}_{F,d} \cup \bar{\mathcal{D}}_{F,d} \cup \bar{\mathcal{E}}_{F,d}.$$

Here $\bar{\mathcal{D}}_{F,d}$ is the closure of $\mathcal{D}_{F,d}$ in $\text{Sym}^d \bar{\Gamma}_F$ and $\bar{\mathcal{E}}_{F,d}$ is a divisor whose irreducible components $\bar{\mathcal{E}}_{F,d}(\infty_i)$ correspond to the points ∞_i in $\bar{\Gamma}_F \setminus \Gamma_F$, namely

$$\bar{\mathcal{E}}_{F,d}(\infty_k) = \left\{ \langle \infty_k, P_2, \ldots, P_d \rangle \mid P_k \in \bar{\Gamma}_F \text{ for } 2 \leq k \leq d \right\}.$$

Each vector field X_{H_i} being a polynomial vector field on $\mathbf{C}^{2d}$, it is holomorphic on $\mathcal{F}_{F,d}$. We determine its behavior along the irreducible components of $\bar{\mathcal{D}}_{F,d}$ and $\bar{\mathcal{E}}_{F,d}$, which may be done by computing the order of vanishing of X_{H_i} at a generic point of each component, which in turn is done by using local coordinates at such a point.

Proposition 3.7 *Every vector field X_{H_i} has a simple pole along all irreducible components of $\bar{\mathcal{D}}_{F,d}$. It has a zero of order ρ_k along $\bar{\mathcal{E}}_{F,d}(\infty_k)$ (i.e., a pole of order $-\rho_k$ if $\rho_k < 0$), where*

$$\rho_k = \begin{cases} \mu_k - \nu_k + d + 1 & \text{if } \nu_k < 0, \\ \mu_k - \nu_k + 1 & \text{if } \nu_k > 0; \end{cases} \tag{3.7}$$

μ_k is the order (of vanishing) of $\frac{\partial F}{\partial y}(x,y)$ at ∞_k, and ν_k is the order of x at ∞_k (resp. the order of $x - x(\infty_k)$ if x is finite at ∞_k).

Proof

We first write down the vector field X_{H_i} at a generic point $(u(\lambda), v(\lambda)) \in \mathcal{F}_{F,d}$; the genericity condition taken here is that for $\phi_{F,d}(u(\lambda), v(\lambda)) = \langle (x_1, y_1), \ldots, (x_d, y_d) \rangle$ all x_i are different and none of the points (x_i, y_i) is a ramification point of x. Varying the point $(u(\lambda), v(\lambda))$ in a small neighborhood, each x_i gives a local coordinate on a neighborhood $U_i \subset \Gamma_F$ of (x_i, y_i) as well as a local coordinate on a neighborhood $U \subset \mathcal{F}_{F,d}$ of $\langle (x_1, y_1), \ldots, (x_d, y_d) \rangle$. Since on the one hand the derivative of $u(\lambda) = \prod_{k=1}^{d}(\lambda - x_k)$ at $\lambda = x_j$ is $X_{H_i}u(x_j) = -\prod_{l \neq j}(x_j - x_l) X_{H_i}x_j$, while on the other hand, direct substitution in (2.12) gives

$$X_{H_i}u(x_j) = \frac{\partial F}{\partial y}(x_j, y_j) \sum_{k=i+1}^{d} u_k x_j^{k-i-1},$$

we find that

$$X_{H_i} x_j = - \prod_{l \neq j} (x_j - x_l)^{-1} \frac{\partial F}{\partial y} (x_j, y_j) \sigma_{d-i-1}(\hat{x}_j), \tag{3.8}$$

where $\sigma_i(\hat{x}_j)$ is the i-th symmetric function in $x_1, \ldots, x_d$, evaluated at $x_j = 0$.

The right hand side of (3.8) has at a generic point of $\bar{\mathcal{D}}_{F,d}$ a simple pole, hence each vector field X_{H_i} has a simple pole on (every component of) $\bar{\mathcal{D}}_{F,d}$. The behavior of X_{H_i} along $\bar{\mathcal{E}}_{F,d}$ is slightly more complicated since it depends on $F(x,y)$, and may even behave differently on each component $\bar{\mathcal{E}}_{F,d}(\infty_k)$. For a generic point in a neighborhood of a point of $\bar{\mathcal{E}}_{F,d}(\infty_k)$, let us introduce coordinates x_i as above. If we denote by μ_k and ν_k the integers introduced in the statement of the proposition, then clearly x_1 is given in a neighborhood of ∞_k in terms of a local parameter t_1 at ∞_k as $x_1 = t_1^{\nu_k}$ ($\nu_k < 0$), or as $x_1 = c_1 + t_1^{\nu_k}$ ($\nu_k > 0$), depending on whether x is infinite in a neighborhood of ∞_k or has a finite value $c_1 \in \mathbf{C}$ at ∞_k; also $\frac{\partial F}{\partial y}(x_1(t), y_1(t)) = t^{\mu_k}(f_1 + \mathcal{O}(t))$ with $f_1 \neq 0$. We define for $2 \leq j \leq d$ local parameters t_j (centered at P_j, which may be assumed to be generic) by $x_j = x(P_j) + t_j$. Direct substitution in (3.8) yields

$$X_{H_i} t_1 = t_1^{\rho_k}(c_i + \mathcal{O}(t_1)),$$
$$X_{H_i} t_j = c_j + \mathcal{O}(t_1), \qquad (j = 2, \ldots, d);$$

where ρ_k is defined in (3.7). We conclude that X_{H_i} has a zero (resp. pole) of order $|\rho_k|$ along $\bar{\mathcal{E}}_{F,d}(\infty_k)$ if $\rho_k \geq 0$ (resp. $\rho_k < 0$). ∎

Thus we have shown that $\mathrm{Sym}^d \bar{\Gamma}_F$ is not a good compactification, since all vector fields X_{H_i} have at least a pole along $\bar{\mathcal{D}}_{F,d}$. This divisor can be contracted in some cases, as we will show in Chapter V. The following example shows that a good compactification does not exist in general.

Example 3.8 Let $F(x,y) = y^3 + f(x)$, where the degree of f is at least three, and let $d = 2$. To show that $\mathcal{F}^{\mathbf{C}}_{F,2}$ has no good compactification we use some results about algebraic surfaces which can be found in [Har] Ch. V. Suppose that $\bar{\mathcal{F}}$ is a good compactification of $\mathcal{F}^{\mathbf{C}}_{F,2}$ then $\bar{\mathcal{F}}$ and $\mathrm{Sym}^2 \bar{\Gamma}_F$ are birational; for surfaces this means that there exists a finite series of monoidal transformations (also known as blow-up's) which transforms $\bar{\mathcal{F}}$ into $\mathrm{Sym}^2 \bar{\Gamma}_F$. Then there exist (Zariski) open subsets $\mathcal{U} \subset \bar{\mathcal{F}}$ and $\mathcal{V} \subset \mathrm{Sym}^2 \bar{\Gamma}_F$ to which all these monoidal transformations restrict as isomorphisms and the vector fields on $\mathcal{U}$ and $\mathcal{V}$ correspond exactly under this isomorphism. In particular $\bar{\mathcal{D}}_{F,2}$ is entirely contained in the complement of $\mathcal{V}$ and must be contracted by one of the monoidal transformations, so at least we know that the genus of $\bar{\mathcal{D}}_{F,2}$ must be 0 (only $\mathbf{P}^1$'s can be contracted).

We may however compute the genus of $\bar{\mathcal{D}}_{F,2}$ directly. Recall that it consists of the points $\langle P_1, P_2 \rangle$ with $x(P_1) = x(P_2)$ for which $P_1 \neq P_2$ or $P_1 = P_2$ is a ramification point, so its smoothness is easy checked. However the map x expresses $\bar{\mathcal{D}}_{F,2}$ as a $3:1$ cover of $\mathbf{P}^1$, ramified at the $n = \deg f$ points (x_i, y_i) for which $f(x_i) = 0$ (and at infinity if n is not divisible by 3). So it has the same ramification divisor as $\bar{\Gamma}_F$, hence genus($\bar{\mathcal{D}}_{F,2}$) = genus($\bar{\Gamma}_F$) > 0, a contradiction.

Example 3.9 In the one-dimensional case ($d = 1$) the level manifolds are punctured Riemann surfaces and have a unique compactification. If the genus of such a Riemann surface exceeds one, then it supports no holomorphic vector fields, so for $d = 1$ good compactifications of $\mathcal{F}_{F,d}$ rarely exist.

As an application of (3.8), let us show how the vector fields can be integrated on these real level sets. Summing up (3.8) over all j (and for any φ) we find that for any fixed integers $i, r < d$,

$$\sum_{j=1}^{d} x_j^r \frac{X_{H_i}^\varphi x_j}{\varphi(x_j, y_j) \frac{\partial F}{\partial y}(x_j, y_j)} = -\sum_{j=1}^{d} \sum_{k=d-j+1}^{d} u_k x_j^{r-i+k-1} \prod_{l \neq j} \frac{1}{x_j - x_l} = -\delta_{i,r}.$$

Therefore the d functions

$$\chi_r = \sum_{i=0}^{d-1} \frac{x_j^r X_{H_i}^\varphi x_j}{\varphi(x_j, y_j) \frac{\partial F}{\partial y}(x_j, y_j)}, \qquad r = 0, \ldots, d-1, \tag{3.9}$$

have linear dynamics in time and lead to the integration of the vector fields X_{H_i} along the (connected components of the) real manifolds $\mathcal{F}_{F,d}^{\mathbf{R}}$.

3.5. The significance of the Poisson structures $\{\cdot, \cdot\}_d^\varphi$

We have constructed for any positive integer d and for any $F(x, y) \in \mathbf{C}[x, y] \setminus \mathbf{C}[x]$ a family of compatible integrable Hamiltonian systems $(\mathbf{C}^{2d}, \{\cdot, \cdot\}_d^\varphi, \mathcal{A}_{F,d})$ which is indexed by $\varphi(x, y) \in \mathbf{C}[x, y] \setminus \{0\}$. For a fixed $F(x, y)$ all vector fields in $\mathrm{Ham}(\{\cdot, \cdot\}_d^\varphi, \mathcal{A}_{F,d})$ are tangent to the d-dimensional fibers of the momentum map $\mathbf{C}^{2d} \to \mathrm{Spec}\,\mathcal{A}_{F,d}$. Since for a fixed φ these vector fields generate at a general point the tangent space, the vector fields obtained for one choice of φ can be written down in terms of another choice of φ. This relation between the vector fields of the different (compatible) integrable Hamiltonian systems is given explicitly by the following proposition.

Proposition 3.10 *For fixed $F(x, y)$ and d let H_i $(i = 0, \ldots, d-1)$ denote the polynomials on $\mathbf{C}^{2d}$, defined in (2.10), and let X_i^1 and X_i^φ denote their Hamiltonian vector fields with respect to $\{\cdot, \cdot\}_d$ and $\{\cdot, \cdot\}_d^\varphi$. Then the transfer matrix T_1^φ, which is defined by*

$$\left(X_{H_{d-1}}^\varphi, \ldots, X_{H_0}^\varphi \right) = \left(X_{H_{d-1}}^1, \ldots, X_{H_0}^1 \right) T_1^\varphi,$$

is given by

$$T_1^\varphi = \varphi(M, v(M)), \quad \text{where} \quad M = \begin{pmatrix} -u_{d-1} & -u_{d-2} & -u_{d-3} & \cdots & -u_0 \\ 1 & 0 & 0 & \cdots & 0 \\ 0 & 1 & 0 & \cdots & 0 \\ \vdots & \ddots & \ddots & \ddots & \vdots \\ 0 & \cdots & 0 & 1 & 0 \end{pmatrix}. \tag{3.10}$$

The general transfer matrices $T_{\varphi_1}^{\varphi_2}$ are immediately computed from (3.10) upon using the cocycle identities

$$T_{\varphi_1}^{\varphi_2} T_{\varphi_2}^{\varphi_1} = 1 \quad \text{and} \quad T_{\varphi_1}^{\varphi_2} T_{\varphi_2}^{\varphi_3} T_{\varphi_3}^{\varphi_1} = 1.$$

Proof

It suffices to express T_1^φ in the neighborhood of a generic point $(u(\lambda), v(\lambda))$ in terms of any local coordinates. At such a generic point all roots x_i of $u(\lambda)$ are different and serve as coordinates; we denote $v(x_i) = y_i$ as before. Then by (3.8),

$$X_H^\varphi = \Delta^\varphi X_H \qquad \text{with} \qquad \Delta^\varphi = \text{diag}(\varphi(x_1, y_1), \ldots, \varphi(x_d, y_d));$$

X_H^φ denotes the matrix with entries $(X_H^\varphi)_{ij} = X_{H_j}^\varphi x_i$ and $X_H = X_H^1$. The above formula implies that T_1^φ is given by

$$\begin{aligned}
T_1^\varphi &= (X_H)^{-1} \Delta^\varphi X_H \\
&\overset{(i)}{=} V \Delta^\varphi V^{-1} \\
&= V \varphi(\Delta^x, v(\Delta^x)) V^{-1} \\
&= \varphi\left(V \Delta^x V^{-1}, v(V \Delta^x V^{-1})\right) \\
&= \varphi(M, v(M)),
\end{aligned}$$

where $M = V \Delta^x V^{-1}$ is easily checked to have the form announced in (3.10). Step (i) requires some extra work (one uses (3.8)); also we have introduced the notation V for the Vandermonde matrix

$$V = \begin{pmatrix}
x_1^{d-1} & x_2^{d-1} & \cdots & x_d^{d-1} \\
x_1^{d-2} & x_2^{d-2} & \cdots & x_d^{d-2} \\
\vdots & \vdots & & \vdots \\
1 & 1 & \cdots & 1
\end{pmatrix}.$$

∎

Remark 3.11   In the special case where $d = 2$ and $\varphi(x, y) = x$ the transfer matrix has the simple form

$$\begin{pmatrix} -u_1 & -u_0 \\ 1 & 0 \end{pmatrix}$$

and we obtain what Caboz et al. call a (ρ, s) bi-Hamiltonian structure (see [CGR]). Our definition of compatible integrable Hamiltonian systems and Proposition 3.10 largely generalize and clarify this concept.

Chapter IV

Interludium: the geometry of Abelian varieties

1. Introduction

For the convenience of the reader who wishes to go on reading the rest of the text we include here a chapter about Abelian varieties. There is nothing new in this chapter, our intention was to give a compact and coherent presentation of the theory of Abelian varieties in a form suitable for applications to integrable Hamiltonian systems. Our exposition is partly algebraic partly analytic, we think that both approaches highlight different aspects of the theory of Abelian varieties, see for example the theorems of Abel, Jacobi and Riemann in Paragraph 4.3. Moreover, for applications to the theory of integrable systems a reasonable amount of understanding of both aspects and their interplay is needed. The main references for the theory of Abelian varieties are [LB], [Kem], [Mum2] and [Mum3]. However the relevant chapters in [ACGH], [GH] and [Mum4] are also highly recommended to learn this subject.

We start by recalling the basic definitions of divisors and line bundles (on a complex manifold) and recall how they are related. The sections of an ample line bundle are used to construct embeddings in projective space and the dimension of this space is computed from the Riemann-Roch Theorem. As an illustration we show that every compact Riemann surface can be embedded in some projective space, hence it is an algebraic curve. It is therefore common not to distinguish between compact Riemann surfaces and algebraic curves, the latter of course being assumed complete, non-singular, irreducible, reduced; we will also use these terms interchangeably, choosing the term compact Riemann surface or algebraic curve according to whether the analytic or the algebraic structure is more relevant.

In Section 3 we give the Riemann conditions which tell which complex tori are Abelian, i.e., can be embedded in projective space. The sections of an embedding line bundle are explicitly described by theta functions and their number is easily computed. A lot is simplified here because every effective divisor on an (irreducible) Abelian variety defines an ample line bundle and the third power of an ample line bundle on an Abelian variety always provides an embedding. In particular in the case of Abelian surfaces everything can be computed very explicitly (except the embedding itself but that is where we will use the theory of integrable systems for!), an ample divisor is then nothing but an embedded curve and its genus relates to the number of sections of the line bundle defined by this divisor.

We also treat in detail the Jacobian of an algebraic curve since it is the most important example of an Abelian variety. There are basically two (very different) ways to define it, there is an algebraic definition (using divisors on the curve) and an analytic/transcendental definition (using integration of differential forms over cycles). The fact that these correspond to the same object is a deep theorem, due to Abel and Jacobi. We could not resist to reproduce a proof of it here. We close Section 3 by discussing the Kummer surface of a Jacobi surface and its 16_6 configuration, which will be used in Section VII.4.

We use the following notation and terminology. On a complex manifold M we will use, aside from $\mathcal{O}$, its sheaf of holomorphic functions[9], and $\mathcal{M}$, its sheaf of meromorphic functions, also the constant sheaves $\mathbf{Z}, \mathbf{Q}, \mathbf{R}$ and $\mathbf{C}$ and the sheaves $\mathcal{O}^*, \mathcal{M}^*, \Omega^p$ defined for each open set U by

$\mathcal{O}^*(U)$ = the multiplicative group of non-zero holomorphic functions on U,

$\mathcal{M}^*(U)$ = the multiplicative group of non-zero meromorphic functions on U,

$\Omega^p(U)$ = the vector space of holomorphic p-forms on U.

We add M as a subscript in the notation when it is not clear from the context that we are talking about a sheaf on M.

---

[9] In this chapter $\mathcal{O}$ will only be used in this sense, so no confusion can arise with the notation $\mathcal{O}(M)$ for the ring of regular functions on an affine variety, which is used in other chapters.

2. Divisors and line bundles

In this section we discuss divisors, line bundles and the way in which they are related. We also explain how the holomorphic sections of a (very ample) line bundle lead to embeddings in projective space. We fix a compact complex manifold M of dimension n, the dimensions $n = 1$ and $n = 2$ being the most important for us.

2.1. Divisors

A subset V of M which is locally given by the zero locus of a holomorphic function is called an *analytic hypersurface*. V is called *reducible* if it is the union of at least two hypersurfaces; otherwise it is called *irreducible*. A *divisor* on M is a finite (formal) sum

$$\mathcal{D} = \sum a_i V_i \qquad (a_i \in \mathbf{Z})$$

of irreducible analytic hypersurfaces of M. Two divisors are added in the obvious way and the resulting group, which is called the *group of divisors on M*, is denoted by $\mathrm{Div}(M)$. If all a_i are positive the divisor is called *effective*. For curves ($n = 1$), the integer $\sum a_i$ is called the *degree* of $\mathcal{D}$ and is noted by $\deg \mathcal{D}$.

If V is an irreducible hypersurface then to any meromorphic function f there can be assigned an integer, "the order of vanishing of f along V" as follows. By definition, V has a local defining function for some neighborhood around any point P of V; if g is such a local defining function then this integer is the largest integer n for which g^n divides f in the ring of germs of holomorphic functions at P. This is well-defined because it turns out that this integer is independent of the point P and the local defining function g. We denote this integer by $\mathrm{ord}_V f$ and say that f has a *zero* (*pole*) of order a if $\mathrm{ord}_V f = a > 0$ ($\mathrm{ord}_V f = -a < 0$). Then a divisor (f) can be assigned to every meromorphic function f by

$$(f) = \sum_V \mathrm{ord}_V f \cdot V,$$

the (finite) sum running over all irreducible analytic hypersurfaces V for which $\mathrm{ord}_V f \neq 0$; it is not hard to show that $\deg(f) = 0$ for any meromorphic function f on a compact Riemann surface.

For a holomorphic top-form $\omega \in \Omega^n(M)$, $\mathrm{ord}_V \omega$ is defined as follows. If $\omega = f_\alpha dz_1 \wedge \ldots \wedge dz_n$ on some coordinate neighborhood U_α, then we define $\mathrm{ord}_V \omega = \mathrm{ord}_V f_\alpha$ for any irreducible analytic hypersurface V intersecting U_α (again this is well-defined because the transition functions lie in $\mathcal{O}^*$). Then the *divisor* of ω is the effective divisor

$$(\omega) = \sum_V \mathrm{ord}_V \omega \cdot V,$$

the (finite) sum running over all irreducible analytic hypersurfaces V for which $\mathrm{ord}_V \omega \neq 0$.

$\mathrm{Div}(M)$ admits a sheaf-theoretic interpretation, namely there is a natural isomorphism

$$\mathrm{Div}(M) \cong H^0(M, \mathcal{M}^*/\mathcal{O}^*).$$

99

To see this let $f = \{f_\alpha\}$ be a global section of $\mathcal{M}^*/\mathcal{O}^*$, then $\mathrm{ord}_V f_\alpha = \mathrm{ord}_V f_\beta$ because

$$\frac{f_\alpha}{f_\beta} \in \mathcal{O}^*(U_\alpha \cap U_\beta)$$

and the divisor associated to this section is naturally taken to be

$$\mathcal{D} = \sum \mathrm{ord}_V f_\alpha \cdot V,$$

where the sum is taken over all irreducible analytic hypersurfaces V for which $V \cap U_\alpha \neq 0$. Conversely, given a divisor, a global section of $\mathcal{M}^*/\mathcal{O}^*$ is defined over small opens by taking the product of the local defining functions of all irreducible analytic hypersurfaces appearing in the divisor (with the coefficients as exponents). Also $\mathrm{Div}(M)$ and $H^0(M, \mathcal{M}^*/\mathcal{O}^*)$ are easily seen to have the same group structure.

2.2. Line bundles

By a *line bundle* $\mathcal{L}$ on M we will always mean a holomorphic vector bundle of rank one over M. For any two overlapping open sets U_α and U_β the *transition functions* $g_{\alpha\beta}$: $U_\alpha \cap U_\beta \to \mathbf{C}^*$ of $\mathcal{L}$ are defined in terms of the *trivializations* $\phi_\alpha : \pi^{-1}(U_\alpha) \to U_\alpha \times \mathbf{C}$ by

$$g_{\alpha\beta}(z) = (\phi_\alpha \circ \phi_\beta^{-1})|_{\mathcal{L}_z} \in \mathbf{C}^*,$$

($\mathcal{L}_z$ is the fiber over $z \in U_\alpha \cap U_\beta$) and take values in $\mathbf{C}^*$; one has that $g_{\alpha\beta} \in \mathcal{O}^*(U_\alpha \cap U_\beta)$ and these transition functions satisfy the so-called *cocycle identities*

$$g_{\alpha\beta} g_{\beta\alpha} = 1,$$
$$g_{\alpha\beta} g_{\beta\gamma} g_{\gamma\alpha} = 1,$$

hence the transition functions represent a Čech 1-cochain which is actually a 1-cocycle because

$$(\delta\{g_{\alpha\beta}\})_{\gamma\delta\epsilon} = g_{\delta\epsilon} g_{\gamma\epsilon}^{-1} g_{\gamma\delta} = g_{\delta\epsilon} g_{\epsilon\gamma} g_{\gamma\delta} = 1,$$

(δ is the coboundary operator). It is a standard result that to any set of functions $\{g_{\alpha\beta}\}$ satisfying the cocycle identities there corresponds a line bundle with these functions as transition functions. Thus, the tensor product of two line bundles is again a line bundle and the set of all line bundles on M up to isomorphism becomes a commutative group, the *Picard group* of M, $\mathrm{Pic}(M)$. The inverse of a line bundle $\mathcal{L}$ is nothing but its dual and will be denoted by $\mathcal{L}^*$.

If other transition functions $\{h_{\alpha\beta}\}$ are given for $\mathcal{L}$, coming from trivializations χ_α, then there are functions $f_\alpha \in \mathcal{O}^*(U_\alpha)$, defined by $\chi_\alpha = f_\alpha \phi_\alpha$, such that

$$h_{\alpha\beta} = \frac{f_\alpha}{f_\beta} g_{\alpha\beta},$$

i.e., $g_{\alpha\beta} h_{\alpha\beta}^{-1}$ is a Čech coboundary. The upshot is that $\mathrm{Pic}(M) = H^1(M, \mathcal{O}^*)$ and both sets have the same group structure.

To every divisor there corresponds a line bundle in the following way. Let $\mathcal{D} \in \text{Div}(M)$ be locally defined by functions $f_\alpha \in \mathcal{M}^*(U_\alpha)$, then the corresponding line bundle $[\mathcal{D}]$ is defined by the transition functions

$$g_{\alpha\beta} = \frac{f_\alpha}{f_\beta},$$

and the definition is clearly independent of the choice of the local defining functions. The map

$$[\cdot] : \text{Div}(M) \to \text{Pic}(M)$$

is a homomorphism of groups, which is surjective if M is algebraic. Its kernel is given by

$$\text{Ker}[\cdot] = \{\mathcal{D} \in \text{Div}(M) \mid \exists f \in \mathcal{M}(M) \text{ for which } (f) = \mathcal{D}\},$$

because, on the one hand, the transition functions for $[(f)]$ are given by

$$g_{\alpha\beta} = f_\alpha / f_\beta = \frac{f|_{U_\alpha}}{f|_{U_\beta}},$$

and on the other hand, if the line bundle $[\mathcal{D}]$ is trivial, then the local defining functions f_α give rise to a function f for which $(f) = \mathcal{D}$, by defining $f|_{U_\alpha} = f_\alpha$ on U_α, which is well-defined when the local defining functions are chosen in such a way that the transition functions $g_{\alpha\beta} = 1$. If two divisors $\mathcal{D}$ and $\mathcal{D}'$ are defined to be linearly equivalent when $\mathcal{D} \sim_l \mathcal{D}'$, if $[\mathcal{D}] = [\mathcal{D}']$, then we see that if M is (in addition) an algebraic variety, then

$$\frac{\text{Div}(M)}{\sim_l} \cong \text{Pic}(M)$$

in a natural way. The map $[\cdot]$ corresponds under the identifications $\text{Pic}(M) \cong H^1(M, \mathcal{O}^*)$ and $\text{Div}(M) \cong H^0(M, \mathcal{M}^*/\mathcal{O}^*)$ to the connecting homomorphism δ^* in the long exact sequence

$$\cdots \to H^0(M, \mathcal{M}^*) \to H^0(M, \mathcal{M}^*/\mathcal{O}^*) \overset{\delta^*}{\to} H^1(M, \mathcal{O}^*) \to \cdots$$

coming from the short exact sequence $0 \to \mathcal{O}^* \to \mathcal{M}^* \to \mathcal{M}^*/\mathcal{O}^* \to 0$.

2.3. Sections of line bundles

Let $\mathcal{L}$ be a line bundle over M with projection $\pi : \mathcal{L} \to M$, vector bundle charts $\phi_\alpha : \pi^{-1}(U_\alpha) \to U_\alpha \times \mathbf{C}$ and transition functions $g_{\alpha\beta}$. A (holomorphic) *section* of $\mathcal{L}$ is a (holomorphic) map $s : M \to \mathcal{L}$ for which $\pi \circ s = 1_M$. The composition

$$\phi_\alpha \circ s|_{U_\alpha} : U_\alpha \to \pi^{-1}(U_\alpha) \to U_\alpha \times \mathbf{C}$$

defines a (holomorphic) map $s_\alpha : U_\alpha \to \mathbf{C}$, for each α. Two of these, say s_α and s_β, with $U_\alpha \cap U_\beta \neq 0$, are related by $s_\alpha = g_{\alpha\beta} s_\beta$. Conversely, a set of holomorphic maps $\{s_\alpha\}$ satisfying $s_\alpha = g_{\alpha\beta} s_\beta$ on each non-empty $U_\alpha \cap U_\beta$ determines a global holomorphic section of the line bundle $\mathcal{L}$. The *sheaf of holomorphic sections* of a line bundle $\mathcal{L}$ will be denoted by $\mathcal{O}(\mathcal{L})$ and the sheaf $\mathcal{O}(\mathcal{L}) \otimes \Omega^p$ will be denoted by $\Omega^p(\mathcal{L})$, the *sheaf of holomorphic differentials with values in $\mathcal{L}$*. We also define the *sheaf of meromorphic sections* of $\mathcal{L}$ as $\mathcal{O}(\mathcal{L}) \otimes \mathcal{M}$; then the meromorphic sections are given by meromorphic functions $\{s_\alpha\}$ satisfying $s_\alpha = g_{\alpha\beta} s_\beta$ on each non-empty

$U_\alpha \cap U_\beta$. Given two meromorphic sections s, t of $\mathcal{L}$, the local defining functions $\{s_\alpha\}$ and $\{t_\alpha\}$ satisfy $s_\alpha/t_\alpha = s_\beta/t_\beta$. Hence we can define a meromorphic function f (the "quotient" of s and t) by $f_\alpha = s_\alpha/t_\alpha$. For example, for $\mathcal{D} \in \mathrm{Div}(M)$, defined locally by $f_\alpha \in \mathcal{M}(U_\alpha)$, we have taken $[\mathcal{D}]$ to be the line bundle with transition functions $g_{\alpha\beta} = f_\alpha f_\beta^{-1}$, so that $f_\alpha = g_{\alpha\beta} f_\beta$. Therefore every divisor $\mathcal{D}$ determines a line bundle $[\mathcal{D}]$ and any set of local defining functions of $\mathcal{D}$ define a meromorphic section s_f of $[\mathcal{D}]$. It is customary to write $\mathcal{O}(\mathcal{D})$ for the sheaf $\mathcal{O}([\mathcal{D}])$.

Just as we did for meromorphic functions, we can associate a divisor to every meromorphic section of a line bundle by first defining, for any irreducible hypersurface $V \subset M$, $\mathrm{ord}_V s = \mathrm{ord}_V s_\alpha$ for any α for which $U_\alpha \cap V \neq 0$, and then setting

$$(s) = \sum_V \mathrm{ord}_V s \cdot V$$

where the sum runs over all irreducible hypersurfaces $V \subset M$. As before, ord_V is well-defined because the transition functions $g_{\alpha\beta} = s_\alpha/s_\beta$ are in $\mathcal{O}^*(U_\alpha \cap U_\beta)$. Taking up the previous example again, the divisor of s_f can be read off from the local defining functions f_α, giving $(s_f) = \mathcal{D}$. More generally, if s is any meromorphic section of $[\mathcal{D}]$, then $(s) \sim_l \mathcal{D}$ because the "quotient" of s_f and s defines a meromorphic function whose divisor is exactly $\mathcal{D} - (s)$. It follows that $[(s)] = [\mathcal{D}]$ and that there is for any divisor $\mathcal{D}' \sim_l \mathcal{D}$, a section s of $[\mathcal{D}]$ for which $(s) = \mathcal{D}'$. Notice that line bundles which come from divisors are exactly those which have (non-zero) meromorphic sections. Since a section s is holomorphic if and only if (s) is effective, it follows that a line bundle is the line bundle of an effective divisor if and only if it has a non-trivial global holomorphic section.

Actually, instead of working with the space $H^0(M, \mathcal{O}(\mathcal{D}))$ of holomorphic sections of a line bundle $[\mathcal{D}]$, one can work with the space $L(\mathcal{D})$ of meromorphic functions on M defined by

$$L(\mathcal{D}) = \{f \in \mathcal{M}(M) \mid (f) + \mathcal{D} \geq 0\}.$$

To see this, fix any meromorphic section s_0 of $[\mathcal{D}]$ for which $(s_0) = \mathcal{D}$. Then every holomorphic section $s \in H^0(M, \mathcal{O}(\mathcal{D}))$ corresponds to a meromorphic function f_s by taking the "quotient", $f_s = s/s_0$, and

$$(f_s) + \mathcal{D} = (s) - (s_0) + \mathcal{D} = (s) \geq 0,$$

so that $f_s \in L(\mathcal{D})$ and the proof of the converse is similar. We have established a non-canonical isomorphism

$$L(\mathcal{D}) \leftrightarrow H^0(M, \mathcal{O}(\mathcal{D})). \tag{2.1}$$

Denoting by $|\mathcal{D}|$ the set of all effective divisors linearly equivalent to $\mathcal{D}$, we have the additional correspondence

$$|\mathcal{D}| \cong \mathbf{P}(L(\mathcal{D})) \cong \mathbf{P}(H^0(M, \mathcal{O}(\mathcal{D})))$$

for compact manifolds M, because $\mathcal{D}' \in |\mathcal{D}|$ implies the existence of a function $f \in L(\mathcal{D})$ such that $\mathcal{D}' = \mathcal{D} + (f)$ and f is uniquely determined up to a constant. Therefore, $|\mathcal{D}|$ is a projective space; any subspaces of it are classically called *linear systems* and $|\mathcal{D}|$ itself is called a *complete linear system*. The common intersection of the divisors in a linear system is called the *base locus* of the system.

2.4. The Riemann-Roch Theorem

The calculation of the dimension of $H^0(M, \mathcal{O}(D))$ is done by using the Riemann-Roch Theorem. We will give this theorem here only for (algebraic) curves, postponing the analogous theorem for surfaces to a later paragraph. To give three equivalent versions of this theorem, we need some terminology. The *holomorphic Euler characteristic* $\chi(\mathcal{L})$ of a line bundle $\mathcal{L}$ over a compact complex manifold M is the integer

$$\chi(\mathcal{L}) = \sum_{p \geq 0} (-1)^p \dim H^p(M, \mathcal{O}(\mathcal{L})),$$

which is a topological invariant of $\mathcal{L}$ and M. In particular the holomorphic Euler characteristic of the trivial line bundle over M is denoted by $\chi(\mathcal{O}_M)$ and is a topological invariant of M. Also, every compact complex manifold M has another distinguished line bundle, its *canonical line bundle* K_M, defined as the line bundle corresponding to the divisor of any holomorphic top-form on M; this line bundle is well-defined because the "quotient" of two holomorphic top-forms defines a meromorphic function on M which establishes the linear equivalence of their divisors. The divisor of any holomorphic top-form is called a *canonical divisor* and will be denoted in the same way as its corresponding line bundle (although this divisor is only uniquely determined up to linear equivalence). Notice that the holomorphic sections of K_M correspond to the holomorphic top-forms on M, $\mathcal{O}(K_M) \cong \Omega_M^n$ (isomorphism of sheaves).

Then the *Riemann-Roch Theorem for curves* can be stated in the following equivalent forms.

Theorem 2.1 Let Γ be a compact Riemann surface of genus g, $\mathcal{D}$ be a divisor on Γ and K_Γ a canonical divisor on Γ. Then the following equivalent formulas hold:

(i) $\chi([\mathcal{D}]) = \chi(\mathcal{O}_\Gamma) + \deg \mathcal{D}$,
(ii) $\dim H^0(\Gamma, \mathcal{O}(\mathcal{D})) = \dim H^0(\Gamma, \Omega^1(-\mathcal{D})) + \deg \mathcal{D} - g + 1$,
(iii) $\dim H^0(\Gamma, \mathcal{O}(\mathcal{D})) = \dim H^0(\Gamma, \mathcal{O}(K_\Gamma - \mathcal{D})) + \deg \mathcal{D} - g + 1$.

Proof
Instead of proving the Riemann-Roch Theorem, we prefer to show the equivalence of the three formulas and to use this theorem to compute the degree of the canonical bundle on a curve.

First, the equivalence of (ii) and (iii) is obvious from the isomorphism $\mathcal{O}(K_\Gamma) \cong \Omega^1$. To show the equivalence between (i) and (ii) we need another fundamental theorem, the *Kodaira-Serre duality Theorem*, which we will formulate in a more general form, since we will also use it for studying sections of line bundles on surfaces.

Theorem 2.2 Let M be a compact complex manifold of dimension n, then the vector spaces $H^q(M, \Omega^p(\mathcal{L}))$ and $H^{n-q}(M, \Omega^{n-p}(\mathcal{L}^*))^*$ are isomorphic.

For the curve Γ, n equals 1 in this theorem. Taking $p = 0$ and a line bundle $\mathcal{L}$ on Γ, we find

$$H^q(\Gamma, \mathcal{O}(\mathcal{L})) = H^q(\Gamma, \Omega^0(\mathcal{L})) \cong H^{1-q}(\Gamma, \Omega^1(\mathcal{L}^*))^*, \qquad (2.2)$$

so that $H^q(\Gamma, \mathcal{O}(\mathcal{L})) = 0$ for $q > 1$. Therefore, the holomorphic Euler characteristic of $\mathcal{L}$, $\chi(\mathcal{L})$, equals

$$\chi(\mathcal{L}) = \dim H^0(\Gamma, \mathcal{O}(\mathcal{L})) - \dim H^1(\Gamma, \mathcal{O}(\mathcal{L})).$$

103

On the other hand, for $q = 1$ (2.2) gives $H^1(\Gamma, \mathcal{O}(\mathcal{L})) \cong H^0(\Gamma, \Omega^1(\mathcal{L}^*))^*$ so that

$$\chi(\mathcal{L}) = \dim H^0(\Gamma, \mathcal{O}(\mathcal{L})) - \dim H^0(\Gamma, \Omega^1(\mathcal{L}^*)).$$

Taking for $\mathcal{L}$ the line bundle $[\mathcal{D}]$ and the trivial line bundle over Γ, this gives respectively

$$\chi([\mathcal{D}]) = \dim H^0(\Gamma, \mathcal{O}(\mathcal{D})) - \dim H^0(\Gamma, \Omega^1(-\mathcal{D})),$$
$$\chi(\mathcal{O}_\Gamma) = \dim H^0(\Gamma, \mathcal{O}_\Gamma) - \dim H^0(\Gamma, \Omega^1) = 1 - \dim H^0(\Gamma, \Omega^1).$$

The equivalence of *(i)* and *(ii)* is established if we can show that $\dim H^0(\Gamma, \Omega^1) = g$. To show it, we consider the short exact sequence of sheaves

$$0 \to \mathbf{C} \to \mathcal{O} \to \Omega^1 \to 0,$$

which gives a long exact sequence in cohomology, namely

$$0 \to H^0(\Gamma, \mathbf{C}) \to H^0(\Gamma, \mathcal{O}) \to H^0(\Gamma, \Omega^1)$$
$$\to H^1(\Gamma, \mathbf{C}) \to H^1(\Gamma, \mathcal{O}) \to H^1(\Gamma, \Omega^1)$$
$$\to H^2(\Gamma, \mathbf{C}) \to 0,$$

because $H^2(\Gamma, \mathcal{O}) = 0$ from Kodaira-Serre duality. Also we know that

$$H^0(\Gamma, \mathbf{C}) \cong H^0(\Gamma, \mathcal{O}) \cong H^1(\Gamma, \Omega^1) \cong H^2(\Gamma, \mathbf{C}) \cong \mathbf{C}$$

and

$$H^1(\Gamma, \mathcal{O}) \cong H^0(\Gamma, \Omega^1).$$

Therefore, counting dimensions in the above exact sequence, we find

$$\dim H^0(\Gamma, \Omega^1) = \frac{1}{2} \dim H^1(\Gamma, \mathbf{C}) = g,$$

proving the claim. ∎

We proceed to calculate $\deg K_\Gamma$, the degree of the canonical bundle on a curve Γ. Taking $\mathcal{D} = K_\Gamma$ in the Riemann-Roch Theorem,

$$\dim H^0(\Gamma, \mathcal{O}(K_\Gamma)) = \dim H^0(\Gamma, \mathcal{O}) + \deg K_\Gamma - g + 1$$
$$= 2 - g + \deg K_\Gamma.$$

Using Kodaira-Serre duality again,

$$\deg K_\Gamma = \dim H^0(\Gamma, \mathcal{O}(K_\Gamma)) + g - 2$$
$$= \dim H^0(\Gamma, \Omega^1(K_\Gamma - K_\Gamma)) + g - 2$$
$$= \dim H^0(\Gamma, \Omega^1) + g - 2$$
$$= 2g - 2,$$

which shows that the degree of the canonical bundle equals $2(g - 1)$.

2.5. Line bundles and embeddings in projective space

One useful application of linear systems is to construct embeddings of compact complex manifolds in projective space, thereby realizing them as algebraic varieties. Let M be a compact complex manifold and $|\mathcal{D}|$ a complete linear system (one proceeds in the same way for general linear systems). If the base locus of $|\mathcal{D}|$ is empty then, for every $p \in M$, the set of sections vanishing at p forms a hyperplane H_p in $\mathbf{P}(H^0(M, \mathcal{O}(\mathcal{D})))$, so we get a map

$$\imath_{[\mathcal{D}]} : M \to \mathbf{P}(H^0(M, \mathcal{O}(\mathcal{D})))^*$$
$$p \mapsto H_p.$$

If $\imath_{[\mathcal{D}]}$ is an embedding then $[\mathcal{D}]$ is called a *very ample* line bundle. If $\mathcal{L}$ is a line bundle and $\mathcal{L}^k$ is very ample for some $k > 0$ then $\mathcal{L}$ is called *ample*. A necessary condition for $\imath_{[\mathcal{D}]}$ to be an embedding is expressed in the following theorem (Kodaira's Theorem).

Theorem 2.3 *Let M be a compact complex manifold and $\mathcal{L}$ a positive line bundle. Then there exists an integer k_0 such that for $k \geq k_0$ the map*

$$\imath_{\mathcal{L}^k} : M \to \mathbf{P}^N$$

is a well-defined embedding of M in $\mathbf{P}^N = \mathbf{P}(H^0(M, \mathcal{O}(\mathcal{L}^k)))^*$. In the language of divisors, if $[\mathcal{D}]$ is positive, then for some $k \in \mathbf{N}$, the functions with a pole of order at most k along $\mathcal{D}$ will provide an embedding of M into projective space. ∎

It is plain from the theorem that the positivity of $\mathcal{L}$ is crucial. A line bundle $\mathcal{L}$ is called *positive* if there exists a metric on $\mathcal{L}$ with curvature form Θ such that $\frac{i}{2\pi}\Theta$ is a positive $(1,1)$-form (see [GH] Ch. 1.2). Actually, the condition that the line bundle is positive turns out to be a topological condition and Kodaira's Theorem can be reformulated as follows.

Theorem 2.4 *A compact complex manifold M is an algebraic variety if and only if it has a closed, positive $(1,1)$-form ω whose cohomology class $[\omega]$ is rational, i.e., $[\omega] \in H^2(M, \mathbf{Q})$. Such a $(1,1)$-form is called a Hodge form.*

As an application of Kodaira's embedding Theorem, we show that every compact Riemann surface is an algebraic curve, thereby justifying our terminology. Let g be any Hermitian metric on the compact Riemann surface Γ and ω the associated $(1,1)$-form $\Im g$. Multiplying ω by a constant if necessary, we may suppose that $\int_\Gamma \omega = 1$, i.e., $[\omega]$ is an integral cohomology class. It follows that Γ can be embedded in projective space by using the meromorphic functions on Γ with poles at one point only. By Chow's Theorem, the embedded curve $\bar{\Gamma}$ is given by the zero locus of a set of homogeneous polynomials. Since the variety of chords of $\bar{\Gamma}$ has dimension three, we can project $\bar{\Gamma}$ to a hyperplane in this projective space and we can repeat this process, until we finally obtain an embedding of Γ in $\mathbf{P}^3$. In general Γ cannot be embedded in $\mathbf{P}^2$; however, since the set of tangents to the embedded curve has only dimension two, we can project the curve birationally to a curve $\tilde{\Gamma}$ whose only singularities are isolated (ordinary) double points, which means that the map $\Gamma \to \tilde{\Gamma}$ is 1 : 1, except in a finite number of points. Conversely, it can be shown that every irreducible algebraic curve $\tilde{\Gamma}$ has a *normalization*, i.e., there exists a compact Riemann surface Γ (which is unique up to biholomorphism) and a map $\nu : \Gamma \to \tilde{\Gamma}$, which is 1 : 1 except at isolated points.

2.6. Hyperelliptic curves

Most compact Riemann surface of genus $g \geq 2$ can be embedded in projective space in a canonical way, namely by using the canonical bundle of the Riemann surface. Suppose Γ has genus $g \geq 2$ and let $\{\omega_1, \ldots, \omega_g\}$ be a basis of $\Omega^1(\Gamma)$. In local coordinates, we can write $\omega_i = f_i(z)dz$, and the canonical mapping $\imath_K$ is given by

$$\imath_K : \Gamma \to \mathbf{P}^{g-1}$$
$$P \mapsto (f_1(P) : \cdots : f_g(P)).$$

The complete linear system has no base points by the Riemann-Roch Theorem, hence $\imath_K$ is an embedding when it is injective and immersive. The compact Riemann surfaces for which the above map is not an embedding are called *hyperelliptic* (compact Riemann surfaces of genus 1 being called *elliptic*). Hyperelliptic curves will be the most interesting for us because of the following theorem.

Theorem 2.5   *Let Γ be a compact Riemann surface of genus $g \geq 2$.*

(1) *Γ is hyperelliptic if and only if it has a non-constant meromorphic function f for which $(f) + P + Q \geq 0$, for some points $P, Q \in \Gamma$;*
(2) *Γ is hyperelliptic if and only if Γ is the normalization of an algebraic curve, given by an affine equation of the form $y^2 = f(x)$, where f is a monic polynomial of degree $2g+1$ or $2g+2$ without multiple roots;*
(3) *If $g = 2$ then Γ is hyperelliptic.*

Proof
    The map $\imath_K$ is $1 : 1$ and immersive if and only if for any two points $P, Q \in \Gamma$, there is a holomorphic differential ω for which $\omega(P) \neq \omega(Q)$ and there is an ω' vanishing once at P exactly. These conditions are equivalent to $\dim H^0(\Gamma, \mathcal{O}(K - P - Q)) < \dim H^0(\Gamma, \mathcal{O}(K - P))$ for any $P, Q \in \Gamma$. This reduces to $\dim H^0(\Gamma, \mathcal{O}(P + Q)) < 2$ upon using $\dim H^0(\Gamma, \mathcal{O}(K - P)) = g - 1$ and

$$\dim H^0(\Gamma, \mathcal{O}(K - P - Q)) = g - 3 + \dim H^0(\Gamma, \mathcal{O}(P + Q)).$$

Since the constants obviously belong to $L(P + Q)$ we conclude that a compact Riemann surface is hyperelliptic if and only if it admits a meromorphic function with only two poles. This proves (1).

    As for (2), Consider the map from Γ to $\mathbf{P}^1$ constructed in (1). It is a $2 : 1$ map and by an elementary count (the *Riemann-Hurwitz formula*) the number of *branch points* of this map (i.e., the number of points where the map is $1 : 1$) equals $2g - 2$. Denoting these points by $x_1, \ldots, x_{2g+2}$ let

$$\Gamma' = \left\{ (x, y) \in \mathbf{C}^2 \mid y^2 = \prod(x - x_i) \right\},$$

where the sum runs over all i for which $x_i \neq \infty$. The projection map $x : \Gamma' \to \mathbf{C}$ is seen to extend to a holomorphic map $\bar{\Gamma} \to \mathbf{P}^1$, when $\bar{\Gamma}$ is obtained from Γ' by adding one or two points "at infinity". Finally $\bar{\Gamma}$ is shown to be a compact Riemann surface isomorphic to Γ.

    To show (3), take Γ of genus 2 and substitute $\mathcal{D} = K$ in the Riemann-Roch Theorem. Since $\deg K = 2g - 2 = 2$, we find

$$\dim H^0(\Gamma, \mathcal{O}(K)) = 1 + 2 - 2 + 1 = 2.$$

We conclude the proof by applying part (1) of the theorem to $K = P + Q$.   ∎

2. Divisors and line bundles

In more general terms, a point P on a compact Riemann surface of genus g is called a *Weierstrass point* if there exists a function which has a pole of order at most g at P and which is holomorphic elsewhere. It follows that a hyperelliptic Riemann surface of genus g has $2g + 2$ Weierstrass points which are the points for which there is a function with a double pole in one point only. Notice that these $2g + 2$ points are intrinsically defined.

Finally, it is easy to check that on a compact hyperelliptic Riemann surface Γ, whose corresponding algebraic curve is given by an equation $y^2 = f(x)$, the g differentials

$$\omega_i = \frac{x^{i-1}dx}{y}$$

are holomorphic. Since they are independent, they form a basis of the vector space of all holomorphic differentials on the Riemann surface. A hyperelliptic Riemann surface has a (holomorphic) involution (which is unique), the *hyperelliptic involution*, which is given by $(x, y) \mapsto (x, -y)$ when an equation for the corresponding algebraic curve is written as $y^2 = f(x)$. Hyperelliptic Riemann surfaces and their Jacobians will be dealt with in detail in Chapter VI. They will also appear frequently in Chapter VII.

3. Abelian varieties

3.1. Complex tori and Abelian varieties

If Λ is any lattice (i.e., a discrete subgroup of maximal rank) in a complex vector space V of dimension g, then the quotient

$$\mathcal{T}^g = V/\Lambda$$

is a compact complex manifold, called a *complex torus*. It is a commutative complex Lie group, but in general it is not an algebraic variety; a complex torus which is at the same time a projective algebraic variety, i.e., which can be embedded in projective space, is called an *Abelian variety*. The conditions on Λ for the derived torus $\mathcal{T}^g$ to be an algebraic variety, which are computed from Theorem 2.4, are expressed by the famous *Riemann conditions*:

Theorem 3.1 $\mathcal{T}^g = V/\Lambda$ *is an Abelian variety if and only if there exists an integral basis* $\{\lambda_1, \ldots, \lambda_{2g}\}$ *of* Λ *and a complex basis* $\{e_1, \ldots, e_g\}$ *of* V *such that the matrix of* Λ *with respect to* $\{e_1, \ldots, e_g\}$ *is given by*

$$\Lambda = (\Delta_\delta \, Z),$$

with $\Delta_\delta = \operatorname{diag}(\delta_1, \ldots, \delta_g)$ *a diagonal matrix whose diagonal elements are positive integers satisfying* $\delta_i | \delta_{i+1}$ *and* Z *a symmetric matrix whose imaginary part,* $\Im(Z)$, *is positive definite. In terms of coordinates* $x_1, \ldots, x_{2g}$ *dual to this basis of* Λ, *the Hodge form* ω *is given by*

$$\omega = \sum_{i=1}^{g} \delta_i dx_i \wedge dx_{i+g}.$$

The Hodge form ω and its cohomology class $[\omega]$ carry extra information about particular embeddings of V/Λ in projective space. Namely, up to an integral multiple, ω is obtained from an embedding

$$\imath : V/\Lambda \to \mathbf{P}^N$$

as $\imath^* \Omega|_{\imath(V/\Lambda)}$, where Ω is the associated $(1,1)$-form associated to the standard Kähler structure of $\mathbf{P}^N$. Different embeddings may lead to classes $[\omega]$ which are different, even up to a multiple; the cohomology class $[\omega]$ of ω on an Abelian variety is called a *polarization* and the pair $(V/\Lambda, [\omega])$ is called a *polarized Abelian variety*. When the embedding is done by using the sections of a (very ample) line bundle $\mathcal{L}$, as explained in Paragraph 2.5, then the polarization is precisely the Chern class of $\mathcal{L}$ and any polarization is the Chern class of an embedding line bundle (Theorem 3.3 below).

The importance of considering polarized Abelian varieties, rather than Abelian varieties, stems also from the fact that their moduli spaces are simpler. They break up in components, given by the polarization type as follows. The integers δ_i in $\omega = \sum \delta_i dx_i \wedge dx_{i+g}$ are invariants of the cohomology class of ω and are called the *elementary divisors* of the polarization and $(V/\Lambda, [\omega])$ is said[10] to have *polarization type* or *type* $(\delta_1, \ldots, \delta_g)$; if the elementary divisors of ω are all one then ω is said to define a *principal polarization* and $(V/\Lambda, [\omega])$ (or just V/Λ) is called a *principally polarized Abelian variety*. Thus the moduli space of all polarized Abelian

---

[10] We will often use the common abbreviation "Abelian variety of type $(\delta_1, \ldots, \delta_g)$" instead of "polarized Abelian variety of type $(\delta_1, \ldots, \delta_g)$".

varieties breaks up in a natural way in components indexed by the polarization type and are thus studied separately for each polarization type.

A holomorphic map between Abelian varieties is a group homomorphism followed by a translation. If such a group homomorphism is surjective with finite kernel then it is called an *isogeny*. Surjectivity is almost automatic in view of the following theorem, known as the *Poincaré reducibility Theorem*.

Theorem 3.2 *If $\mathcal{T}^g$ is an Abelian variety which contains a non-trivial Abelian subvariety A then there exists an Abelian subvariety B of $\mathcal{T}^g$ such that $A \cap B$ is a finite subgroup and such that there exists a surjective homomorphism $A \oplus B \to \mathcal{T}^g$ whose kernel equals $A \cap B$: up to an isogeny such an Abelian variety is a product of Abelian varieties.*

Abelian varieties which contain a non-trivial subtorus are called *reducible Abelian varieties* and it follows e.g. from an easy dimension count that a general Abelian variety is irreducible. Interesting isogenies, used below, are obtained by the following: any polarized Abelian variety is isogenous to a principally polarized Abelian variety (but not in a unique way). Another interesting isogeny is an isogeny between a polarized Abelian variety and its dual which will be described in the next paragraph.

3.2. Line bundles on Abelian varieties

The Riemann conditions give us necessary and sufficient conditions for a complex torus to be an Abelian variety and Kodaira's Theorem says that the embedding can be done using the sections of a positive line bundle. The positive line bundles on an Abelian variety can be described very explicitly and a basis of the space of holomorphic sections can be written down. To show this, let $\mathcal{T}^g = V/\Lambda$ be a complex torus and $\mathcal{L}$ a line bundle on $\mathcal{T}^g$. Then $\pi^*\mathcal{L}$ is trivial ($\pi : V \to \mathcal{T}^g$) because V is contractible. Hence, there exists a global trivialization

$$\phi : \pi^*\mathcal{L} \to V \times \mathbf{C}.$$

Then $(\pi^*\mathcal{L})_z = (\pi^*\mathcal{L})_{z+\lambda}$ and since ϕ_z maps $(\pi^*\mathcal{L})_z$ to $\mathbf{C}$ we get

$$\mathbf{C} \leftarrow (\pi^*\mathcal{L})_z = (\pi^*\mathcal{L})_{z+\lambda} \to \mathbf{C},$$

giving a linear automorphism $\mathbf{C} \to \mathbf{C}$, i.e., multiplication by a non-zero number $e_\lambda(z)$. The functions $\{e_\lambda \in \mathcal{O}^*(V)\}_{\lambda \in \Lambda}$ are called the *multipliers* of $\mathcal{L}$ and satisfy

$$e_{\lambda'}(z + \lambda)\, e_\lambda(z) = e_\lambda(z + \lambda')\, e_{\lambda'}(z) = e_{\lambda+\lambda'}(z).$$

Conversely, multipliers which satisfy these relations define a unique line bundle with these multipliers.

Line bundles can be constructed using multipliers $\{e_\lambda(z)\}$ of a simple character. From the exponential sequence

$$0 \to \mathbf{Z} \to \mathcal{O} \overset{\exp}{\to} \mathcal{O}^* \to 0$$

we get

$$\cdots \to H^1(\mathcal{T}^g, \mathcal{O}) \to H^1(\mathcal{T}^g, \mathcal{O}^*) \overset{c_1}{\to} H^2(\mathcal{T}^g, \mathbf{Z}) \to H^2(\mathcal{T}^g, \mathcal{O}) \to \cdots$$

where $c_1(\mathcal{L})$ is called the *Chern class* of the line bundle $\mathcal{L}$. If ω is a positive integral form of type $(1,1)$ then

$$\omega = \sum \delta_\alpha dx_\alpha \wedge dx_{n+\alpha},$$

with respect to the basis $\{x_1, \ldots, x_{2g}\}$ dual to some basis $\{\lambda_1, \ldots, \lambda_{2g}\}$ of Λ. Setting $e_\lambda = \lambda_\alpha/\delta_\alpha$ and letting $z_1, \ldots, z_g$ be linear coordinates dual to $e_1, \ldots, e_g$, we get the following theorem.

Theorem 3.3 *The line bundle $\mathcal{L} \to \mathcal{T}^g$ with multipliers*

$$e_{\lambda_\alpha} = 1,$$
$$e_{\lambda_{n+\alpha}} = e^{-2\pi i z_\alpha},$$

(for $\alpha = 1 \ldots, g$) has Chern class $c_1(\mathcal{L}) = [\omega]$.

Up to a translation in $\mathcal{T}^g$ every line bundle is uniquely determined by its Chern class and this Chern class is $c_1(\mathcal{L}) = \left[\frac{i}{2\pi}\Theta\right]$ so it is a positive integral form of type $(1,1)$.

The fact that the line bundle is given by simple multipliers allows us to construct explicitly its holomorphic sections; they can be seen as functions on $\mathbf{C}^g$ which are periodic in g directions and "quasi-periodic" in g other directions. The number of independent holomorphic sections is given by

$$\dim H^0(\mathcal{T}^g, \mathcal{O}(\mathcal{L})) = \prod \delta_\alpha, \tag{3.1}$$

where $(\delta_1, \ldots, \delta_g)$ are the elementary divisors of the polarization $c_1(\mathcal{L})$. For a line bundle defining a principal polarization, for example, there is only one section which, as a quasi-periodic function on $\mathbf{C}^g$, is given by *Riemann's theta function*

$$\vartheta(z) = \sum_{l \in \mathbf{Z}^n} e^{\pi i \langle l, Zl \rangle} e^{2\pi i \langle l, z \rangle} \qquad (\Lambda = (I\ Z)). \tag{3.2}$$

Its divisor of zeros, Θ, is determined uniquely by $\mathcal{L}$, hence up to a translation by $[c_1(\mathcal{L})]$ and is called the *Riemann theta divisor*.

The group of all line bundles of degree 0 on a polarized Abelian variety $\mathcal{T}^g$ is a complex torus, called its *dual* and denoted by $\hat{\mathcal{T}}^g$. If $\mathcal{T}^g$ corresponds to a period matrix (Δ_δ, Z) then a "dual" basis can be picked such that the matrix defining the lattice defining $\hat{\mathcal{T}}^g$ is given by

$$\left(\delta_n \Delta_\delta^{-1}, \delta_n \Delta_\delta^{-1} Z \Delta_\delta^{-1}\right). \tag{3.3}$$

In this representation it is easy to check the Riemann conditions, which show that the dual is indeed an Abelian variety. For $\mathcal{L}$ a fixed positive line bundle on $\mathcal{T}^g$ one defines an isogeny between $\mathcal{T}^g$ and its dual by $v \mapsto \mathcal{L}^{-1} \otimes \tau_v^* \mathcal{L}$. The degree of this isogeny is $\prod \delta_i^2$.

If $\mathcal{T}^g$ is irreducible then the line bundle of any effective divisor is ample; moreover the third power of any ample line bundle on an Abelian variety is very ample, hence gives an embedding in projective space (these two properties are particular for Abelian varieties, for general algebraic varieties both are false; the last property is due to Lefschetz). For example, if $\mathcal{L}$ defines a principal polarization on an irreducible Abelian variety $\mathcal{T}^g$, then $\mathcal{L}^3$ induces a polarization of type $(3, 3, \ldots, 3)$, and hence every irreducible principally polarized Abelian variety can be embedded in $\mathbf{P}H^0(\mathcal{T}^g, \mathcal{O}(\mathcal{L}^3))$ which is by (3.1) isomorphic to $\mathbf{P}^{3^g-1}$.

3.3. Abelian surfaces

Since we are mainly interested in two dimensional a.c.i. systems, the Abelian varieties which we will encounter are often Abelian surfaces. In what follows we give some useful techniques to study these surfaces.

As we have seen in the case of curves, varieties are often studied by examining the (mero-morphic) functions on them, more specifically by examining the divisors of these functions; we will call an effective divisor C on an algebraic surface S an *(embedded) curve on S*. The curve C is said to be *smooth* if it is a submanifold of S (taken with multiplicity 1) and *irreducible* if it is not the union of two effective divisors.

A fundamental result here is that the canonical bundle of a curve on a surface and the canonical bundle of the surface itself are intimately related, as is expressed in the following *adjunction formula*.

Theorem 3.4 *If S is an algebraic surface and C a smooth curve on S, then the canonical bundles K_S and K_C of S and C are related by*

$$K_C = (K_S \otimes [C])|_C.$$

Since we have shown that $\deg(K_C) = 2g - 2$ if C has genus g, we can calculate the genus of C by

$$g = \frac{1}{2} \deg(K_S \otimes [C])|_C + 1.$$

Now on S there is a natural non-degenerate *intersection pairing*

$$\cdot : H_2(S, \mathbf{Z}) \times H_2(S, \mathbf{Z}) \to \mathbf{Z},$$

which counts the (signed) number of intersection points of arbitrary transversely meeting 2-cycles representing the homology classes. The pairing also gives a natural definition of the intersection of divisors by taking the intersection of their fundamental classes in $H_2(S, \mathbf{Z})$. Under the natural isomorphism $H_2(S, \mathbf{R}) \to H^2(S, \mathbf{R}) \cong H_{DR}^2$ which derives from it, each divisor $\mathcal{D}$ corresponds to a two-form $\eta_\mathcal{D} \in H_{DR}^2(S)$, its *Poincaré dual*, and the intersection of cycles can be shown to be Poincaré dual to the wedge product of forms, i.e., if Ω denotes the top-form corresponding to the natural orientation, then $\eta_\mathcal{D} \wedge \eta_{\mathcal{D}'} = (\mathcal{D} \cdot \mathcal{D}')\Omega$. This suggests to define the intersection $\mathcal{L} \cdot \mathcal{L}'$ for two line bundles $\mathcal{L}$ and $\mathcal{L}'$ as the cup product of their first Chern classes, thought of as an element of $\mathbf{Z}$,

$$c_1(\mathcal{L}) \cup c_2(\mathcal{L}') = (\mathcal{L} \cdot \mathcal{L}')\Omega.$$

Using the fact that the first Chern class of $[\mathcal{D}]$ is Poincaré dual to $\mathcal{D}$ (which can be shown by computing the curvature form of a metric connection on $[\mathcal{D}]$) we see that

$$([\mathcal{D}] \cdot [\mathcal{D}'])\Omega = c_1([\mathcal{D}]) \cup c_1([\mathcal{D}']) = \eta_\mathcal{D} \wedge \eta_{\mathcal{D}'} = (\mathcal{D} \cdot \mathcal{D}')\Omega.$$

Hence, $[\mathcal{D}] \cdot [\mathcal{D}'] = \mathcal{D} \cdot \mathcal{D}'$, and the two definitions correspond under the basic correspondence between line bundles and divisors. It follows that we can also define the intersection of a line bundle $\mathcal{L}$ and a divisor $\mathcal{D}$ by putting

$$\mathcal{L} \cdot \mathcal{D} = \mathcal{L} \cdot [\mathcal{D}] = c_1(\mathcal{L})(\mathcal{D}).$$

Using the intersection pairing we can calculate the genus of the smooth curve C on S more explicitly as

$$g = \frac{1}{2}\deg(K_S \otimes [C])|_C + 1 = \frac{1}{2}(K_S + C) \cdot C + 1 = \frac{K_S \cdot C + C \cdot C}{2} + 1.$$

For arbitrary curves on S we define the *virtual genus* $\pi(C)$ by this formula. Hence, $\pi(C)$ is the genus of any smooth curve homologous to C. Let $\phi : \Gamma \to C$ be a normalization of C and let $\{P_1, \ldots, P_s\}$ denote the *singularities* of C, i.e. the points where C is not smooth. If we denote by k_i the *multiplicity* of C in P_i (i.e., the number of sheets in the projection, in a small coordinate disc in S around P_i, of C onto a generic disc), then

$$g(\Gamma) \leq \frac{K_S \cdot C + C \cdot C}{2} + 1 - \sum_{i=1}^{s} \frac{k_i(k_i - 1)}{2}, \tag{3.4}$$

and the equality holds if and only if all P_i are *ordinary singularities* (i.e., the k_i tangents at P_i are different).

We are now ready to give and prove the *Riemann-Roch Theorem for line bundles on a surface*.

Theorem 3.5 *Let $\mathcal{L}$ be a line bundle on a (smooth) surface S with canonical bundle K_S. Then*

$$\chi(\mathcal{L}) = \chi(\mathcal{O}_S) + \frac{\mathcal{L} \cdot \mathcal{L} - \mathcal{L} \cdot K_S}{2}.$$

Proof
   We give the proof for line bundles of the form $\mathcal{L} = [\mathcal{D}]$, where $\mathcal{D}$ is a smooth, irreducible curve on S. Starting from the short exact sequence

$$0 \to \mathcal{O}_S \to \mathcal{O}_S(\mathcal{L}) \to \mathcal{O}_D(\mathcal{L}) \to 0,$$

and expressing the fact that the alternating sum of the dimensions of the vector spaces appearing in the associated long exact sequence is zero, we obtain

$$\chi(\mathcal{L}) = \chi(\mathcal{O}_S) + \chi(\mathcal{O}_D(\mathcal{L})).$$

Now $\chi(\mathcal{O}_D(\mathcal{L}))$ is the Euler characteristic of $\mathcal{L}$ restricted to the curve $\mathcal{D}$, and it can be calculated using the Riemann-Roch Theorem for curves, giving

$$\chi(\mathcal{O}_D(\mathcal{L})) = \chi(\mathcal{O}_D) + \deg(\mathcal{L}|_D) = 1 - g + \mathcal{L} \cdot \mathcal{L} = \frac{\mathcal{L} \cdot \mathcal{L} - \mathcal{L} \cdot K_S}{2},$$

which leads to the proposed formula for $\chi(\mathcal{L})$. ∎

3. Abelian varieties

It remains to compute the holomorphic Euler characteristic of the trivial bundle over S. This result is given by *Noether's formula*:

Theorem 3.6 *Let S be a complex surface and K_S its canonical bundle. Then*

$$\chi(\mathcal{O}_S) = \frac{1}{12}(K_S \cdot K_S + \chi(S)),$$

where $\chi(S)$ denotes the Euler-Poincaré characteristic of S.

In the case of an Abelian surface $\mathcal{T}^2$ containing an arbitrary curve C, the adjunction formula and the Riemann-Roch formula become extremely simple:

Theorem 3.7 *Let $\mathcal{T}^2$ be an Abelian surface and C a curve on $\mathcal{T}^2$. Then*

$$\pi(C) = \frac{C \cdot C}{2} + 1 = \chi([C]) + 1.$$

If $[C]$ is positive and induces a polarization of type (δ_1, δ_2) on S, then

$$\pi(C) = \frac{C \cdot C}{2} + 1 = \dim L(C) + 1 = \delta_1 \delta_2 + 1. \tag{3.5}$$

Moreover, for a general Abelian surface the intersection of two line bundles $[C]$ and $[D]$ is deduced from these formulas by replacing C and D by linear equivalent multiples of one divisor.

Proof

Let (z_1, z_2) be the coordinates on $\mathcal{T}^2$ coming from $\mathbf{C}^2$. Then the two-form

$$\omega = dz_1 \wedge dz_2$$

has no zeros. Hence its canonical bundle vanishes, $K = 0$. Because $\mathcal{T}^2$ has a natural cell-decomposition with $\binom{4}{i}$ cells of dimension i, $(i = 0, \ldots, 4)$, its Euler-Poincaré characteristic equals $1 - 4 + 6 - 4 + 1 = 0$, leading to the first string of formulas.

We now show that $\chi([D]) = \dim L(D)$ for positive line bundles, the equality $\dim L(D) = \delta_1 \delta_2$ being given by (3.1). Clearly, it suffices to show that $\dim H^i(\mathcal{T}^2, \mathcal{O}(D)) = 0$ for $i \geq 1$. For this purpose, we use the *Kodaira vanishing Theorem*, which states that $H^p(M, \Omega^q(\mathcal{L})) = 0$ for any positive line bundle $\mathcal{L}$ over a compact complex n-dimensional manifold M, if $p + q > n$. For Abelian surfaces, $\Omega^2(\mathcal{L}) = \mathcal{O}(K)(\mathcal{L}) = \mathcal{O}(\mathcal{L})$, because $K = 0$ and the Kodaira vanishing Theorem reduces for Abelian surfaces to

$$H^p(\mathcal{T}^2, \mathcal{O}(\mathcal{L})) = 0 \qquad \text{for } p > 0.$$

The last claim follows from the fact that the Néron-Sevieri group of a general Abelian surface is isomorphic to $\mathbf{Z}$. ∎

4. Jacobi varieties

There are two very different ways to define the Jacobian of a non-singular curve, the equivalence of the two definitions being given by two fundamental theorems, Abel's Theorem and the Jacobi inversion Theorem. We prefer to give both definitions here because of the importance of both of them in application to integrable Hamiltonian systems. We start by giving the algebraic definition, then we give the analytic/transcendental definition and finally prove their equivalence. It is also shown that the Jacobian of a curve of genus g is a principally polarized Abelian variety of dimension g.

4.1. The algebraic Jacobian

We fix a curve (compact Riemann surface) Γ of genus g. For the *algebraic* definition, recall that we constructed the group $\text{Pic}(\Gamma)$ of all line bundles on Γ, and showed that this group is isomorphic to $\text{Div}(\Gamma)/\sim_l$, the group of all divisors modulo linear equivalence. Since $\deg(f) = 0$ for any meromorphic function f, it follows that $\mathcal{D} \sim_l \mathcal{D}'$ implies $\deg \mathcal{D} = \deg \mathcal{D}'$, hence deg induces a homomorphism

$$\deg_\sim : \frac{\text{Div}(\Gamma)}{\sim_l} \to \mathbf{Z}.$$

We define the *Jacobian of* Γ, $\text{Jac}(\Gamma)$, to be $\text{Ker} \deg_\sim$. Said differently, $\text{Jac}(\Gamma) = \text{Div}^0(\Gamma)/\sim_l$, the group of divisors of divisors of degree zero modulo linear equivalence. The map $\deg_\sim$ allows us to define the *degree* of a line bundle $\mathcal{L} = [\mathcal{D}]$ by $\deg \mathcal{L} = \deg_\sim \mathcal{D}$. Defining $\text{Pic}^i(\Gamma)$ as the group of (isomorphism classes of) line bundles of degree i on Γ we have, with the above definition of the Jacobian, that $\text{Jac}(\Gamma)$ is canonically isomorphic to the group $\text{Pic}^0(\Gamma)$. Of course, $\text{Pic}^0(\Gamma)$ is isomorphic to $\text{Pic}^i(\Gamma)$ for any other integer i but the isomorphism is not canonical. For reasons that will become clear later some authors prefer to *define* $\text{Jac}(\Gamma)$ as $\text{Pic}^{g-1}(\Gamma)$.

4.2. The analytic/transcendental Jacobian

For the *analytic* definition, choose any basis $\vec{\omega} = {}^t(\omega_1, \ldots, \omega_g)$ of the space of holomorphic differentials on Γ and let Λ denote the discrete subgroup of $\mathbf{C}^g$ consisting of all vectors in $\mathbf{C}^g$ of the form $\oint_\gamma \vec{\omega}$, with γ running through $H_1(\Gamma, \mathbf{Z})$. Since $\{\omega_1, \ldots, \omega_g, \bar{\omega}_1, \ldots, \bar{\omega}_g\}$ generate $H^1_{DR}(\Gamma) = H^{1,0} \oplus H^{0,1}$ (the first the Rham group of Γ and its splitting in the holomorphic and anti-holomorphic part), Λ is actually a lattice (called the *period lattice* of $\text{Jac}(\Gamma)$). This shows that $\mathbf{C}^g/\Lambda$ is a complex torus, the *analytic* Jacobian of Γ. More intrinsically, the analytic Jacobian of Γ is defined by

$$\text{Jac}(\Gamma) = \frac{H^0(\Gamma, \Omega^1)^*}{H_1(\Gamma, \mathbf{Z})},$$

where $H_1(\Gamma, \mathbf{Z})$ is viewed as a subgroup of $H^0(\Gamma, \Omega^1)^*$ via the natural injective homomorphism Ψ which maps $\gamma \in H_1(\Gamma, \mathbf{Z})$ to the linear map

$$\Psi(\gamma) : H^0(\Gamma, \Omega^1) \to \mathbf{C}$$

$$\omega \mapsto \oint_\gamma \omega.$$

4. Jacobi varieties

Before showing how the algebraic and the analytic Jacobian are related we want to show that the analytic Jacobian is a (principally polarized) Abelian variety. Choosing a basis of $H_1(\Gamma, \mathbf{Z})$, for example a symplectic basis $\{A_1, \ldots, , A_g, B_1, \ldots, B_g\}$, i.e., a basis for which $A_i \cdot A_j = B_i \cdot B_j = 0$ and $A_i \cdot B_j = \delta_{ij}$, the lattice Λ is then conveniently represented as the column space (over $\mathbf{Z}$) of the matrix

$$
\begin{pmatrix}
\oint_{A_1} \omega_1 & \cdots & \oint_{A_g} \omega_1 & \oint_{B_1} \omega_1 & \cdots & \oint_{B_g} \omega_1 \\
\vdots & & \vdots & \vdots & & \vdots \\
\oint_{A_1} \omega_g & \cdots & \oint_{A_g} \omega_g & \oint_{B_1} \omega_g & \cdots & \oint_{B_g} \omega_g
\end{pmatrix}.
$$

The first $g \times g$ block is called the *matrix of A-periods* and the last $g \times g$ block the *matrix of B-periods*. For a single 1-form its *i*-th period $(1 \le i \le 2g)$ is its integral over the cycle A_i if $i \le g$, otherwise it is its integral over the cycle B_{i-g}. The following theorem states that the (analytic) Jacobian of Γ is a principally polarized Abelian variety.

Theorem 4.1 *Let $\{A_1, \ldots, A_g, B_1, \ldots, B_g\}$ be a symplectic basis of $H_1(\Gamma, \mathbf{Z})$. For any basis of $H^1(\Gamma, \Omega^1)$ the matrix of A-periods is non-singular and hence a basis of the latter space can be chosen such that the matrix of A-periods is the identity matrix. In this basis the matrix of B-periods is symmetric and its imaginary part is positive definite. Thus $\mathrm{Jac}(\Gamma)$ is a principally polarized Abelian variety.*

The main ingredient in the proof (given below) is the following proposition (known as the *reciprocity law for differentials of the first and third kind*).

Proposition 4.2 *Let Γ be a compact Riemann surface of genus g, equipped with a symplectic basis $\{A_1, \ldots, A_g, B_1, \ldots, B_g\}$ of $H_1(\Gamma, \mathbf{Z})$. Let ω be a holomorphic 1-form and let η be a meromorphic 1-form whose poles are simple; we call these poles $s_1, \ldots, s_n$. Also $s_0 \in \Gamma$ denotes an arbitrary fixed point. If Π_i (resp. Π_i') denotes the i-th period of ω (resp. of η) then*

$$
\sum_{k=1}^{g} \Pi_k \Pi_{g+k}' - \Pi_{g+k} \Pi_k' = 2\pi i \sum_{j=1}^{n} \mathrm{Res}_{s_j} \eta \cdot \int_{s_0}^{s_j} \omega. \tag{4.1}
$$

Proof

Representatives for the A_i and B_i may be chosen such that none of them pass through the poles of η and by cutting Γ along the cycles the surface may be represented by a polygon $\tilde{\Gamma}$ (with $4g$ sides) as in the following figure.

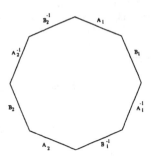

Figure 6

The boundary pieces which correspond to the cycles A_i and B_i will simply be called γ_i resp. γ_{i+g} and accordingly the cycles are sometimes denoted by the uniform notation $\delta_i,\ i = 1, \ldots, 2g$, (thus with this notation γ_i is just the boundary piece corresponding to the cycle δ_i). Let

$$\pi : \tilde{\Gamma} \to \mathbf{C} : s \mapsto \int_{s_0}^{s} \omega,$$

then π is holomorphic and $\pi\eta$ is a meromorphic 1-form with poles in the points s_i. Since these poles are simple we have by the residue theorem

$$\oint_{\partial\tilde{\Gamma}} \pi\eta = 2\pi i \sum_j \mathrm{Res}_{s_j} \pi\eta = 2\pi i \sum_j \mathrm{Res}_{s_j} \eta \int_{s_0}^{s_j} \omega,$$

which gives the right hand side of (4.1). Also

$$\oint_{\partial\tilde{\Gamma}} \pi\eta = \sum_{j=1}^{2g} \int_{\gamma_j + \gamma_j^{-1}} \pi\eta = \sum_{j=1}^{2g} \oint_{\delta_j} \left(\pi(P^+) - \pi(P^-)\right) \eta(P).$$

Here P^+ and P^- are the two points corresponding to P which lie on γ_j resp. γ_j^{-1}; we used that $\eta(P^+) = \eta(P^-)$. To finish, look at the above figure and compute

$$\pi(P^+) - \pi(P^-) = \int_{P^-}^{P^+} \omega = -\int_{B_j} \omega = -\Pi_{j+g}$$

if $P \in \gamma_j,\ 1 \le j \le g$. Similarly for $P \in \gamma_{j+g},\ g+1 \le j \le 2g$ one finds

$$\pi(P^+) - \pi(P^-) = \int_{P^-}^{P^+} \omega = \int_{A_j} \omega = \Pi_j.$$

From this we find the left hand side of (4.1):

$$\oint_{\partial\tilde{\Gamma}} \pi\eta = \sum_{j=1}^{g} \oint_{\delta_j} -\Pi_{j+g}\eta(P) + \oint_{\delta_{j+g}} \Pi_j\eta(P) = \sum_{k=1}^{g} \Pi_k\Pi'_{g+k} - \Pi_{g+k}\Pi_k. \qquad (4.2)$$

∎

Proof of Theorem 4.1

For a holomorphic 1-form η the right hand side in (4.1) vanishes and gives the simple relation

$$\sum_{k=1}^{g} \Pi_k \Pi'_{g+k} - \Pi_{g+k} \Pi_k = 0. \tag{4.3}$$

Denoting the complex conjugate of η by $\bar{\eta}$ one has

$$\iint_{\tilde{\Gamma}} \omega \wedge \bar{\eta} = \iint_{\tilde{\Gamma}} d\pi \wedge \bar{\eta} = \iint_{\tilde{\Gamma}} d(\pi \cdot \bar{\eta}) = \oint_{\partial \tilde{\Gamma}} \pi \cdot \bar{\eta},$$

which, upon using (4.2) to

$$\iint_{\tilde{\Gamma}} \omega \wedge \bar{\eta} = \sum_{k=1}^{g} \Pi_k \bar{\Pi}'_{g+k} - \Pi_{g+k} \bar{\Pi}'_k. \tag{4.4}$$

For a holomorphic 1-form $\omega = f(z)dz$

$$\omega \wedge \bar{\omega} = -2i|f(z)|^2 dx \wedge dy,$$

which leads for $\omega \neq 0$ in combination with (4.4) to

$$0 > \Im \iint_{\Gamma} \omega \wedge \bar{\omega} = 2\Im \sum_{k=1}^{g} \Pi_k \bar{\Pi}_{g+k}. \tag{4.5}$$

A first conclusion is that if all A-periods of a holomorphic 1-form ω are zero then $\omega = 0$. This implies that the matrix of A-periods of the basis $\{\omega_1, \ldots, \omega_g\}$ is non-singular: otherwise some non-trivial combination of the elements of this basis would have all its A-periods equal to zero.

By changing our basis of the space of holomorphic differentials (if necessary) we may therefore assume that the ω_i are *normalized* in the sense that

$$\oint_{A_j} \omega_i = \delta_{ij}.$$

The matrix of periods now takes the form $(I \; Z)$, where Z is the matrix of B-periods. For the B-periods of two 1-forms ω_i and ω_j one finds from (4.3) that $\Pi'_{g+i} = \Pi_{g+j}$, i.e.,

$$\oint_{B_i} \omega_j = \oint_{B_j} \omega_i$$

and Z is a symmetric matrix.

Finally $\Im Z$ is positive definite. Let $0 \neq \omega = \sum_{i=1}^{g} c_k \omega_k$ with $c_k \in \mathbf{R}$. Then

$$\Pi_k = \oint_{A_k} \sum_{j=1}^{g} c_j \omega_j = c_k,$$

$$\Pi_{k+g} = \oint_{B_k} \sum_{j=1}^{k} c_j \omega_j = \sum_{j=1}^{k} c_j Z_{jk},$$

117

which leads by substitution in (4.5) to

$$\sum_{k=1}^{g} c_k \left(\Im Z_{kj} \right) \bar{c}_j < 0.$$

This shows that the period matrix has the desired form and we conclude from Theorem 3.1 that $\mathrm{Jac}(\Gamma)$ is a principally polarized Abelian variety whose dimension is the genus of Γ. ∎

Thus the Jacobian of a curve of genus g is a principally polarized Abelian variety whose dimension equals g. The following converse also holds: every irreducible principally polarized Abelian variety of dimension 2 or 3 is the Jacobian of a curve of genus 2 or 3. In higher dimensions this is no longer true (as can be checked by an easy dimension count), and there is the famous *Schotky problem* which asks for a characterization of those matrices Z for which $(\mathrm{Id}_g \, Z)$ is a Jacobi variety (see [Mul1], [Mum3] and [Shi]).

The Riemann theta function, defined in (3.2), can be used to construct (all) meromorphic functions on $\mathrm{Jac}(\Gamma)$. Namely from the definition one checks the following quasi-periodicity of this entire function on $\mathbf{C}^g$: if $m \in \mathbf{Z}^g$ then

$$\vartheta(z + m) = \vartheta(z),$$
$$\vartheta(z + Zm) = e^{-\pi i \langle m, Zm \rangle} e^{-2\pi i \langle m, z \rangle} \vartheta(z). \tag{4.6}$$

From this it is clear that for all $1 \le i, j \le g$ the meromorphic function

$$u(z) = \frac{\partial^2}{\partial z_i \partial z_j} \log \vartheta(z)$$

is periodic, $u(z+m) = u(z+Zm) = u(z)$, hence it descends to a meromorphic function on the quotient with a double pole on the theta divisor. Another way to construct such functions, which will also be used later, is by using *theta functions with characteristics*. For $a, b \in \mathbf{Q}^n$ (called *characteristics*) one defines

$$\vartheta \begin{bmatrix} a \\ b \end{bmatrix} (z) = \sum_{l \in \mathbf{Z}^g} e^{\pi i \langle l+a, Z(l+a) \rangle} e^{2\pi i \langle l+a, z+b \rangle}. \tag{4.7}$$

Formulas (4.6) now become

$$\vartheta \begin{bmatrix} a \\ b \end{bmatrix} (z + m) = e^{2\pi i \langle a, m \rangle} \vartheta \begin{bmatrix} a \\ b \end{bmatrix} (z),$$
$$\vartheta \begin{bmatrix} a \\ b \end{bmatrix} (z + Zm) = e^{-\pi i \langle m, Zm \rangle} e^{-2\pi i \langle m, z \rangle} e^{-2\pi i \langle b, m \rangle} \vartheta \begin{bmatrix} a \\ b \end{bmatrix} (z). \tag{4.8}$$

From these formulas it is easy to see that if a_i, b_i and a_i', b_i' are characteristics (for $i = 1, \ldots, n$) such that $\sum (a_i - a_i') \in \mathbf{Z}^g$ and $\sum (b_i - b_i') \in \mathbf{Z}^g$ then

$$\prod_{i=1}^{n} \frac{\vartheta \begin{bmatrix} a_i \\ b_i \end{bmatrix} (z)}{\vartheta \begin{bmatrix} a_i' \\ b_i' \end{bmatrix} (z)} \tag{4.9}$$

is a meromorphic function on $\mathrm{Jac}(\Gamma)$.

4.3. Abel's Theorem and Jacobi inversion

We now show that the algebraic and the analytic/transcendental Jacobians correspond to the same object, namely we show that the *Abel-Jacobi map*

$$\mathrm{Ab} : \mathrm{Div}^0(\Gamma) \to \mathbf{C}^g/\Lambda : \sum_{i=1}^g P_i - Q_i \mapsto \sum_{i=1}^g \int_{Q_i}^{P_i} \vec{\omega} \pmod{\Lambda},$$

with $\vec{\omega}$ and Λ as defined above, is surjective with kernel consisting of the principal divisors. We start with the latter which is the content of *Abel's Theorem*.

Theorem 4.3 *For any divisor $\mathcal{D}$ of degree 0, $\mathrm{Ab}(\mathcal{D}) = 0$ if and only if $\mathcal{D}$ is the divisor of a meromorphic function.*

Proof

If $\mathcal{D} = (f)$ then $d\log f$ is a well-defined meromorphic 1-form with simple poles on the support of $\mathcal{D} = \sum_{k=1}^n (P_k - Q_k)$ and residue 1 (resp. -1) at P_k (resp. at Q_k). Denote by $\omega_{P_k Q_k}$ the unique meromorphic 1-form with the same poles and residues as $d\log f$ but with its A-periods zero. Then their difference is a holomorphic 1-form

$$d\log f - \sum_{k=1}^n \omega_{P_k Q_k} = \sum_{j=1}^g c_j \omega_j,$$

$c = (c_1, \ldots, c_g)$ is just the vector of A-periods of $d\log f$. Another application of Proposition 4.2, now with $\omega = \omega_j$ and $\eta = \omega_{P_k Q_k}$ leads to

$$\oint_{B_j} \omega_{P_k Q_k} = 2\pi i \int_{Q_k}^{P_k} \omega_j.$$

This allows us to compute explicitly the integrals appearing in the Abel map:

$$\sum_{k=1}^n \int_{Q_k}^{P_k} \omega_j = \frac{1}{2\pi i} \sum_{k=1}^n \oint_{B_j} \omega_{P_k Q_k}$$

$$= \frac{1}{2\pi i} \oint_{B_j} d\log f - \frac{1}{2\pi i} \sum_{l=1}^g c_l \oint_{B_j} \omega_l$$

$$= n_j - \sum_{l=1}^n m_l Z_{jl},$$

in which the integers m_j and n_j are up to a factor $2\pi i$ the A- resp. B-periods of $d\log f$.

This shows one direction of Abel's Theorem. For the other direction, if $\mathrm{Ab}(\mathcal{D}) = 0$ one is tempted to define a function on Γ by

$$f(P) = \exp\left(2\pi i \int_{P_0}^P \eta\right)$$

where η is a 1-form with residue $\frac{1}{2\pi i}$ in the points P_k and $\frac{-1}{2\pi i}$ in Q_k. Indeed, if f is well-defined then $(f) = \mathcal{D}$ and we are done. The problem is thus to find an η with all its periods integral. Any such η can however be written as

$$\eta = \eta_0 + \sum_{l=1}^{g} c_l \omega_l$$

where η_0 has its A-periods equal to zero. We must take c_k integral in order to have integral A-periods for η. As for the B-periods,

$$\oint_{B_j} \eta = \sum_{k=1}^{n} \int_{Q_k}^{P_k} \omega_j + \sum_{l=1}^{g} c_l Z_{lj},$$

where we computed the B-periods of η_0 from Proposition 4.2 (with $\omega = \omega_j$ and $\eta = \eta_0$). Since $\mathrm{Ab}(\mathcal{D}) = 0$ the first term has the form $n_j + (Zm)_j$ where n_i, $m_i \in \mathbf{Z}$ (and $m = (m_1 \ldots, m_g)$). Thus it suffices to choose $c_j = -m_j$ to obtain a 1-form η which has all its periods integral. ∎

To show surjectivity, the *Jacobi inversion Theorem*, it is better to define other maps which are quite similar to the Abel-Jacobi map. To do this we fix any $P_0 \in \Gamma$ and we define for any $d \in \mathbf{N}$

$$\mathrm{Ab}_d : \mathrm{Sym}^d \Gamma \to \mathbf{C}^g/\Lambda : \langle P_1, \ldots, P_d \rangle \mapsto \sum_{j=1}^{d} \int_{P_0}^{P_j} \vec{\omega} \quad (\mathrm{mod}\ \Lambda). \tag{4.10}$$

Theorem 4.4 *The map Ab_g is surjective.*

Proof

$\mathrm{Sym}^g \Gamma$ and $\mathbf{C}^g/\Lambda$ are both compact connected complex manifolds and Ab_g is a holomorphic map. Hence the image of Ab_g is a compact subvariety of $\mathbf{C}^g/\Lambda$. Its differential in a general point $\langle P_1, \ldots, P_g \rangle$ is given by

$$\begin{pmatrix} \omega_1(P_1) & \cdots & \omega_g(P_1) \\ \vdots & & \vdots \\ \omega_1(P_g) & \cdots & \omega_g(P_g) \end{pmatrix}$$

where we have written ω_j locally as $\omega_j = f_j(z)dz$ and $\omega_j(P)$ is a (bad) notation for $f_j(P)$. By Riemann-Roch there exists no holomorphic differential with its zeros on a general set $\{P_1, \ldots, P_g\}$ of points, hence this matrix is invertible. This shows that the image of Ab_g has dimension g, hence Ab_g is surjective. ∎

The link between the algebraic and the analytic/transcendental Jacobian is pursued further in the following theorem, attributed to Riemann.

Theorem 4.5 *There is a constant $\vec{\Delta} \in \mathbf{C}^g$ (called Riemann's constant) such that*

$$\vartheta(Z) = 0 \iff \exists P_1, \ldots, P_{g-1} \in \Gamma : Z = \mathrm{Ab}\left\{ \sum_{i=1}^{g-1} (P_i - P) \right\} - \vec{\Delta} \quad (\mathrm{mod}\ \Lambda). \tag{4.11}$$

The important condition in the right-hand side is that the sum runs over $g - 1$ points only. Because of this theorem many people prefer to define the Jacobian by $\mathrm{Pic}^{g-1}(\Gamma)$ (rather than $\mathrm{Pic}^0(\Gamma)$) since with the former definition the theta divisor is canonical determined (while otherwise it is defined only up to a translation which depends on the choice of a point on Γ).

4.4. Jacobi and Kummer surfaces

Everything discussed above is easily specialized to *Jacobi surfaces*, i.e., Jacobians of genus two curves. They are especially important when working with Abelian surfaces since, as we explained, every Abelian surface is isogenous to a principally polarized Abelian surface and these are (under the assumption of irreducibility) Jacobians of genus two curves (recall also that every Riemann surface of genus two is hyperelliptic, hence can it can easily be described explicitly). The theta divisor of a Jacobi surface is by Riemann's Theorem nothing but a copy of the curve, embedded in its Jacobian; conversely if in some Abelian surface a (non-singular) curve of genus two is found, then it defines by (3.7) a principal polarization on the surface which necessitates it to be the Jacobian of this curve. Since the third power of any ample line bundle on an Abelian variety is very ample, we may take the third power of the line bundle corresponding to the theta divisor to construct an embedding of a Jacobi surface in $\mathbf{P}^8$.

Associated to an Abelian variety $\mathcal{T}^g$ is a singular variety, its *Kummer variety*, which is defined as the quotient surface $K = \mathcal{T}^g/(-1)$, where -1 is the involution given by $(z_1, \ldots, z_g) \mapsto (-z_1, \ldots, -z_g)$ in linear coordinates coming from the universal covering space $\mathbf{C}^g$ of $\mathcal{T}^g$. Clearly the Kummer variety has 2^{2g} singular points which correspond to the two-torsion points (also called *half-periods*) of $\mathcal{T}^g$. We will consider in this text only the case $g = 2$, the case of *Kummer surfaces*. Let A be an Abelian surface of type (δ_1, δ_2). Its Kummer surface $K = A/(-1)$ has sixteen singular points $e_1, \ldots, e_{16}$. The desingularization of K can be described as follows. Let $p : \tilde{A} \to A$ be the blow-up of A at all its half periods and denote the corresponding exceptional divisors by E_i. $(-1)_A$ extends to an involution $(-1)_{\tilde{A}}$ on $\tilde{A}$ and the quotient $\tilde{K} = \tilde{A}/(-1)_{\tilde{A}}$ is a *K-3 surface* (see [Beal], Proposition VIII.11). $\tilde{K}$ is the desingularisation (minimal resolution) of K and we have the following commutative diagram.

Associated to A there are also several *intermediate Kummer surfaces* which are desingularizations of K at some but not all singular points. A similar construction can be performed when the Abelian surface admits an automorphism, different from the (-1)-involution, leading to a *generalized Kummer surface*. See Paragraph VII.3 for an example.

In the case of the Kummer surface of a Jacobi surface, taking $\mathcal{L} = [\Theta]$, all sections of $\mathcal{L}^2$ are even and the singular surface can be embedded in $\mathbf{P}^3$ by using these sections. Being two-dimensional the image is given by a single equation. To compute the degree of this equation, which is the degree of the hypersurface, we use the fact that this degree is given by $\int_K \Omega$, where Ω is associated $(1,1)$-form of the standard Kähler structure on $\mathbf{P}^3$. Clearly this is twice the volume of K, which itself is half the volume of the Jacobi surface (embedded with the polarization of type $(2, 2)$). For $\omega = 2dx_1 \wedge dx_2 + 2dx_3 \wedge dx_4$ we get $\int \omega^2 = 8$, hence the Jacobi surface has degree 8, its volume is 4, the volume of K is 2 and the degree of K is 4. Explicit equations for this quartic polynomial, in terms of an equation for the underlying algebraic curve, will be given in Section VII.4.

Another classical result about the Kummer surface of a Jacobi surface is that it has a 16_6 configuration. Namely for Γ a curve of genus two, the (-1)-involution $\imath$ on $\mathrm{Jac}(\Gamma)$ has, apart from 16 fixed points also 16 invariant *theta curves* (i.e., translates of the theta divisor); each of these 16 curves contains 6 of the fixed points and through any of the 16 fixed points pass 6 of the invariant theta curves. The configuration can be described completely as follows (see [Hud]). Let $W_1, \ldots, W_6$ be the Weierstrass points on Γ, then the points

$$W_{ij} = \int_{W_i}^{W_j} \vec{\omega} \pmod{\Lambda}$$

are half-periods of $\mathrm{Jac}(\Gamma)$ since $2W_i \sim_l 2W_j$. There are sixteen half-periods in total since the equalities $W_{ij} = W_{ji}$ and $W_{ii} = W_{jj}$ hold for all $i, j = 1, \ldots, 6$; apart from these they are all different. The sixteen invariant theta curves Γ_{ij} in $\mathrm{Jac}(\Gamma)$ are the translates $W_{ij} + \Gamma_{kk}$ of the single curve $\Gamma_{11} = \cdots = \Gamma_{66}$, which can e.g. be taken as $\{\mathrm{Ab}(P - W_1) \mid P \in \Gamma\}$. From this it is clear that Γ_{11}, hence all Γ_{ij}, pass through six points W_{kl} and also that each point belongs to six lines Γ_{ij}. Since the whole configuration is invariant under $\imath$ it goes down to the Kummer surface in $\mathbf{P}^3$ and gives there a 16_6 configuration of points and planes, classically called *Kummer's configuration*. The sixteen points are *nodes* (singular points) and the sixteen planes the lines belong to are *tropes* (singular planes) of the Kummer surface.

The 16_6 configuration is best visualized by the *incidence diagram*, which consists of a pair of square diagrams, such as the following.

W_{11}	W_{12}	W_{23}	W_{13}	Γ_{11}	Γ_{12}	Γ_{23}	Γ_{13}
W_{45}	W_{36}	W_{16}	W_{26}	Γ_{45}	Γ_{36}	Γ_{16}	Γ_{26}
W_{46}	W_{35}	W_{15}	W_{25}	Γ_{46}	Γ_{35}	Γ_{15}	Γ_{25}
W_{56}	W_{34}	W_{14}	W_{24}	Γ_{56}	Γ_{34}	Γ_{14}	Γ_{24}

Namely, the points incident with a line at position (m, n) in the second square diagram are those six points in the m-th row and n-th column, but not in both, of the first square diagram. Dually, the same applies for the lines incident with a point. The 24^2 incidence diagrams obtained by permuting the rows or columns of both square diagrams in an incidence diagram (in the same way) are defined to be the same as the original incidence diagram (we will see in Section VII.4 that there are 20 incidence diagrams which are different in this sense). These incidence diagrams will be used in Section VII.4 to describe Abelian surfaces of type $(1,4)$.

5. Abelian surfaces of type (1,4)

Closely related to Jacobi surfaces are Abelian surfaces of type $(1,4)$. Their geometry is discussed here in detail because they will appear in the two examples discussed in Chapter VII. The results in this section are taken from [BLS]. As in that paper we will without further mention always restrict ourselves to those Abelian surfaces of type $(1,4)$ which are not isomorphic to a product of elliptic curves as polarized Abelian surfaces. Let $\mathcal{L}$ be a line bundle of type $(1,4)$ on an Abelian surface $\mathcal{T}^2$. It follows from (3.7) that $\dim H^0(\mathcal{T}^2, \mathcal{O}(\mathcal{L})) = 4$ and $\mathcal{L}$ induces a rational map $\phi_{\mathcal{L}} : \mathcal{T}^2 \to \mathbf{P}^3$.

5.1. The generic case

In the generic case, the image of this map $\mathcal{O} = \phi_{\mathcal{L}}(\mathcal{T}^2) \subset \mathbf{P}^3$ is an octic and $\phi_{\mathcal{L}}$ is birational on its image. Let $K(\mathcal{L})$ be the kernel of the isogeny

$$I_{\mathcal{L}} : \mathcal{T}^2 \to \hat{\mathcal{T}}^2$$
$$a \mapsto t_a \mathcal{L} \otimes \mathcal{L}^{-1}$$

between $\mathcal{T}^2$ and its dual $\hat{\mathcal{T}}^2$ (defined as the set of all line bundles on $\mathcal{T}^2$ of degree 0; t_a is the translation by $a \in \mathcal{T}^2$), then $K(\mathcal{L})$ is a group of translations, isomorphic to $\mathbf{Z}/4\mathbf{Z} \oplus \mathbf{Z}/4\mathbf{Z}$. Picking any such isomorphism, let σ and τ be generators of the subgroups corresponding to this decomposition. Then homogeneous coordinates $(y_0 : y_1 : y_2 : y_3)$ for $\mathbf{P}^3$ can be picked, such that σ, τ and the (-1)-involution $\imath$ on $\mathcal{T}^2$ (defined as $\imath(z_1, z_2) = (-z_1, -z_2)$ for $(z_1, z_2) \in \mathbf{C}^2/\Lambda$) act as follows (see [LB]):

$$\begin{aligned}
\sigma(y_0 : y_1 : y_2 : y_3) &= (y_2 : y_3 : y_0 : -y_1), \\
\tau(y_0 : y_1 : y_2 : y_3) &= (y_1 : y_0 : iy_3 : iy_2), \\
\imath(y_0 : y_1 : y_2 : y_3) &= (y_0 : y_1 : y_2 : -y_3),
\end{aligned} \tag{5.1}$$

(strictly speaking it may be necessary to replace τ by 3τ; it is easily checked that these coordinates exist only for (σ, τ) and $(3\sigma, 3\tau)$ or for $(\sigma, 3\tau)$ and $(3\sigma, \tau)$). [BLS] show that the octic $\mathcal{O}$ is given in these coordinates by

$$\begin{aligned}
&\lambda_0^2 y_0^2 y_1^2 y_2^2 y_3^2 + \lambda_1^2(y_0^4 y_1^4 + y_2^4 y_3^4) + \lambda_2^2(y_0^4 y_2^4 + y_1^4 y_3^4) + \lambda_3^2(y_0^4 y_3^4 + y_1^4 y_2^4) + \\
&2\lambda_1\lambda_2(y_0^2 y_1^2 + y_2^2 y_3^2)(y_1^2 y_3^2 - y_0^2 y_2^2) + 2\lambda_1\lambda_3(y_0^2 y_3^2 - y_1^2 y_2^2)(y_0^2 y_1^2 - y_2^2 y_3^2) + \\
&2\lambda_2\lambda_3(y_1^2 y_2^2 + y_0^2 y_3^2)(y_1^2 y_3^2 + y_0^2 y_2^2) = 0,
\end{aligned} \tag{5.2}$$

for some $(\lambda_0 : \lambda_1 : \lambda_2 : \lambda_3) \in \mathbf{P}^3 \setminus S$ where S is some divisor of $\mathbf{P}^3$ which we will determine later (Paragraph VII.4.4). Notice that for any $\epsilon_i = \pm 1$, the coordinates $(\epsilon_0 y_0 : \epsilon_1 y_1 : \epsilon_2 y_2 : \epsilon_0 \epsilon_1 \epsilon_2 y_3)$ will also satisfy (5.1) and these are the only coordinates with this property. It is also seen that, if (σ, τ) is replaced by $(3\sigma, 3\tau)$, then the coordinates $(y_0 : y_1 : y_2 : y_3)$ are replaced by $(y_0 : y_1 : y_2 : -y_3)$. Since the equation of $\mathcal{O}$ depends only on y_i^2 these choices do not affect the equation (5.2), so there is associated to a decomposition $K(\mathcal{L}) = K_1 \oplus K_2$ (where K_1 and K_2 are cyclic of order 4) an equation for $\mathcal{O}$. [BLS] also show that the polarized Abelian surface as well as the decomposition of $K(\mathcal{L})$ can be recovered from (5.2) and that every octic of the type (5.2) (with $(\lambda_0 : \lambda_1 : \lambda_2 : \lambda_3) \notin S$) is the image $\phi_{\mathcal{L}}(\mathcal{T}^2)$ of some $(1,4)$-polarized Abelian surface $(\mathcal{T}^2, \mathcal{L})$.

If we denote by $\tilde{\mathcal{A}}^0_{(1,4)}$ the moduli space of (isomorphism classes of) $(1,4)$-polarized Abelian surfaces for which $\phi_{\mathcal{L}}$ is birational, equipped with a decomposition of $K(\mathcal{L})$ as above, then it follows that

$$\tilde{\mathcal{A}}^0_{(1,4)} \cong \frac{\mathbf{P}^3 \setminus S}{\lambda_0 \sim -\lambda_0}. \tag{5.3}$$

Moreover, if we denote by K the subgroup of $K(\mathcal{L})$ of two-torsion elements,

$$K = \{0, 2\sigma, 2\tau, 2\tau + 2\sigma\},$$

then $\mathcal{T}^2/K$ is a principal polarized Abelian surface, which is the Jacobian of a curve of genus two; we call $\mathcal{T}^2/K$ the *canonical Jacobian* associated to $\mathcal{T}^2$. Recall that for a two-dimensional Jacobian J its Kummer surface is the image of $\phi_{[2\Theta]} \subset \mathbf{P}^3$, where Θ is the theta divisor of J. Then it is seen from (5.1) that an equation for the Kummer surface of $\mathcal{T}^2/K$ is given by the quartic Q in $\mathbf{P}^3$, obtained by replacing y_i^2 by z_i in the equation (5.2) for $\mathcal{O}$ and there is a natural projection $\bar{p} : \mathcal{O} \to Q$. In fact, choosing the origin of $\mathcal{T}^2$ such that $\mathcal{L}$ becomes symmetric, $\mathcal{L}$ is the pull-back of a line bundle $\mathcal{N}$ on $\mathcal{T}^2/K$ of type $(1,1)$ via the canonical projection

$$p : \mathcal{T}^2 \to \mathcal{T}^2/K,$$

and $\phi_{\mathcal{N}^2}$ induces the Kummer mapping; [BLS] prove that the following diagram commutes

$$
\begin{array}{ccc}
\mathcal{T}^2 & \xrightarrow{\phi_{\mathcal{L}}} & \mathcal{O} \\
\downarrow{\scriptstyle p} & & \downarrow{\scriptstyle \bar{p}} \\
\mathcal{T}^2/K & \xrightarrow[\phi_{\mathcal{N}^2}]{} & Q
\end{array}
\tag{5.4}
$$

5.2. The non-generic case

If $\phi_{\mathcal{L}}$ is not birational, then it is $2 : 1$ and $\phi_{\mathcal{L}}(\mathcal{T}^2)$ is a quartic in $\mathbf{P}^3$, given by one of the equations

$$\lambda_1(y_0^2 y_1^2 + y_2^2 y_3^2) + \lambda_2(y_1^2 y_3^2 - y_0^2 y_2^2) = 0,$$
$$\lambda_1(y_2^2 y_3^2 - y_0^2 y_1^2) + \lambda_3(y_1^2 y_2^2 - y_0^2 y_3^2) = 0,$$
$$\lambda_2(y_1^2 y_3^2 + y_0^2 y_2^2) + \lambda_3(y_1^2 y_2^2 + y_0^2 y_3^2) = 0,$$

depending on the choice of the decomposition; in this case the Abelian surface as well as the decomposition of $K(\mathcal{L})$ can only partly be recovered from these equations and $\mathcal{T}^2/K$ is

5. Abelian surfaces of type (1,4)

a product of elliptic curves (in particular $\mathcal{T}^2$ is isogeneous to a product of elliptic curves). Squaring each of these equations we find equation (5.2) respectively with

$$\begin{cases} \lambda_0^2 = 2(\lambda_2^2 + \lambda_3^2) \\ \lambda_1 = 0 \end{cases} \quad \lambda_2 \lambda_3 \neq 0,\ \lambda_2^2 - \lambda_3^2 \neq 0,$$

$$\begin{cases} \lambda_0^2 = -2(\lambda_1^2 + \lambda_3^2) \\ \lambda_2 = 0 \end{cases} \quad \lambda_1 \lambda_3 \neq 0,\ \lambda_1^2 - \lambda_3^2 \neq 0, \qquad (5.5)$$

$$\begin{cases} \lambda_0^2 = 2(\lambda_1^2 - \lambda_2^2) \\ \lambda_3 = 0 \end{cases} \quad \lambda_1 \lambda_2 \neq 0,\ \lambda_1^2 + \lambda_2^2 \neq 0.$$

Summarizing, in the first case (the generic case), $\phi_{\mathcal{L}}(\mathcal{T}^2)$ is an octic, $\mathcal{T}^2/K$ is a Jacobian and $\mathcal{T}^2$ as well as the decomposition of $K(\mathcal{L})$ can be reconstructed from the octic; in the other case $\phi_{\mathcal{L}}(\mathcal{T}^2)$ is a quartic, $\mathcal{T}^2/K$ is a product of elliptic curves and $\mathcal{T}^2$ cannot be reconstructed from the quartic. The rational map $\phi_{\mathcal{L}}$ provides us with a natural surjective map

$$\psi^0 : \mathcal{A}_{(1,4)}^0 \to \left((\mathbf{P}^3 \setminus S) \bigcup (\text{three rational curves in } S \text{ missing eight points}) \right) / (\lambda_0 \sim -\lambda_0),$$

where $\mathcal{A}_{(1,4)}^0$ denotes the moduli space of (isomorphism classes of) $(1,4)$-polarized Abelian surfaces together with a decomposition of $K(\mathcal{L})$ (as above). The map ψ^0 extends the bijection (5.3) defined on the dense subset $\tilde{\mathcal{A}}_{(1,4)}^0$ of $\mathcal{A}_{(1,4)}^0$ and maps the (two-dimensional) complement of $\tilde{\mathcal{A}}_{(1,4)}^0$ to the three rational curves, which are thought of as lying inside the boundary of $\psi^0\left(\tilde{\mathcal{A}}_{(1,4)}^0\right)$, i.e., in S; the generic point of S however does not correspond to Abelian surfaces, but to surfaces which can be interpreted as degenerations of Abelian surfaces (see [BLS]).

Chapter V

Algebraic completely integrable Hamiltonian systems

1. Introduction

In many integrable Hamiltonian systems of interest the general level set of the momentum map are isomorphic to an affine part of an Abelian variety and the flow of the integrable vector fields is linearized by this isomorphism. These two properties lead to the definition of an algebraic completely integrable Hamiltonian system (a.c.i. system). We will discuss three quite different definitions of an a.c.i. system, which have been proposed by different authors, and we extract from it a definition which is consistent with our approach to integrable Hamiltonian systems. All constructions of integrable Hamiltonian systems easily specialize to the case of a.c.i. systems, except in the case of the quotient, which requires a real proof. The definitions and these properties will be considered in Section 2.

Painlevé analysis is an important tool for studying an (irreducible) a.c.i. system, since it allows us to determine the nature of the general level sets of its momentum map and to construct an explicit embedding of the completed general level set (which is isomorphic to an Abelian variety) in projective space. Although the nature of these Abelian varieties can often also be deduced from a Lax representation of the a.c.i. system (see e.g. [Aud3]), Painlevé analysis is at present the only available method to construct an embedding of these Abelian varieties in projective space. Since several such embeddings will be constructed and used, we will explain what Painlevé analysis is about in Section 3. We wish to point out that Painlevé analysis can, by a theorem of Adler and van Moerbeke (see [AM7]), in principle be used to prove that a certain integrable Hamiltonian system is a.c.i., but applying this theorem requires even for the simplest systems a considerable amount of work. In Section 4 we recall from [Van2] our algorithm which allows one to linearize explicitly two-dimensional a.c.i. systems starting from the differential equations for one of the integrable vector fields. It is used in this text to construct morphisms of two-dimensional a.c.i. systems.

In Section 5 we will explain briefly how Lax equations appear and what is their relevance for a.c.i. systems, in particular when dealing with Lax equations with a spectral parameter. Lax equations have by now been obtained for all integrable Hamiltonian systems which were classically known, however some of these have been obtained after a lot of effort, reflecting the fact that it is not clear how to write down Lax equations for a given integrable Hamiltonian system, even if the underlying geometry has been completely revealed. For the general integrable Hamiltonian systems introduced by us in Chapter III for example it is not clear how to write Lax equations for the integrable vector fields (apart from the very special case treated in Paragraph III.2.4). Lax equations constitute a complete chapter in the theory of integrable systems, however we will not treat them as being basic in their study since very often they only appear at the end. Of course, once obtained, Lax equations beautifully exhibit many aspects of the integrable Hamiltonian system in a unifying way; moreover, a careful analysis of the underlying algebra leads in many cases to generalizations and to a quantification of the integrable system.

2. A.c.i. systems

We start by discussing three definitions of algebraic complete integrability. Often in the literature this term is being used meaning that the general fiber of the integrable Hamiltonian system under consideration is an Abelian variety, an affine part of an Abelian variety or just a dense subset of an Abelian variety, the flow of the integrable vector fields being in any case linear on it. Indeed that is the main idea and will be present in the different definitions of a.c.i. systems given below.

Let us start with the Adler-van Moerbeke definition (see [AM7]). They consider a polynomial integrable Hamiltonian system (i.e., the Poisson bracket is polynomial as well as the functions in involution) on $\mathbf{R}^n$ (fixing a basis of the integrable algebra, but that is not relevant here) which they complexify in the natural way. They suppose the following conditions about the general level sets of its momentum map and the flows on them to be verified:

1) The general fiber of the momentum map is (isomorphic to) an affine part of an Abelian variety;

2) The flow of the integrable vector fields on the general fiber of the momentum map is linear;

3) The coordinate functions on $\mathbf{C}^n$ restrict to the general level sets as regular functions which extend to meromorphic functions on the corresponding Abelian varieties;

4) The divisor to be adjoined to the general level set is minimal in the sense that for each irreducible component of this divisor there is at least one of these meromorphic functions which has a pole on it.

Next we give Mumford's definition (see [Mum5] pp. 3.51 – 3.54). He starts from a real symplectic manifold (M, ω) of dimension $2n$ and a smooth function H on it. His requirement for algebraic complete integrability is the existence of a smooth real algebraic variety $(M^{\mathbf{C}}, \tau)$ such that:

1) M is a component of the real part of $(M^{\mathbf{C}}, \tau)$;

2) $M^{\mathbf{C}}$ is endowed with a (co-)symplectic structure $\hat{\omega}$ which restricts to ω along M;

3) There exists on $M^{\mathbf{C}}$ a proper map which is submersive onto a Zariski open subset of $\mathbf{C}^n$ and whose components are in involution, H being dependent on these component functions (restricted to M).

The conditions are sufficient to imply that the fibers are Abelian varieties or extensions of these by $\mathbf{C}^{*m}$ and that the flow of the integrable vector fields is linear.

Third, let us recall Knörrer's definition (see [Knö]). He considers an (irreducible) complex manifold M of dimension $2n$ with a holomorphic symplectic structure ω, a holomorphic function H on M and n holomorphic functions in involution (and in involution with H) which define a submersive map onto an open dense subset B of $\mathbf{C}^n$,

$$F : F^{-1}(B) \subset M \to B \subset \mathbf{C}^n.$$

For algebraic complete integrability he demands the existence of

1) a bundle $\pi : A \to B$ of Abelian varieties;

2) a divisor $\mathcal{D} \subset M$;

3) an isomorphism $\phi : F^{-1}(B) \to A \setminus \mathcal{D}$;

4) a vector field Y on $A \setminus \mathcal{D}$ which restricts to a linear vector field on the fibers of π;

so that the diagram

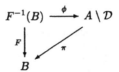

is commutative and such that the vector field X_H is ϕ-related to Y.

Let us comment on these definitions and deduce from it the definition which is appropriate in the case of integrable Hamiltonian systems on affine Poisson spaces. We don't discuss here the fact that all three choose a specific basis of the integrable algebra since we discussed this aspect already in Chapter II (and it is not really relevant here). The last two definitions are situated on symplectic manifolds, while the first is on Poisson manifolds, a difference which is easy to overcome. The first two definitions start from a real system and complexify it, the last one is completely situated in the complex. Also here, there is nothing to be worried about; from the point of view of physics and when one wants to study the real topology of integrable Hamiltonian systems, one may want to impose a reality condition, but it is clear that the condition is not essential in the definition. So we think it should not be encoded in the definition but put as a restriction when the applications which one has in mind ask so.

We now come to the more essential points in the definitions which all relate to the nature of the general level sets of the momentum map.

• The Adler-van Moerbeke definition can in this respect be reformulated saying that the general fiber of the momentum map (on $\mathbf{C}^n$), with its algebraic structure which comes from the ambient space $\mathbf{C}^n$, is isomorphic to an affine part of an Abelian variety; the added condition that the missing divisor is minimal relates to the possibility to perform the compactification of this general fiber into an Abelian variety by using the Laurent solutions to the differential equations describing one of the vector fields (see Section 3 below). We think that one should not insist on putting this condition in the general definition.

• D. Mumford's definition does not demand anything about the nature of the general fiber of the momentum map but his assumptions imply what it is: it is an Abelian variety or an extensions of an Abelian variety by $\mathbf{C}^{*m}$. The key is that he asks properness of the morphism (which plays the role of the momentum map). As such, this is very strict: in the examples we know of one rarely finds complete Abelian varieties (but $\mathbf{C}^{*m}$ are found in several examples). After completion however, as is discussed below, a proper map and hence an a.c.i. system in this sense[11] is found. An advantage of his definition is that he allows fibers which are more general than Abelian varieties.

• H. Knörrer's definition asks that the level manifolds be isomorphic to affine parts of Abelian varieties, but even as a family. Whereas the Adler-van Moerbeke definition asks that every generic level manifold can be compactified into an Abelian variety, his definition asks for a so-called partial compactification of the momentum map (at least over a Zariski open subset); this would then lead to a compactification of each general fiber by restriction. With these remarks in mind we will now define two notions of a.c.i. systems, one notion being stronger (but harder to verify) than the other.

---

[11] Strictly speaking the symplectic or Poisson structure may not extend to the completion.

Definition 2.1 Let $(M, \{\cdot\,,\cdot\}, \mathcal{A})$ be an integrable Hamiltonian system on an affine Poisson variety and denote its momentum map by $\pi_{\mathcal{A}} : M \to \mathrm{Spec}(\mathcal{A})$. Then $(M, \{\cdot\,,\cdot\}, \mathcal{A})$ is called an *algebraic completely integrable Hamiltonian system* or an *a.c.i. system* if there exists a Zariski open subset $B \subset \mathrm{Spec}(\mathcal{A})$ and a bundle of Abelian groups $\pi : A \to B$ such that for each $b \in B$ there exists a divisor $\mathcal{D}_b \subset \pi^{-1}(b)$ and an isomorphism $\phi_b : \pi_{\mathcal{A}}^{-1}(b) \to \pi^{-1}(b) \setminus \mathcal{D}_b$, such that the restriction of each vector field in $\mathrm{Ham}(\mathcal{A})$ to $\pi_{\mathcal{A}}^{-1}(b)$ is ϕ_b-related to a linear[12] vector field on $\pi^{-1}(b)$.

If there exists instead a bundle of Abelian groups $\pi : A \to B$, where $B \subset \mathrm{Spec}\,\mathcal{A}$ is a Zariski open subset, a divisor $\mathcal{D}$ on A and an isomorphism ϕ such that

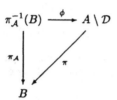

is a commutative diagram and such that each vector field in $\mathrm{Ham}(\mathcal{A})$ is ϕ-related to a vector field on $A \setminus \mathcal{D}$ which restricts to a linear vector field on each level set of π, then we say that $(M, \{\cdot\,,\cdot\}, \mathcal{A})$ is a *completable a.c.i. system*.

Here are some remarks about this definition. Clearly a completable a.c.i. system is an a.c.i. system: one defines $\phi_b = \phi_{|\pi_{\mathcal{A}}^{-1}(b)}$ and $\mathcal{D}_b = \mathcal{D}_{|\pi^{-1}(b)}$. Whether every a.c.i. system is completable is a question which has been studied in more general terms in algebraic geometry. In simple terms the way they put this question is demanding for a partial[13] compactification of a morphism whose fibers over closed points are (isomorphic to) affine parts of algebraic groups; this partial compactification should be such that the compactified fibers are isomorphic to the corresponding algebraic groups. Their solution is to construct the points to be added as a sum[14] of two points in the affine part; more precisely they construct the fibred product of $\pi_{\mathcal{A}}^{-1}(B)$ with itself and define an equivalence relation by which two pairs of points are equivalent if they belong to the same fiber and they have the same sum (the sum to be taken in the completed[15] fiber). The question is now if this leads to an algebraic quotient. In general the answer is no, it works only up to a base extension. The problem of completing an a.c.i. system comes from the monodromy of the base space; in a real setting the monodromy is an obstruction for the global existence of action-action variables, as was shown by Duistermaat (see [Dui]). As Michèle Audin pointed out to me, these ideas are very close.

It turns out that many a.c.i. systems are actually completable (more precisely: a counterexample is not known). One way to verify this is to compute an embedding of the general fiber of the momentum map in some projective space $\mathbf{P}^n$; this can be done explicitly by using

---

[12] By a *linear vector field* on an Abelian group G we mean a G-invariant vector field.

[13] In a partial compactification only the fibers are compactified, not the base space.

[14] In our case the Abelian varieties will appear without origin, i.e., as homogeneous spaces and one cannot assume the existence of a section over B which picks an origin on each of these Abelian varieties. Then the construction is modified by picking the fibred cube instead of the fibred square, using a similar equivalence relation.

[15] It is given that we can complete each level separately.

the Laurent solutions to the differential equations describing one of the integrable vector fields as we will explain in Section 3. In all examples we have seen the embedding of $\mathcal{F}_b = \pi_{\mathcal{A}}^{-1}(b)$ ($b \in B$, a Zariski open subset of Spec $\mathcal{A}$) in projective space is given by functions which are regular on M and independent of the chosen fiber, i.e., there exist $f_1, \ldots, f_N \in \mathcal{O}(M)$ such that the map $\phi : M \to \mathbf{P}^N$ given by

$$p \mapsto (1 : f_1(p) : \cdots : f_N(p))$$

has the following property: for any $b \in B$ the restriction of ϕ to $\mathcal{F}_b$ is an embedding and the closure of its image is an Abelian group (isomorphic to $\pi^{-1}(b)$). In this case we obviously have an embedding $\bar{\phi} : \pi_{\mathcal{A}}^{-1}(B) \to \mathbf{P}^N \times B$ given by

$$p \mapsto ((1 : f_1(p) : \cdots : f_N(p)), \pi_{\mathcal{A}}(p))$$

and the closure of $\bar{\phi}(\pi_{\mathcal{A}}^{-1}(B))$ is the desired partial compactification. The same construction applies when the functions $f_1, \ldots, f_N$ depend regularly on B, i.e., if $f_i \in \mathcal{O}(B \times M)$ for $i = 1, \ldots, N$.

From now on we will almost exclusively (i.e., except in Section VII.4.5) deal with irreducible a.c.i. systems, defined as follows.

Definition 2.2 An a.c.i. system is called *irreducible* if the general fiber of the momentum map is an affine part of an irreducible Abelian variety.

Example 2.3 If one of the vector fields of an a.c.i. system is super-integrable then the a.c.i. system is not irreducible. Namely suppose that its general level sets are Abelian varieties, then the flow of the super-integrable vector field, being linear on the Abelian variety on the one hand and being contained in a subvariety of lower dimension on the other hand, must evolve on an Abelian subvariety, hence the general fiber of the momentum map would not be an irreducible Abelian variety. Whether an a.c.i. system which is not irreducible admits a super-integrable vector field is not known.

Example 2.4 The product of two a.c.i. systems is an a.c.i. system which is never irreducible. In this case every integrable vector field of the original systems leads to a super-integrable vector field on the product.

On each fiber $\pi_{\mathcal{A}}^{-1}(b)$ of an irreducible a.c.i. system the divisor $\mathcal{D}_b$ induces a polarization $c_1([\mathcal{D}_b])$ since any effective divisor on an irreducible Abelian variety is ample. Thus one may think of the general fiber of the momentum map of an irreducible a.c.i. system as being a *polarized* Abelian variety. If the a.c.i. system is moreover completable then the polarization type of this general level set is constant since it is discrete; probably the assumption that the system is completable is superfluous here but we don't have a proof of this. In any case, if the polarization type of the general level sets of an irreducible a.c.i. system is constant then we call it the *(polarization) type* of the a.c.i. system.

If a divisor is removed from a (completable) a.c.i. system as in Proposition II.2.35 then the resulting integrable Hamiltonian system is also a (completable) a.c.i. system. There are two very distinct possibilities.

- If the function f whose zero divisor is removed belongs to $\mathcal{A}$ then some level sets are left out, the others remain intact. Of course the ones which are left out are not general, so we still have an a.c.i. system;

– If f does not belong to $\mathcal{A}$ its zero divisor cuts off from every level set a divisor (a dramatic change), of course the general level set is still an affine part of an Abelian variety (or Abelian group in general).

For an a.c.i. system which depends on parameters as in Proposition II.3.24 we easily see that the big integrable Hamiltonian system which is given by the latter proposition is also a.c.i.; clearly it contains as its level sets all level sets of all the a.c.i. systems obtained by freezing the parameters.

Some care has to be taken when restricting a (completable) a.c.i. system to a level set of the Casimirs. By Proposition II.3.19 one gets on a *general* level set an integrable Hamiltonian system; in order for it to be a (completable) a.c.i. system one has to verify in addition that the general level set of this restriction is contained in the collection of general fibers (which are known to be affine parts of Abelian groups) of the original system. For special level sets of the Casimirs we may not even get an integrable Hamiltonian system.

Less obvious is the useful property that the quotient construction also leads to a.c.i. systems as given in the following proposition[16].

Proposition 2.5 *Let $(M, \{\cdot, \cdot\})$ be a Poisson variety with a Poisson action by a finite group G. If $(M, \{\cdot, \cdot\}, \mathcal{A})$ is a (completable) a.c.i. system with an affine part of an Abelian variety as its general level set and for each $g \in G$ one has $g \cdot \mathcal{A} \subset \mathcal{A}$ then $(M/G, \{\cdot, \cdot\}_0, \mathcal{A}^G)$ is a (completable) a.c.i. system (with an affine part of an Abelian variety as its general level set) and the quotient map π is a morphism ($\{\cdot, \cdot\}_0$ is the quotient bracket on M/G).*

Proof

In view of Proposition II.3.25 it suffices to identify the general fiber of the momentum map as being isomorphic to an affine part of an Abelian variety. Clearly the action descends to an action on $\text{Spec}\,\mathcal{A}$, (denoted in the same way) namely for each $g \in G$ one has the following commutative diagram.

$$
\begin{array}{ccc}
M & \xrightarrow{\;g\cdot\;} & M \\
{\scriptstyle \pi_{\mathcal{A}}}\downarrow & & \downarrow{\scriptstyle \pi_{\mathcal{A}}} \\
\text{Spec}\,\mathcal{A} & \xrightarrow[\;g\cdot\;]{} & \text{Spec}\,\mathcal{A}
\end{array}
$$

Since $(M, \{\cdot, \cdot\}, \mathcal{A})$ is a.c.i. there is a Zariski open subset $B \subset \text{Spec}\,\mathcal{A}$ and a bundle $\pi : T \to B$ whose fibers are Abelian varieties which compactify the fibers of $\pi_{\mathcal{A}}$ over B. Since G is finite, B may be assumed stable for the action of G; also by passing to a smaller group if necessary we may assume that the action is effective.

Let $b \in B$ be a general point. Since each $g\cdot$ maps level sets (of $\pi_{\mathcal{A}}$) to level sets and since all $g\cdot$ are invertible, each $g\cdot$ restricts to an isomorphism $\pi_{\mathcal{A}}^{-1}(c) \to \pi_{\mathcal{A}}^{-1}(g \cdot b)$ which after composing with ϕ_b and $\phi_{g\cdot b}$ leads to an isomorphism $\pi^{-1}(b)\backslash \mathcal{D}_b \to \pi^{-1}(g\cdot b)\backslash \mathcal{D}_{g\cdot b}$ of the affine parts of the corresponding Abelian varieties; since $g\cdot$ is a Poisson map and ϕ_b and $\phi_{g\cdot b}$ linearize

---

[16] The proposition is not valid if the action is generated by a quasi-automorphism (of finite order). If one considers for example the $\mathbf{Z}_2$ action generated by a (-1)-involution then the general level set of the quotient is an affine part of the Kummer variety of the Abelian variety; obviously also the vector fields do not descend to the quotient.

the Hamiltonian vector fields, each $g\cdot$ extends to an isomorphism $g\cdot : \pi^{-1}(b) \to \pi^{-1}(g \cdot b)$ hence extends to T. The upshot is that the action of G on M induces an action on T (such that ϕ is equivariant). As we have seen in Paragraph IV.3.1 this implies that each map $g\cdot : T \to T$ restricts to each fiber as a group homomorphism followed by a translation.

At first suppose that for any $g \in G$ the level set of π over any closed point is mapped to another level set of π then we easily pass to the quotient, giving a bundle $T/G \to B/G$ whose fibers are isomorphic to the original fibers of $\pi : T \to B$, hence are Abelian varieties and we are done. If for any $g \in G$ the level set of π over a general point is mapped to another level set of π then we may replace B and T by Zariski open subsets which put us in the previous situation leading again to an a.c.i. system. In this case, if the original a.c.i. system is completable then obviously the quotient is also completable.

The situation is very different when the general fiber is stable for the action of G, because in this case we get non-trivial quotients of these level sets. In this case all points of Spec $\mathcal{A}$ are fixed (for the induced action), hence all fibers are stable and $\mathcal{A} \subset \mathcal{O}(M)^G$. Since for any $g \in G$ the morphism $g\cdot$ is Poisson, all vector fields X_f, $f \in \mathcal{A}$ are preserved by the action; by linearity of these vector fields the action of each $g \in G$ on the level sets of π is by translation over an integral part of a period. The quotient of an Abelian variety by a finite group of translations is again an Abelian variety and we may form the quotient of T by the action of G, obtaining a new bundle T/G of Abelian varieties over B. Thus in this case the quotient is also a.c.i., but the Abelian varieties which appear in this quotient system are not isomorphic but merely isogeneous to the Abelian varieties which appear in the original a.c.i. system (in particular the quotient system will most often have a different polarization type). Again completability of the a.c.i. system (if present) is not lost. Notice also that in this last case the group G acts as a group of translations, hence G is commutative if its action is effective.

We are left with the case in which some elements of G map the level sets over every (or the generic) closed point to another level set, while some other elements fix all these level sets. In this case one may take the quotient in two steps, since the subgroup of G which corresponds to the latter elements is a normal subgroup of G. ∎

The following converse of the above proposition is not true. Let $(M, \{\cdot, \cdot\})$ be a Poisson variety with a Poisson action by a finite group G and suppose that $(M, \{\cdot, \cdot\}, \mathcal{A})$ is an integrable Hamiltonian system such that for each $g \in G$ one has $g \cdot \mathcal{A} \subset \mathcal{A}$. If the quotient $(M/G, \{\cdot, \cdot\}_0, \mathcal{A}^G)$ (which we know to be an integrable Hamiltonian system) happens to be an a.c.i. system, then it does *not* follow that $(M, \{\cdot, \cdot\}, \mathcal{A})$ is itself a.c.i. See Paragraph VII.6.2 for a counterexample.

Notice that in the first case treated in the proof of the proposition the map $g\cdot : B \to B$ is a covering map over the moduli space of Abelian varieties which appear as level sets in the a.c.i. system. The quotient B/G is an intermediate object between B and this moduli space.

3. Painlevé analysis for a.c.i. systems

The differential equations describing an integrable vector field of an irreducible a.c.i. system possess families of Laurent solutions (see [AM7]). In slightly different terms this was already known to Kowalevski; her original idea was taken up and extended by Adler and van Moerbeke to give necessary and sufficient conditions for algebraic complete integrability in terms of Laurent solutions (see [AM7]). We restrict ourselves here to part of their result[17].

Proposition 3.1   *Let $(M, \{\cdot, \cdot\}, \mathcal{A})$ be an irreducible a.c.i. system. Then for any integrable vector field X_H, $H \in \mathcal{A}$ the space of Laurent solutions has dimension $\dim M - 1$.*

Proof

Pick a general fiber $\mathcal{F}$ of the momentum map $\pi : M \to \operatorname{Spec} \mathcal{A}$. By assumption $\mathcal{F}$ is isomorphic to $\mathcal{T} \setminus \mathcal{D}$ where $\mathcal{T}$ is an Abelian variety and $\mathcal{D}$ is a divisor on $\mathcal{T}$; we denote this isomorphism by ϕ. Let Z be any point on $\mathcal{T}$ and let us choose a system of generators $z_1, \ldots, z_m$ of $\mathcal{O}(M)$; upon restriction to $\mathcal{F}$ they lead to a system of generators of $\mathcal{O}(\mathcal{F})$, which we still denote by z_i. The functions $z_i \circ \phi^{-1}$ provide a system of generators of $\mathcal{O}(\mathcal{T} \setminus \mathcal{D})$ and in terms of these the differential equations for the vector fields $\phi_* X_H$ ($H \in \mathcal{A}$) are linear, hence if $Z \in \mathcal{T} \setminus \mathcal{D}$ then one finds a solution for $z_i \circ \phi^{-1}$, hence also for z_i, which is holomorphic in t. It is just the description of the integral curve of the vector field X_H starting from the point $\phi^{-1}(Z) \in M$. Of course the dimension of the space of such solutions is $\dim M$, since we have precisely one solution for every point of M.

Next, suppose that $Z \in \mathcal{D}$ is such that one or several of the functions $z_i \circ \phi^{-1}$ have a pole at Z; notice that the functions $z_i \circ \phi^{-1}$ which are regular on $\mathcal{T} \setminus \mathcal{D}$ uniquely extend to meromorphic functions which have their poles on at least one irreducible component of $\mathcal{D}$. The divisor of e.g. $z_1 \circ \phi^{-1}$ can be written (uniquely) as

$$(z_1 \circ \phi^{-1}) = \sum_{i=1}^{k} n_i \mathcal{D}_i - \sum_{i=1}^{l} m_i \mathcal{D}_i' \qquad (m_i, n_i \in \mathbf{N} \setminus \{0\}),$$

where all $\mathcal{D}_i$ and $\mathcal{D}_i'$ are different and irreducible. If $z_1 \circ \phi^{-1}$ is not holomorphic around Z then Z belongs to one or more $\mathcal{D}_i'$, but it may belong as well to some of the $\mathcal{D}_i$. In any case, if we pick for each irreducible component of $\mathcal{D}$ a local defining function around Z, say f_i for $\mathcal{D}_i$ and g_i for $\mathcal{D}_i'$ (if Z does not belong to some divisor then the local defining function may be taken as the constant function 1), then $z_1 \circ \phi^{-1}$ is written around Z as

$$z_1 \circ \phi^{-1} = f \frac{f_1^{n_1} f_2^{n_2} \cdots f_k^{n_k}}{g_1^{m_1} g_2^{m_2} \cdots g_l^{m_l}},$$

with f holomorphic around Z and $f(Z) \neq 0$.

We may take linear coordinates $x_1 = t, x_2, \ldots, x_n$ for the torus, and think of the local defining functions as being expressed in terms of these. The t-axis cannot be contained in any of the divisors $\mathcal{D}_i$ or $\mathcal{D}_i'$ since otherwise the general fiber would contain a subtorus, i.e.,

---

[17] Since we are talking in this section about the solutions to some differential equations (in other words the integral curves of a vector field) we will talk here about holomorphic and meromorphic functions rather than regular and rational maps.

be reducible, contrary to our assumptions. It follows that all these functions can (again up to a nonvanishing holomorphic function) be written as a (Weierstrass) polynomial in t (by the Weierstrass Preparation Theorem) and we see that the zero or pole z_1 has in Z (as a function of t) depends on the components of the divisor of z_1 to which Z belongs but also on the singularity these divisors have in Z (since then the first few terms in the series vanish) and on the contact the vector field X_H has with these divisors (for the same reason).

Proceeding in this way for all functions z_i we find a Laurent solution to the differential equations, which starts from Z. Since Z is an arbitrary point of a divisor of the general fiber, the space of Z from which there starts a Laurent solution is of dimension $\dim M - 1$.

Notice that this divisor is contained in but is not necessarily equal to the divisor which needs to be adjoined to the level in order to complete it into an Abelian variety. If there is for every component of $\mathcal{D}$ at least one of the functions $z_i \circ \phi^{-1}$ which has a pole on this component, then a Laurent solution starts from an arbitrary point of $\mathcal{D}$ and we have an exact bijection of the space of Laurent solutions and the points to be added to the general level sets in order to complete them into Abelian varieties. This is the case in all examples that we will consider. ∎

The Laurent series organize themselves naturally in families as follows: for every z_i, fix an intersection of some divisors (contained in the divisor of poles of (z_i)), fix an order of singularity and an order of tangency of the vector field. On this set all z_i are written as Laurent series depending on a number of free parameters, equal to the dimension of this set (corresponding to the starting point of the series which can be chosen in it) and in a dense subset the order of pole each expansion experiences is fixed. The pole may however become less severe in an analytic subset, obtained from the intersection with one of the divisors on which z_i has a zero; in such a case the leading coefficient of the Laurent series must be (dependent on) a free parameter, so that it can in particular take the value 0. Thus, there is always at least one family of Laurent solutions which depends on $\dim M - 1$ free parameters (called a *principal balance*). If there is for every component of $\mathcal{D}$ at least one of the functions $z_i \circ \phi^{-1}$ which has a pole on this component, then there are as many principal balances as irreducible components in $\mathcal{D}$. There are also always families of Laurent solutions which depend on fewer free parameters, known as *lower balances*.

The geometric study of the Laurent solutions of one of the integrable vector fields of an integrable Hamiltonian system is called the *Painlevé analysis* of this system.

Remark 3.2 The different sets which correspond to the different balances do *not* give a stratification of the Abelian variety in general; indeed, if, for example, z_1 and z_2 both have a pole on some smooth divisor and the intersection of these divisors is singular, then the singular locus of this intersection will in general not show up as a separate family of Laurent solutions, leading to a singular stratum. An example where the Laurent solutions do lead to a (family of) stratification(s) of all hyperelliptic Jacobians will be given in Paragraph VI.4.2.

Finding all Laurent solutions in a direct way is in general a difficult task. One encounters the following problems.

1) It is not clear by looking at the differential equations with which exponents to start in order to find a (all) solution(s);

2) For a given choice of exponents one needs to solve a nonlinear system of algebraic equations (called the *indicial equations*) for the leading term, which may be

very difficult, especially when the number of variables is indefinite (see e.g., Paragraph VI.4.2);

3) The presence of free parameters (giving information about the dimension of the corresponding subset) can in favourable cases be detected by computing the eigenvalues of a matrix, depending on these leading terms, but this is again very difficult when the number of variables, hence the size of the matrix, is indefinite;

4) One also has to show the convergence of all Laurent solutions;

5) It is not obvious to figure out how the different sets the different families of Laurent solutions correspond to are related (see [AM7]).

The main use that we will make of Painlevé analysis in the remaining chapters is not to detect a.c.i. systems (see e.g. [Hai2]) or to prove algebraic complete integrability (see [AM7]), but to construct explicit projective embeddings of the Abelian varieties whose affine parts appear as the fibers of the momentum map.

Proposition 3.3 *Let $\mathcal{T}$ be an Abelian variety and let $\mathcal{D}$ some divisor such that $\mathcal{N} = \mathcal{T} \setminus \mathcal{D}$ is an affine variety; consider also a linear vector field Y on $\mathcal{T}$. If for every irreducible component of $\mathcal{D}$ the space of Laurent solutions, which corresponds to a general point of it, is explicitly known (in the form of the first few terms), then an explicit embedding of $\mathcal{T}$ in projective space can be computed concretely from it.*

Proof

We know that any irreducible component $\mathcal{D}_i$ of $\mathcal{D}$ is ample since $\mathcal{T}$ is irreducible; moreover $3\mathcal{D}_i$ is very ample and we may construct an embedding of the Abelian variety by using $3\mathcal{D}_i$. For any irreducible component of $\mathcal{D}$ these Laurent solutions express a system of generators $z_1, \ldots, z_m$ of $\mathcal{O}(M)$ (restricted to $\mathcal{N}$) as Laurent series in t, hence can be used to express any element of $\mathcal{O}(M)$ (restricted to $\mathcal{N}$) in terms of t. What is important here is that the pole which the Laurent solution of a function z_i on $\mathcal{N}$ has in t coincides with the pole the extension of f to $\mathcal{T}$ has on the divisor which corresponds to the Laurent solution; this follows as above by writing z_i (or $z_i \circ \phi^{-1}$) as a quotient of holomorphic functions which in turn are expressed in terms of Weierstrass polynomials, but now this is done at a general point of the divisor. This allows to look for elements of $\mathcal{O}(M)$ which lead to independent meromorphic functions on $\mathcal{T}$ which have a certain pole at $\mathcal{D}'$ but are holomorphic on the other divisors in $\mathcal{D}$. This is a finite proces: we can first look for functions of degree one, then of degree two and so on and by Formula (IV.3.1) we know when we have found a complete set of (independent) functions and these provide the embedding. ∎

Amplification 3.4 One can also construct an embedding by using some combination of the different components in the divisor, taking these components only with multiplicity one or two. It is easy to figure out which set of multiplicities suffices when the Néron-Severi group of the Abelian variety is trivial (i.e., is equal to $\mathbf{Z}$). Notice that since this is the case for a generic Abelian variety this is a mild assumption. Then every component $\mathcal{D}_i$ is algebraically equivalent to a multiple of some fixed divisor and it suffices that, under this algebraic equivalence, the divisor which one picks to construct the embedding is equivalent to (at least) three times this divisor. One fixes such a choice of embedding divisor $\mathcal{D}'$ and determines as before the Laurent solutions which correspond to the irreducible components which belong to this divisor as well as the Laurent solutions which correspond to the remaining components in $\mathcal{D}$; from it a concrete embedding can be constructed.

4. The linearization of two-dimensional a.c.i. systems

Let $(M, \{\cdot, \cdot\}, \mathcal{A})$ be a two-dimensional a.c.i. system and let us denote for any $c \in \operatorname{Spec} \mathcal{A}$ the fiber of the momentum map over c by $\mathcal{F}_c$. If c is general then $\mathcal{F}_c$ completes into an Abelian surface by adding one or several (possibly singular) curves. The following algorithm, proposed in [Van2] leads to an explicit linearization (i.e., integration) of any integrable vector field X_H, $H \in \mathcal{A}$. (steps (1) and (2) are due to Adler and van Moerbeke, see [AM9]).

(1) Compute the first few terms of the Laurent solutions to the differential equations, and use these to construct an embedding of the general fiber $\mathcal{F}_c$ in projective space (Proposition 3.3).

(2) Deduce from the embedding the structure of the divisors $\mathcal{D}_c$ to be adjoined to $\mathcal{F}_c$ in order to complete $\mathcal{F}_c$ into an Abelian surface $\mathcal{T}_c^2$. At this point the type of polarization induced by each irreducible component of $\mathcal{D}_c$ can also be determined (see Section IV.3.3).

(3) a) If one of the components of $\mathcal{D}_c$ is a smooth curve Γ_c of genus two, compute the image of the rational map
$$\phi_{[2\Gamma_c]} : \mathcal{T}_c^2 \to \mathbf{P}^3$$
which is a singular surface in $\mathbf{P}^3$, the Kummer surface $\mathcal{K}_c$ of $\operatorname{Jac}(\Gamma_c)$.

b) Otherwise, if one of the components of $\mathcal{D}_c$ is a $d : 1$ unramified cover $\mathcal{C}_c$ of a smooth curve Γ_c of genus two, $p : \mathcal{C}_c \to \Gamma_c$, the map p extends to a map $\bar{p} : \mathcal{T}_c^2 \to \operatorname{Jac}(\Gamma_c)$. In this case, let $\mathcal{E}_c$ denote the (non-complete) linear system $\bar{p}^*|2\Gamma_c| \subset |2\mathcal{C}_c|$ which corresponds to the complete linear system $|2\Gamma_c|$ and compute now the Kummer surface $\mathcal{K}_c$ of $\operatorname{Jac}(\Gamma_c)$ as the image of
$$\phi_{\mathcal{E}_c} : \mathcal{T}_c^2 \to \mathbf{P}^3.$$

c) Otherwise, change the divisor at infinity so as to arrive in case a) or b). This can always be done for any irreducible Abelian surface.

(4) Choose a Weierstrass point W on the curve Γ_c and coordinates $(z_0 : z_1 : z_2 : z_3)$ for $\mathbf{P}^3$ such that $\phi_{[2\Gamma_c]}(W) = (0 : 0 : 0 : 1)$ in case (3) a) and $\phi_{\mathcal{E}_c}(W) = (0 : 0 : 0 : 1)$ in case (3) b). Then this point will be a singular point (node) for $\mathcal{K}_c$ and $\mathcal{K}_c$ has an equation
$$p_2(z_0, z_1, z_2) z_3^2 + p_3(z_0, z_1, z_2) z_3 + p_4(z_0, z_1, z_2) = 0,$$
where the p_i are polynomials of degree i. After a projective transformation which fixes $(0 : 0 : 0 : 1)$ we may assume that
$$p_2(z_0, z_1, z_2) = z_1^2 - 4z_0 z_2.$$

(5) Finally, let x_1 and x_2 be the roots of the quadratic equation $z_0 x^2 + z_1 x + z_2 = 0$, whose discriminant is $p_2(z_0, z_1, z_2)$, with the z_i expressed in terms of the original variables $q_1, \ldots, q_4$. Then the differential equations describing the vector field X_H are rewritten by direct computation in the classical Weierstrass form
$$\frac{dx_1}{\sqrt{f(x_1)}} + \frac{dx_2}{\sqrt{f(x_2)}} = \alpha_1 dt,$$
$$\frac{x_1 dx_1}{\sqrt{f(x_1)}} + \frac{x_2 dx_2}{\sqrt{f(x_2)}} = \alpha_2 dt,$$

where α_1 and α_2 depend on c (i.e., on the torus) only. From it, the symmetric functions $x_1 + x_2 (= -z_1/z_0)$ and $x_1 x_2 (= z_2/z_0)$ and hence also all functions in $\mathcal{O}(M)$ can be written in terms of the Riemann theta function associated to the curve $y^2 = f(x)$.

We will show below (in Section VII.5) on a non-trivial example that this algorithm is very effective and easy to apply. Other worked-out examples can be found in [Van2]. We wish to make the remark that it is also shown in [Van2] how a Lax representation and action-angle variables for the system derive from the above linearization.

5. Lax equations

An interesting way to construct integrable Hamiltonian systems is by means of Lax equations. In many cases they turn out to be even a.c.i. We give a sketch of how this works in the case of finite dimensional integrable Hamiltonian systems. The literature on Lax equations is immense, original references are [Lax], [Fla1], [Fla2], [AM2], [AM3], [RS1] and [RS2]; the approach we describe here is due to Semenov-Tian-Shansky (see [Sem]). We also wish to note here Garnier's paper [Gar], since it is the first paper (as early as 1919) in which Lax equations (with spectral parameter!) were written down.

Let $\mathfrak{g}$ be a Lie algebra with Lie bracket $[\cdot,\cdot]$ and R an endomorphism of $\mathfrak{g}$. The new bracket

$$[X,Y]_R = \frac{1}{2}\left([RX,Y]+[X,RY]\right)$$

satisfies the Jacobi identity (hence defines another Lie bracket on $\mathfrak{g}$) if and only if

$$[B_R(X,Y),Z]+[B_R(Y,Z),X]+[B_R(Z,X),Y]=0, \tag{5.1}$$

with

$$B_R(X,Y)=[RX,RY]-R\left([RX,Y]+[X,RY]\right).$$

If so, then R is said to define a structure of a *double Lie algebra* on $\mathfrak{g}$. A particular class of solutions to (5.1) is found by looking for solutions to $B_R(X,Y)=0$, i.e.,

$$[RX,RY]=R\left([RX,Y]+[X,RY]\right).$$

the so-called *classical Yang-Baxter equation*; more general solutions are obtained via solutions to the *modified classical Yang-Baxter equation*

$$B_R(X,Y)=-c[X,Y], \tag{5.2}$$

where c is a constant. If for example $\mathfrak{g}$ is a direct sum (as a vector space) of two Lie subalgebras, $\mathfrak{g}=\mathfrak{g}_++\mathfrak{g}_-$ then the corresponding projection operators $P_+ : \mathfrak{g} \to \mathfrak{g}_+$ and $P_- : \mathfrak{g} \to \mathfrak{g}_-$ lead to a solution of this modified Yang-Baxter equation (for $c=1$), by taking

$$R=P_+-P_-, \tag{5.3}$$

as is easy to verify. Notice that in this case the formula for the R-bracket takes the following simple form

$$[X,Y]_R=[P_+X,P_+Y]-[P_-X,P_-Y]. \tag{5.4}$$

Having two Lie structures on $\mathfrak{g}$ we also have two Lie-Poisson structures on $\mathfrak{g}^*$, denoted by $\{\cdot,\cdot\}$ and $\{\cdot,\cdot\}_R$. Then the relevance of double Lie algebras for integrable Hamiltonian systems comes from the following proposition:

Proposition 5.1 Cas($\{\cdot,\cdot\}$) *is involutive for the R-bracket* $\{\cdot,\cdot\}_R$.

Proof
Recall from Example II.2.8 the explicit formula

$$\{f,g\}(\xi)=\xi\left(\left[\widehat{df(\xi)},\widehat{dg(\xi)}\right]\right)$$

5. Lax equations

for the Lie-Poisson bracket ($f, g \in \mathcal{O}(\mathfrak{g}^*)$, $\xi \in \mathfrak{g}^*$). From it, it follows that

$$f \in \operatorname{Cas} \mathcal{O}(\mathfrak{g}^*) \quad \Rightarrow \quad \forall \xi \in \mathfrak{g}^*, \forall x \in \mathfrak{g} \quad \xi\left(\left[\widehat{df(\xi)}, x\right]\right) = 0.$$

Therefore, if $f, g \in \operatorname{Cas} \mathcal{O}(\mathfrak{g}^*)$ then

$$\{f, g\}_R(\xi) = \xi\left(\left[R\widehat{df(\xi)}, \widehat{dg(\xi)}\right]\right) + \xi\left(\left[\widehat{df(\xi)}, R\widehat{dg(\xi)}\right]\right) = 0,$$

showing that they are in involution with respect to the R-bracket. ∎

Thus the algebra generated by $\operatorname{Cas}(\{\cdot, \cdot\})$ and $\operatorname{Cas}(\{\cdot, \cdot\}_R)$ is a good candidate of being integrable on ($\mathfrak{g}^*, \{\cdot, \cdot\}_R$); in order for $\operatorname{Cas}(\{\cdot, \cdot\})$ to be big enough to imply integrability one often has to restrict the phase space to a Poisson subvariety. For many choices of $\mathfrak{g}$ and R one finds indeed interesting integrable Hamiltonian systems in this way; for example if $\mathfrak{g}$ is a simple Lie algebra then the root space decomposition of $\mathfrak{g}$ leads to a natural choice for R similar to (5.3) and one finds a large family of integrable Hamiltonian systems, the *generalized Toda lattices*.

If $\mathfrak{g}$ admits a nondegenerate, invariant bilinear form then the phase space $\mathfrak{g}^*$ can be identified with $\mathfrak{g}$ and the differential equations which describe the integrable vector fields $X_f = \{\cdot, f\}_R$ on $\mathfrak{g}$ (for $f \in \operatorname{Cas}(\{\cdot, \cdot\})$) have the following peculiar form

$$X_f L = [L, M_f] \quad \text{where} \quad M_f = -\frac{1}{2} R(\widehat{df(L)}). \tag{5.5}$$

Such an equation is called a *Lax equation* and an integrable Hamiltonian system on a Poisson subspace of the dual of a double Lie algebra will be said to be in *Lax form* or to be of *Lax type* (to be distinguished from Lax representations, defined below). The determination of the integral curves of (5.5) can be reduced to the Riemann problem (see [RS1]).

One may however also consider the loop algebra $\mathfrak{g}[\lambda, \lambda^{-1}]$ of a Lie algebra $\mathfrak{g}$, which inherits a Lie structure from $\mathfrak{g}$ and has a natural decomposition

$$\mathfrak{g}[\lambda, \lambda^{-1}] = \mathfrak{g}[\lambda] \oplus \lambda^{-1}\mathfrak{g}[\lambda^{-1}],$$

with corresponding R-bracket given by (5.3). It also has a nondegenerate, invariant bilinear form if $\mathfrak{g}$ has one. In this case the above Lax equations (5.5) retain their form but L and M_f are now dependent on λ; this λ is called a *spectral parameter* and the Lax equations are also said to be *Lax equations with a spectral parameter*. Of course the loop algebra is infinite dimensional but in concrete examples a finite-dimensional Poisson subspace is usually considered. An example will be discussed in detail in Section VI.3.

Integrable Hamiltonian systems of Lax type for which the Lie algebra is a loop algebra often turn out to be a.c.i. Still assuming that $\mathfrak{g}$ has a nondegenerate, invariant bilinear form, we may identify $\mathfrak{g}^*$ with $\mathfrak{g}$ and the spectral invariants of $\mathfrak{g}$ are Casimirs of $\{\cdot, \cdot\}$ when viewed as functions on $\mathfrak{g}^*$. Applied to the case of a loop algebra, $L(\lambda) \in \mathfrak{g}[\lambda, \lambda^{-1}]$ and the characteristic polynomial $\det(L(\lambda) - z\operatorname{Id})$ defines the affine part of an algebraic curve $\Gamma_{L(\lambda)}$ in $\mathbf{C} \times \mathbf{C}^*$, namely the curve

$$\Gamma_{L(\lambda)} : \det(L(\lambda) - z\operatorname{Id}) = 0$$

141

which is called the *spectral curve*. Thus there is associated to each matrix $L(\lambda)$ an algebraic curve and $L(\lambda)$ moves under the flow of X_f in such a way that the curve which is associated to it is constant; what moves under this flow is a line bundle of degree 0 on this spectral curve, which is constructed from the eigenvectors of $L(\lambda)$ (for the precise construction, see [Gri]) and thus the flow of $L(\lambda)$ can be seen as a flow on $\mathrm{Pic}^0(\Gamma_{L(\lambda)})$, i.e., on the Jacobian of $\Gamma_{L(\lambda)}$ (more precisely on the Jacobian of the Riemann surface obtained by compactifying $\Gamma_{L(\lambda)}$ over 0 and ∞). Linearity of this flow is not guaranteed, although for many examples of interest this vector field *is* linear and the above eigenvector map is said to linearize the integrable Hamiltonian system; the Lax equations constructed in Paragraph III.2.4 for example do not have this property except in the special case considered in Paragraph VI.4.2. This can for example be checked by the necessary and sufficient conditions for the eigenvector map of a vector field (which is defined by a Lax equation) to linearize, which is given in [Gri].

The name Lax equation is also often used in the following sense (but it is not explained in this way). Let $(M, \{\cdot, \cdot\}_1, \mathcal{A}_1)$ be an integrable Hamiltonian system and $(N, \{\cdot, \cdot\}_2, \mathcal{A}_2)$ be another integrable Hamiltonian system which is assumed to be of Lax type (as defined above). Then a finite morphism $\phi : (M, \{\cdot, \cdot\}_1, \mathcal{A}_1) \to (N, \{\cdot, \cdot\}_2, \mathcal{A}_2)$ is called a *Lax representation* of the integrable Hamiltonian system $(M, \{\cdot, \cdot\}_1, \mathcal{A}_1)$. Notice that even if the integrable Hamiltonian system on N is a.c.i. the one on M needs not be, moreover the Lax equations (5.5) lead to similar equations for (some of) the integrable vector fields on M, namely we may compose the Lax equation (5.5) with ϕ to obtain

$$X_f L \circ \phi = [L \circ \phi, M_f \circ \phi], \qquad (5.6)$$

which represents the integrable vector field $X_{\phi^* f}$ on M. Notice that not all integrable vector fields X_g with $g \in \mathcal{A}_1$ are of this form since $\phi^* \mathcal{A}_2 \neq \mathcal{A}_1$ (in general). Also the vector fields (5.6) are not vector fields on an affine subspace of the dual of a Lie algebra. Special care has to be taken when determining information about the level sets of $\mathcal{A}_1$, which are covers of (Zariski open subsets of) the level sets of $\mathcal{A}_2$; even when the degree is one, these need not be isomorphic.

Chapter VI

The Mumford systems

1. Introduction

In this chapter we introduce the odd and the even Mumford systems. The odd Mumford system (sometimes also called the *Jacobi-Mumford system*) was first constructed by Mumford in [Mum5], who attributes the idea of the construction to Jacobi. We will not repeat Mumford's construction here, but we will show how this system (and its generalization to $n \times n$-matrices) is obtained naturally as a (finite-dimensional) affine space of commuting differential operators (of fixed degree), with commuting vector fields coming from the KP vector fields (Section 2). It will be shown in Section 3 that this affine space, which may be thought of as a subspace of the loop algebra $\widetilde{\mathfrak{gl}}_q$ of $\mathfrak{gl}_q$, carries a multi-Hamiltonian structure, which comes from a natural family of compatible R-brackets on $\widetilde{\mathfrak{gl}}_q$. Moreover the commuting vector fields turn out to be multi-Hamiltonian and all ingredients of the system (its functions in involution, its compatible Poisson structures and its integrable vector fields) are neatly related by a symmetry vector field $\mathcal{W}$. For $q = 2$ we get the odd Mumford system; in fact, following Mumford, we will take the matrices of the odd Mumford system traceless, which corresponds to fixing a level set of some of the Casimirs.

There is another way to construct the odd Mumford system, namely it can be obtained as a special case of the hyperelliptic case, discussed in Paragraph III.2.4, with a very special choice of Poisson structure (namely the easiest one, which corresponds to $\varphi = 1$) and the underlying space $\mathbf{C}^{2d}$ should have a dimension which is twice the genus of the hyperelliptic curve. We will explain this in Paragraph 4.2 by showing how this construction leads to the *even* Mumford system, a variant of the odd Mumford system which was first introduced in [Van2].

In Paragraph 4.3 we will exhibit some interesting features of the even and the odd Mumford systems. Namely we will show how Painlevé analysis leads to a stratification of the Jacobians which appear as the fibers of the momentum map; notice that the stratification which is obtained in the even case is very different from the one which is induced in the odd case. These stratifications are also obtained from the natural stratification of the Sato Grassmannian via (an extension of) the Krichever map, but this will not be discussed here (see [Van4]).

In the final section (Section 5) we show how to construct for any smooth curve in $\mathbf{C}^2$ an integrable Hamiltonian system which has a level set (of the momentum map) which is isomorphic to an affine part of the Jacobian of this curve. Moreover, the restriction of the integrable vector fields to this level are linearized by this isomorphism. For a large and most important class of curves this leads to an a.c.i. system. The construction which we give here is a modification of the one given in Chapter III, in particular the construction is also completely explicit. A generalization, which is not a.c.i. was given in Chapter III and is further generalized in [Van5] to arbitrary families of curves. For a generalization of the construction in Section 2 to matrix differential operators, see [KV]. A variant of the (even) Mumford system in which the polynomials have parities (odd/even) has recently be constructed in [FV].

144

2. Genesis

2.1. The algebra of pseudo-differential operators

We start by reviewing the basic definitions and properties of (formal) pseudo-differential operators. Let $\mathcal{D}$ denote the non-commutative algebra $\mathbf{C}[[x]][\partial]$ of differential operators, the multiplication being given by juxtaposition and applying the commutation rule $[\partial, a(x)] = a'(x)$; here $a(x)$ is any formal power series, $a(x) \in \mathbf{C}[[x]]$, and $a'(x)$ its (formal) derivative. $\mathcal{D}$ has as a distinguished maximal commutative subalgebra the algebra $\mathcal{D}^c = \mathbf{C}[\partial]$ of constant coefficient differential operators. $\mathcal{D}$ is contained in the larger algebra $\Psi = \mathbf{C}[[x]]((\partial^{-1}))$ of *pseudo-differential operators*, the (associative) rule of multiplication being formally derived from the commutation rule $[\partial, a(x)] = a'(x)$, i.e.,

$$\partial^{-1} a(x) = \sum_{i=0}^{\infty} (-1)^i a^{(i)}(x) \partial^{-i-1}.$$

An element $Q \in \Psi$, $Q = \sum_{i=-\infty}^{q} a_i \partial^i$, is said to have *order* q if $a_q \neq 0$ and is said to be *monic (of order q)* when $a_q = 1$; if, in addition, $a_{q-1} = 0$ then Q is called *normalized (of order q)*. The subalgebra $\mathbf{C}((\partial^{-1}))$ of Ψ consisting of constant coefficient pseudo-differential operators is denoted by Ψ^c. The following properties of differential operators are easily established.

Proposition 2.1

(1) *Every monic pseudo-differential operator Q of order q has a unique inverse Q^{-1} in Ψ. In particular the monic pseudo-differential operators of order zero form a group, called the Volterra group and denoted by* Volt. *Its subgroup of constant coefficient operators is denoted by* Volt^c.

(2) *Every normalized differential operator Q of order $q > 0$ has a unique monic q-th root $Q^{1/q}$ in Ψ. This root $Q^{1/q}$ is normalized and has order 1.*

(3) *Every normalized differential operator Q of order $q > 0$ is conjugated to ∂^q, i.e., $Q = T^{-1} \partial^q T$ for some $T \in$ Volt which is unique up to left multiplication by an element of* Volt^c.

Finally we recall the definition of the Sato Grassmannian Gr. Let us denote by δ Dirac's delta function, thought of as a zeroth order differential operator. It has the fundamental property that for any $Q \in \Psi$ there exists a unique $Q^c \in \Psi^c$ such that $Q\delta = Q^c\delta$. The left coset

$$V := \mathbf{C}((\partial^{-1}))\delta = \Psi^c\delta$$

is a left Ψ-module in a natural way: for $P \in \Psi$ and for $Q \in \mathbf{C}((\partial^{-1})) \subset \Psi$ we define $P \cdot (Q\delta) = (PQ)\delta$. A distinguished subspace of V is defined by

$$H = \mathbf{C}[\partial]\delta = \mathcal{D}^c\delta.$$

The multiplication allows us to associate to each element in Ψ a subspace of $\mathbf{C}((\partial^{-1}))\delta$ in a natural way, namely given $Q \in \Psi$ define $W_Q \subset \mathbf{C}((\partial^{-1}))\delta$ by $W_Q = Q \cdot H$. We call the set of all W_T which correspond to elements T in the Volterra group the *Sato Grassmannian*,

$$\text{Gr} := \{W_T \mid T \in \text{Volt}\}.$$

If $T \in \mathrm{Volt}$ the linear space W_T has a basis, similar to the standard basis of H, namely it has a basis whose elements have the form

$$T \cdot \partial^i = (\partial^i + \mathcal{O}(\partial^{i-1}))\delta, \qquad (i \geq 0). \tag{2.1}$$

It follows that the map $\mathrm{Volt} \to \mathrm{Gr}$ given by $P \mapsto W_P$ is injective hence bijective. The following proposition gives a useful characterization of differential operators in terms of this map (see [Mul3]).

Proposition 2.2 *Let $Q \in \Psi$. Then $Q \in \mathcal{D}$ if and only if $W_Q \subset H$.*

Since no confusion is possible we remove δ from the elements of W_Q, i.e., we identify V with $\mathbf{C}((\partial^{-1}))$.

2.2. The matrix associated to two commuting operators

In this section we assume that P and Q are monic differential operators which commute, $[P, Q] = 0$. We will assume that one of them, say Q, is normalized. Finally we assume that the orders p and q of P and Q are positive and coprime. In this paragraph we show how to associate a unique element $M \in \mathfrak{gl}_q[\lambda]$ to (P, Q); it will be shown in the next section how to reconstruct (P, Q) from M.

In a first step we associate to the pair (P, Q) a pair $(\tilde{P}, W)$ of elements $\tilde{P} \in \Psi^c$ and $W \in \mathrm{Gr}$ (this pair is unique up to an equivalence specified below). Since Q is normalized there exists by Proposition (2.1) an element $T \in \mathrm{Volt}$ such that $Q = T^{-1}\partial^q T$. Choosing such an element T we define $W = W_T = T \cdot H \in \mathrm{Gr}$. If we let $\tilde{P} = TPT^{-1}$ then $\tilde{P} \in \Psi$ is monic of order p and $[\partial^q, \tilde{P}] = 0$. We claim that the latter implies that $\tilde{P} \in \Psi^c$. To show this it is sufficient to show that if $a(x) \in \mathbf{C}[[x]]$ is such that $[\partial^q, a(x)] = 0$ for some $q > 0$ then $a(x)$ is constant. Since

$$0 = [\partial^q, a(x)] = \partial^q a(x) - a(x)\partial^q = qa'(x)\partial^{q-1} + \mathcal{O}(\partial^{q-2}),$$

we find indeed that $a'(x) = 0$.

The pair $(\tilde{P}, W)$ clearly depends on the choice of T: if T is replaced by $T^c T$ where $T^c \in \mathrm{Volt}^c$ then $(\tilde{P}, W)$ is replaced by

$$T^c \cdot (\tilde{P}, W) = (T^c \tilde{P}(T^c)^{-1}, T^c \cdot W) = (\tilde{P}, T^c \cdot W),$$

since constant coefficient differential operators commute. The above equation defines an action of Volt^c on $\Psi^c \times \mathrm{Gr}$; we say that two elements in $\Psi^c \times \mathrm{Gr}$ are *equivalent* when they correspond under this action. We now give a characterizing property of the pair $(\tilde{P}, W)$ and we use it to show that the pair (P, Q) can be reconstructed from it.

Proposition 2.3 *The element $W \in \mathrm{Gr}$ is stable under the action of ∂^q and $\tilde{P}$,*

$$\begin{aligned}
\partial^q \cdot W &\subset W, \\
\tilde{P} \cdot W &\subset W,
\end{aligned} \tag{2.2}$$

Conversely, suppose that $(\tilde{P}, W) \in \Psi^c \times \mathrm{Gr}$ satisfies (2.2), $\tilde{P}$ being monic. The pair $(\tilde{P}, W)$ is associated to a unique pair (P, Q) of commuting differential operators, such that P is monic and Q is normalized.

Proof

The verification of (2.2) is easy, for example

$$\partial^q \cdot W = T \cdot (Q \cdot H) \subset T \cdot H = W,$$

in which the inclusion $Q \cdot H \subset H$ holds because $Q \in \mathcal{D}$. Let us show how to reconstruct (P, Q) from $(\tilde{P}, W)$. Since $W \in \mathrm{Gr}$ it is of the form $W = T \cdot H$ for a unique $T \in \mathrm{Volt}$. If we define

$$Q = T^{-1} \partial^q T,$$
$$P = T^{-1} \tilde{P} T,$$

then Q is normalized of order q while P is monic of order p; also $[P, Q] = 0$. The crucial property is that (2.2) implies that P and Q are differential operators. We have that $W_Q \subset H$ and $W_P \subset H$ because e.g. for P one computes

$$W_P = P \cdot H = T^{-1} \cdot (\tilde{P} \cdot W) \subset T^{-1} \cdot W = H.$$

Using this, the fact that P and Q are differential operators follows from Lemma (2.2). Clearly, if we replace $(\tilde{P}, W)$ by an equivalent pair then the same pair (P, Q) is obtained. ∎

Notice that in the above proposition the orders of P and Q need not be coprime.

The next step is to associate to the pair $(\tilde{P}, W)$ (up to equivalence) a matrix $M \in \mathfrak{gl}_q[\lambda]$ from which $(\tilde{P}, W)$ can be reconstructed. Since $\tilde{P} \in \Psi^c$ and $\tilde{P} \cdot W \subset W$ it follows that $\tilde{P}$ is an endomorphism of W, hence can be represented by a semi-infinite matrix (with entries in $\mathbf{C}$) by choosing any basis of W. This matrix becomes a periodic matrix (with period q) when a periodic basis $(E_0, E_1, \dots,)$ is chosen, i.e., a basis such that for any basis element E_i the element $\partial^q E_i$ is also a basis element. Note that the existence of a periodic basis follows from the inclusion $\partial^q W \subset W$. We choose q vectors in W of the form

$$E_i = \partial^i + \mathcal{O}(\partial^{i-1}), \qquad (0 \le i \le q-1),$$

and extend them to a basis of W by introducing, for arbitrary $i \ge 0$, the vectors $E_{i+q} = \partial^q E_i$. Periodic matrices can be rewritten as square matrices at the price of allowing entries which are polynomials in $\lambda = \partial^q$. Explicitly, if we define the matrix[18] $M = (m_{ij}) \in \mathfrak{gl}_q[\lambda]$ by

$$\tilde{P} \cdot E_i = \sum_{j=0}^{q-1} m_{ij} E_j, \qquad (0 \le i \le q-1), \tag{2.3}$$

---

[18] The rows and columns of the matrix M and of the matrices e_{ij}, introduced below, are labelled from 0 to $q-1$.

then the elements of M have the following degree constraints:

$$
\begin{pmatrix}
\leq [p/q] & & & [p/q] & & \leq [p/q]-1 \\
& \ddots & & & \ddots & \\
& & \leq [p/q] & & & [p/q] \\
[p/q]+1 & & & \leq [p/q] & & \\
& \ddots & & & \ddots & \\
\leq [p/q]+1 & & [p/q]+1 & & & \leq [p/q]
\end{pmatrix} \tag{2.4}
$$

since

$$
\tilde{P} \cdot E_i = E_{p+i} + \text{ lower order in } \partial
$$
$$
= \lambda^{[\frac{p+i}{q}]} E_{(p+i) \bmod q} + \text{ lower order in } \partial.
$$

It is implicit in (2.4) that when the degree is exact (no inequality signs) then the corresponding polynomial is monic. We denote the affine space of matrices of the form (2.4) by $\mathcal{M}^{p,q}$. Let $N_q^- \subset GL(q)$ denote the subgroup of lower triangular matrices, which acts on $\mathcal{M}^{p,q}$ by conjugation.

Proposition 2.4 *The above procedure associates to the pair (P, Q) a well-defined element of $\mathcal{M}^{p,q}/N_q^-$.*

Proof

Two bases of the form (2.1) are related by an element of N_q^- hence their matrices are N_q^--conjugated. When $(\tilde{P}, W)$ is replaced by any other representative $T^c \cdot (\tilde{P}, W)$ then precisely the same matrix M is obtained when using the basis of $T^c \cdot W$ which is obtained by multiplying all elements of the basis of W by T^c. Notice that such a basis is always of the form (2.1). ∎

As it turns out the quotient space $\mathcal{M}^{p,q}/N_q^-$ is in a natural way isomorphic to an affine subspace $\tilde{\mathcal{M}}^{p,q}$ of $\mathcal{M}^{p,q}$. To show this we need to introduce some notation which is motivated by (2.4). For $0 \leq i, j \leq q-1$ let e_{ij} denote the $(q \times q)$-matrix with a 1 at position (i, j) and zeros elsewhere and decompose $\mathfrak{gl}_q$ as $\mathfrak{gl}_q = \bigoplus_{\gamma=1-q}^{q-1} \mathfrak{g}_\gamma$, where $\mathfrak{g}_\gamma$ is the subspace of $\mathfrak{gl}_q$, generated by the matrices e_{ij} for which $j - i = \gamma$. When $Z \in \mathfrak{g}_\gamma$ we also write $\deg Z = \gamma$. The projection $\mathfrak{gl}_q \to \mathfrak{g}_\gamma$ will be denoted by Π_γ. Let $d = p \bmod q$, $0 < d < q$, and let S and R be the elements of $\mathfrak{gl}_q$, given by

$$
S = \begin{pmatrix} 0 & 0 \\ I_{q-d} & 0 \end{pmatrix}, \qquad R = \begin{pmatrix} 0 & I_d \\ 0 & 0 \end{pmatrix}.
$$

We have that $\deg S = -d$ and that $\deg R = q - d$. With this notation $\mathcal{M}^{p,q}$ consists of all matrices $\sum_{i=0}^{[p/q]+1} M_i \lambda^i$ for which

$$
M_{[p/q]+1} \in S + \bigoplus_{\gamma < -d} \mathfrak{g}_\gamma,
$$
$$
M_{[p/q]} \in R + \bigoplus_{\gamma < q-d} \mathfrak{g}_\gamma.
$$

The affine subspace $\tilde{\mathcal{M}}^{p,q}$ of $\mathcal{M}^{p,q}$ consists of those matrices in $\mathcal{M}^{p,q}$ for which

$$M_{[p/q]+1} = S, \qquad M_{[p/q]} = \begin{pmatrix} 0 & I_d \\ \star & \star \end{pmatrix},$$

where the stars denote arbitrary matrices of the appropriate size. We also introduce an intermediate space $\mathcal{M}_{RS}^{p,q}$, which is also an affine subspace of $\mathcal{M}^{p,q}$. The subspace $\mathcal{M}_{RS}^{p,q}$ consists of all matrices in $\mathcal{M}^{p,q}$ for which

$$M_{[p/q]+1} = S, \qquad M_{[p/q]} \in R + \bigoplus_{\gamma < q-d} \mathfrak{g}_\gamma.$$

Let us denote the Lie algebra of N_q^- by $\mathfrak{n}_q^-$ and let $\mathfrak{g}_S$ denote the isotropy algebra of S,

$$\mathfrak{g}_S = \{ X \in \mathfrak{n}_q^- \mid [X, S] = 0 \},$$

which is the Lie algebra of $G_S = \exp \mathfrak{g}_S$, the stabilizer of S. We have that the (adjoint) action of G_S on $\mathcal{M}^{p,q}$ leaves $\mathcal{M}_{RS}^{p,q}$ invariant. Notice that $\mathfrak{g}_S$ is given explicitly as the algebra of strictly lower triangular matrices M for which $M_{i+d,j+d} = M_{ij}$ for all $0 \leq j < i \leq q-1$.

Proposition 2.5 $\mathcal{M}^{p,q}/N_q^-$ is isomorphic $\mathcal{M}_{RS}^{p,q}/G_S$ which, in turn, is isomorphic to the $q(p+q-2d)$-dimensional affine space $\tilde{\mathcal{M}}^{p,q}$.

Proof

We show that every element in $\mathcal{M}^{p,q}$ is N_q^--conjugate to a unique element of $\tilde{\mathcal{M}}^{p,q}$. The proof then follows from the fact that the corresponding map $\mathcal{M}^{p,q} \to \tilde{\mathcal{M}}^{p,q}$ is regular. Let $M = \sum_{i=0}^{[p/q]+1} M_i \lambda^i \in \mathcal{M}^{p,q}$. We first show that $M_{[p/q]+1}$ is N_q^--conjugate to S. Take any element $\xi \in \mathfrak{g}_{-1}$ and let $g = \exp \xi \in N_q^-$. Then

$$\text{Ad}_g \, M_{[p/q]+1} = \exp \text{ad}_\xi \, M_{[p/q]+1} = M_{[p/q]+1} + [\xi, M_{[p/q]+1}] + \ldots,$$

which after projection on $\mathfrak{g}_{-d-1}$ becomes

$$\Pi_{-d-1}\left(\text{Ad}_g \, M_{[p/q]+1}\right) = \Pi_{-d-1}\left(M_{[p/q]+1}\right) + [\xi, S].$$

Since the linear map $\text{ad}_S : \mathfrak{g}_{-1} \to \mathfrak{g}_{-d-1}$ is surjective we can pick g (i.e. ξ) such that $\Pi_{-d-1}\left(\text{Ad}_g \, M_{[p/q]+1}\right) = 0$. Repeating this procedure with $\xi \in \mathfrak{g}_{-\gamma}$ where $\gamma = 2, 3, \ldots$ and using the surjectivity of $\text{ad}_S : \mathfrak{g}_{-\gamma} \to \mathfrak{g}_{-d-\gamma}$ we find that $M_{[p/q]+1}$ is N_q^--conjugate to S. Since, by definition, $G_S \subset N_q^-$ is the stabilizer of S and since the map $\mathcal{M}^{p,q} \to \mathcal{M}_{RS}^{p,q}$ can be picked regular it follows that $\mathcal{M}^{p,q}/N_q^-$ is isomorphic to $\mathcal{M}_{RS}^{p,q}/G_S$.

We next show that $M_{[p/q]}$ is G_S-conjugate to a unique element of the form $\begin{pmatrix} 0 & I_d \\ \star & \star \end{pmatrix}$. Notice that if $M \in \mathcal{M}_{RS}^{p,q}$ then $M_{[p/q]} \in R + \bigoplus_{\gamma < q-d} \mathfrak{g}_\gamma$. Fix any $\delta \in \{1, \ldots, q-1\}$ and consider the space of matrices of the form

$$\begin{pmatrix}
\star & \cdots & \star & \star_1 & 0 & \cdots & 0 & 1 & & 0 \\
\vdots & \ddots & & \ddots & \ddots & \ddots & & & \ddots & \ddots \\
\star & \cdots & \star & \cdots & \star & \star_s & 0 & \cdots & 0 & 1 \\
\star & & & \cdots & & & & \cdots & & \star \\
\vdots & & & & & & & & & \vdots \\
\star & & & \cdots & & & & \cdots & & \star
\end{pmatrix} \tag{2.5}$$

where the diagonal with the stars $\star_1, \ldots, \star_s$ and the diagonal with the I_n are precisely a distance δ apart. The number s is given in terms of δ by $s = \min\{d, q - \delta\}$. It is easy to see that the adjoint action of $G_\delta = \exp(\mathfrak{g}_{-\delta} \cap \mathfrak{g}_S)$ on $\mathfrak{gl}_q$ leaves the space (2.5) invariant. Also, at level $q-d-\delta$ (the diagonal where the $\star_i$ in (2.5) are) the adjoint action induces for any element $g = \exp \xi \in G_\delta$ the affine map (translation) $\mathfrak{g}_{q-d-\delta} \to \mathfrak{g}_{q-d-\delta} : Z \mapsto Z + [R, \xi]$. In turn this map induces on $\mathbf{C}^s$ the translation by $\Pi[R, \xi]$ where Π is the linear map $\mathfrak{gl}_q \to \mathbf{C}^s$ which maps the matrix (2.5) to $(\star_1, \ldots, \star_s)$. We denote this map by χ_ξ, thus $\chi_\xi(z) = z + \Pi[R, \xi]$ for $z \in \mathbf{C}^s$. We wish to show that there exists for any $z \in \mathbf{C}^s$ a unique $\xi \in \mathfrak{g}_{-\delta} \cap \mathfrak{g}_S$ such that $\chi_\xi(z) = 0$. Equivalently, that for any $z \in \mathbf{C}^s$ the affine map $\chi_z : \mathfrak{g}_{-\delta} \cap \mathfrak{g}_S \to \mathbf{C}^s : \xi \mapsto \chi_\xi(z)$ is a bijection. Taking the corresponding linear map this means that we need to show that

$$\chi : \mathfrak{g}_{-\delta} \cap \mathfrak{g}_S \to \mathbf{C}^s : \xi \mapsto \Pi[R, \xi]$$

is an isomorphism. Since both spaces have dimension $s = \min\{d, q - \delta\}$ it suffices to show that χ is injective. Let $\xi = \sum_{\gamma=1}^{s} \xi_\gamma e_{\gamma+\delta,\gamma}$ be in the kernel of χ. If $\delta \geq q - d$ then $s = q - \delta > d - \delta$ and

$$\chi(\xi) = (0, 0, \ldots, 0, \xi_1, \ldots, \xi_{d-\delta}) - (\xi_1, \ldots, \xi_s)$$

so that χ is injective. When $\delta < q - d$ the proof is more delicate and depends on the fact that p and q are relatively prime (indeed if p and q have a common divisor then χ is not injective for some values of δ). Then

$$\chi(\xi) = (0, \ldots, 0, \xi_1, \xi_2, \ldots, \xi_{d-\delta}) - (\xi_{q-d-\delta+1}, \ldots, \xi_{q-\delta}),$$

where ξ_1 appears at position $\delta + 1$, the length of these vectors being $s = d$. If ξ is in the kernel of χ then $\xi_{q-d-\delta+1} = \cdots = \xi_{q-d} = 0$ and $\xi_k = \xi_{q-d+k}$ for $k = 1, \ldots, d - \delta$. But remember that $\xi_k = \xi_{k+d}$ because $\xi \in \mathfrak{g}_S$. This means that the indices of ξ may be thought of as lying in $\mathbf{Z}_d$. Now the fact that p and q are relatively prime implies that d and q are coprime, hence also $q - d$ and q. The fact that $\xi_k = \xi_{q-d+k}$ for $k = 1, \ldots, d - \delta$ then implies that all ξ_γ are equal, hence they are all equal to 0. Thus χ is injective in all cases and we can make all $\star_j$ in (2.5) equal to zero by using a unique element of G_δ; doing this consecutively for $\delta = 1, 2, \ldots, q - 1$ leads to the desired result. ∎

We call $M \in \tilde{\mathcal{M}}^{p,q}$ the *matrix* of (P, Q) or of $(\tilde{P}, W)$.

2.3. The inverse construction

We will now show that every element $M \in \tilde{\mathcal{M}}^{p,q}$ (with p and q coprime) is the matrix of a pair (P, Q) of commuting differential operators such that P is monic of order p and Q is normalized of order q. By Proposition 2.3 it suffices to show that M is the matrix of a pair $(\tilde{P}, W) \in \Psi^c \times \mathrm{Gr}$ satisfying $\partial^q \cdot W \subset W, \tilde{P} \cdot W \subset W$ and $\tilde{P}$ monic. Equivalently we need to show that there exist a monic element $\tilde{P} \in \Psi^c$ of order p and q vectors $E_0, \ldots, E_{q-1}$ in W such that $\mathrm{ord}\, E_i = i$, and such that

$$\tilde{P} \cdot E_i = \sum_{j=0}^{q-1} m_{ij} E_j. \tag{2.6}$$

In order to do this we expand $\tilde{P}$ and E_i in terms of ∂,

$$\tilde{P} = \partial^p + p_1 \partial^{p-1} + p_2 \partial^{p-2} + \cdots,$$
$$E_i = \partial^i + g_1^i \partial^{i-1} + g_2^i \partial^{i-2} + \cdots,$$

$$(2.7)$$

and we define for $r \in \mathbf{Z}$ an element $M^{(r)} \in \mathfrak{gl}_q$ by

$$M^{(r)} = \left(m_{ij}^{(r)} \right)_{0 \le i,j \le q-1} \quad \text{where} \quad m_{ij} = \sum_r m_{ij}^{(r)} \partial^{p+i-j-r}.$$

$$(2.8)$$

Lemma 2.6

(1) $M^{(r)} = 0$ for $r < 0$;

(2) $m_{ij}^{(0)} = \begin{cases} 1 & \text{if } p+i-j = 0 \bmod q, \\ 0 & \text{if } p+i-j \ne 0 \bmod q; \end{cases}$

(3) $M^{(0)}$ is the matrix of a cyclic permutation σ of $\{0, 1, \ldots, q-1\}$.

Proof

If $r < 0$ then

$$p + i - j - r > q \left[\frac{p+i-j}{q} \right] \ge \operatorname{ord} m_{ij}$$

and so $m_{ij}^{(r)}$, which is the coefficient of $\partial^{p+i-j-r}$, is zero. The same inequality holds for $r = 0$ when q does not divide $p+i-j$ so that for such values of i and j we also have $m_{ij}^{(0)} = 0$. If q divides $p+i-j$ then m_{ij} is monic of order $p+i-j$ hence $m_{ij}^{(0)} = 1$. It implies that $M^{(0)}$ is a permutation matrix. The fact that this permutation is *cyclic* follows from the fact that p and q are coprime; indeed, this permutation corresponds to the translation over p in $\mathbf{Z}_q$. ∎

If we plug (2.7) and (2.8) into (2.6) then we find that for any $\gamma \ge 0$ and for $0 \le i \le q-1$

$$\sum_{r=0}^{\gamma} p_r g_{\gamma-r}^i = \sum_{r=0}^{\gamma} \sum_{j=0}^{q-1} m_{ij}^{(r)} g_{\gamma-r}^j,$$

$$(2.9)$$

which can also be written as

$$\sum_{r=0}^{\gamma} \begin{pmatrix} p_r g_{\gamma-r}^0 \\ p_r g_{\gamma-r}^1 \\ \vdots \\ p_r g_{\gamma-r}^{q-1} \end{pmatrix} = \sum_{r=0}^{\gamma} M^{(r)} \begin{pmatrix} g_{\gamma-r}^0 \\ g_{\gamma-r}^1 \\ \vdots \\ g_{\gamma-r}^{q-1} \end{pmatrix},$$

$$(2.10)$$

for any $\gamma \ge 0$. We show that (2.10) can be solved recursively for p_j and g_j^i. Since $p_0 = g_0^i = 1$ for $i = 0, \ldots, q-1$, and since $M^{(0)}$ is a permutation matrix, the equation (2.10) is satisfied identically for $\gamma = 0$. Let us assume that we have constructed p_r and g_r^i for all $r < \gamma$ and $i \in \{0, \ldots, q-1\}$. Then equation (2.10) can be written as

$$\begin{pmatrix} p_\gamma \\ p_\gamma \\ \vdots \\ p_\gamma \end{pmatrix} = (M^{(0)} - I_q) \begin{pmatrix} g_\gamma^0 \\ g_\gamma^1 \\ \vdots \\ g_\gamma^{q-1} \end{pmatrix} + \text{(known stuff)}$$

where "(known stuff)" involves only the previous g_r^i and p_r. Summing up the q rows of this equation we find p_γ because the sum of the rows of any permutation matrix equals the sum of the rows of the identity matrix. We take $g_\gamma^0 = 1$ and we use (2.10) to solve for $g_\gamma^1, \ldots, g_\gamma^{q-1}$. To see how this is done, rewrite (2.10) as follows,

$$(M^{(0)} - I_q) \begin{pmatrix} g_\gamma^0 \\ g_\gamma^1 \\ g_\gamma^2 \\ \vdots \\ g_\gamma^{q-1} \end{pmatrix} = \text{(known stuff)}.$$

Recall that $M^{(0)}$ is a permutation matrix which corresponds to a permutation σ of the set $\{0, 1, \ldots, q-1\}$. Since σ is a cyclic permutation of order q we can solve for the g_γ^i in the following order: solve first for $i = \sigma(0)$, then for $i = \sigma^2(0)$ and so on. Notice that the last equation is precisely the equation defining p_γ, showing that the solution exists and is unique once the vector g_γ^0 has been chosen. Clearly the freedom in choice for g_γ^0 corresponds to the left action of Volt^c. Notice that it is only in the very last step that we used that p and q are coprime.

Summarizing we have shown the following proposition.

Proposition 2.7 *Let $M \in \tilde{\mathcal{M}}^{p,q}$, where p and q are coprime. Then there exists a pair $(\tilde{P}, W) \in \Psi^c \times \text{Gr}$ such that $\tilde{P} \cdot W \subset W$ and $\partial^q \cdot W \subset W$ and such that M is the matrix of $\tilde{P}$. The pair $(\tilde{P}, W)$ is unique up to the left action of Volt^c. The correspondence which associates to the pair (P, Q) its matrix is a bijection between $\tilde{\mathcal{M}}^{p,q}$ and the space of all pairs (P, Q) of commuting differential operators with P monic of order p and Q normalized of order q.*

2.4. The KP vector fields

In this section we will realize the KP vector fields, which are a natural collection of commuting vector fields on the Sato Grassmannian Gr, as a collection of commuting vector fields on the affine space $\tilde{\mathcal{M}}^{p,q}$. In order to write down the KP vector fields on the Grassmannian, let us first show that the tangent space at a point W of the finite-dimensional Grassmannian $G = G(k, n)$ of k planes in $\mathbf{C}^n$ is naturally given by $\text{Hom}(W, \mathbf{C}^n/W)$. To see this, we consider G as the homogeneous space $GL_n/\text{Stab}(W)$, where $GL_n = GL(n, \mathbf{C})$ and $\text{Stab}(W) \subset GL_n$ is the stabilizer of W, with Lie algebra

$$\mathfrak{stab}(W) = \{\phi \in \mathfrak{gl}_n \mid \phi(W) \subset W\}.$$

Then $T_W G = T_W(GL_n/\text{Stab}(W)) = \mathfrak{gl}_n/\mathfrak{stab}(W) = \text{Hom}(W, \mathbf{C}^n/W)$; for the last equality one associates to a representative ϕ of an element of $\mathfrak{gl}_n/\mathfrak{stab}(W)$ the composite map

$$W \longrightarrow \mathbf{C}^n \overset{\phi}{\longrightarrow} \mathbf{C}^n \longrightarrow V/W$$

In the case of the Sato Grassmannian (which is infinite-dimensional) we *define* the tangent space at a point $W \in \text{Gr}$ to be given by $\text{Hom}(W, \Psi^c/W)$, where we consider W in the last equation as a subspace of Ψ^c. In this language the i-th KP vector field is given by $V_i : W \to \Psi^c/W : w \mapsto \partial^i w \bmod W$.

2. Genesis

We first transfer these vector fields to the space of pairs (P, Q) of commuting scalar differential operators with P monic of order p and Q normalized of order q. To do this we use the bijection Volt $\to$ Gr $: T \mapsto W_T$, which gives the following commuting vector fields on Volt

$$\frac{dT}{dt_i} = T(T^{-1}\partial^i T)_-,$$

(see [Mul3]). If we have a constant coefficient pseudo-differential operator $\tilde{U}$ and we define $U = U(t) = T^{-1}\tilde{U}T$ then

$$\frac{dU}{dt_i} = [T^{-1}\tilde{U}T, (T^{-1}\partial^i T)_-] = [U, Q_-^{i/q}] = [Q_+^{i/q}, U].$$

Applying this for $\tilde{U}$ given by ∂^q and by $\tilde{P}$ we find the following Lax representation for the KP vector fields on the above space of commuting operators (P, Q),

$$\frac{dQ}{dt_i} = [Q_+^{i/q}, Q], \qquad \frac{dP}{dt_i} = [Q_+^{i/q}, P].$$

We proceed to write these vector fields down on $\mathcal{M}^{p,q}$. More precisely, we will write down the vector fields that correspond to the constant coefficient scalar differential operators $\left[\tilde{P}^i/\lambda^k\right]_+$; each KP vector field is then a linear combination of these vector fields. We fix i, k and denote the derivative in the direction of the vector field corresponding to $\left[\tilde{P}^i/\lambda^k\right]_+$ by a dot. We choose a periodic basis of W and denote the column vector containing its first q elements by $\vec{e}$. By the above interpretation of the tangent space at W to the Grassmannian, we can write

$$\dot{\vec{e}} = \left[\tilde{P}^i/\lambda^k\right]_+ \vec{e} - A(\partial^q)\vec{e}, \tag{2.11}$$

where A is the polynomial matrix (in $\lambda = \partial^q$) such that the order of the i-th component of the right hand side of (2.11) is smaller than i. Also equation (2.3), which is the defining equation of the matrix $M \in \mathcal{M}^{p,q}$ with respect to the chosen periodic basis, can be rewritten as

$$\tilde{P}\vec{e} = M\vec{e}. \tag{2.12}$$

If we differentiate (2.12) then we find $\tilde{P}\dot{\vec{e}} = \dot{M}\vec{e} + M\dot{\vec{e}}$, which is easily rewritten as

$$\dot{M}\vec{e} = (MA - \tilde{P}A)\vec{e} = [M, A]\vec{e},$$

(one uses that elements of $\mathcal{D}^c$ commute among themselves and with matrices which are independent of x, such as M and A). From the last equality we can conclude that $\dot{M} = [M, A]$, because $\dot{M}$, M and A are polynomials in $\lambda = \partial^q$ (rather than in ∂). We claim that A can be taken as $(M^i/\lambda^k)_+$. To prove this we must show that the i-th component of

$$\left(\tilde{P}^i/\lambda^k\right)_+ \vec{e} - \left(M^i/\lambda^k\right)_+ \vec{e}$$

has order smaller than i (for $i = 1, \ldots, q$). Since $\tilde{P}$ commutes with M we have that

$$\left(\tilde{P}^i/\lambda^k\right)\vec{e} = \left(M^i/\lambda^k\right)\vec{e}$$

153

and it suffices to show that the order of the i-th component of

$$\left(\tilde{P}^i/\lambda^k\right)_- \vec{e} - \left(M^i/\lambda^k\right)_- \vec{e}$$

is smaller than i. Since the i-th component of $\vec{e}$ has order i this is clearly the case for the first term; in the second term every component has negative order because M depends on $\lambda = \partial^q$ only. Thus we have shown that the KP vector fields lead to the following commuting vector fields on $\mathcal{M}^{p,q}$:

$$\frac{dM}{dt_{ij}} = [M, [M^i/\lambda^{j+1}]_+].\tag{2.13}$$

Given $M \in \mathcal{M}^{p,q}$ we have shown in Section 2.2 that there exists a unique $g \in N_q^-$ such that $Y = gMg^{-1} \in \tilde{\mathcal{M}}^{p,q}$. Differentiating $Y = gMg^{-1}$ we find for any i, j the following commuting vector fields on $\tilde{\mathcal{M}}^{p,q}$:

$$\frac{dY}{dt_{ij}} = [Y, [Y^i/\lambda^{j+1}]_+ - \dot{g}g^{-1}].\tag{2.14}$$

In the next section we will show that these vector fields are Hamiltonian with respect to a family of compatible Poisson structures on $\tilde{\mathcal{M}}^{p,q}$.

3. Multi-Hamiltonian structure and symmetries

In the previous section we have constructed a family of commuting vector fields on the affine space $\tilde{\mathcal{M}}^{p,q} \subset \mathfrak{gl}_q[\lambda]$. In this section we will show that these vector fields are multi-Hamiltonian. We will do this by using the loop algebra $\widetilde{\mathfrak{gl}}_q$ of $\mathfrak{gl}_q$.

3.1. The loop algebra $\widetilde{\mathfrak{gl}}_q$

The loop algebra $\widetilde{\mathfrak{gl}}_q$ of $\mathfrak{gl}_q$ is defined by

$$\widetilde{\mathfrak{gl}}_q = \mathfrak{gl}_q((\lambda^{-1})) = \mathfrak{gl}_q[\lambda] \oplus \lambda^{-1}\mathfrak{gl}_q[[\lambda^{-1}]].$$

Elements of the loop algebra will be denoted by capital letters; for an element $X = X(\lambda) = \sum x_i \lambda^i \in \widetilde{\mathfrak{gl}}_q$ we write $X = X_+ + X_-$ according to the above (vector space) decomposition. We define an inner product on $\mathfrak{gl}_q$ by $\langle x, y \rangle = \mathrm{Trace}\, xy$, which is non-degenerate and ad-invariant, $\langle x, [y, z] \rangle = \langle [x, y], z \rangle$. According to Example II.2.8, $\mathcal{O}(\mathfrak{gl}_q)$ carries a natural Poisson bracket, which we will denote by $\{\cdot, \cdot\}$. The inner product $\langle \cdot, \cdot \rangle$ leads to an ad-invariant, non-degenerate inner product $\langle \cdot, \cdot \rangle$ on $\widetilde{\mathfrak{gl}}_q$ via

$$\langle X(\lambda), Y(\lambda) \rangle = \mathrm{Res}_{\lambda=0} \langle X(\lambda), Y(\lambda) \rangle,$$

where the right hand side is a shorthand for $\sum_{i+j=-1} \langle x_i, y_j \rangle = \sum_{i \in \mathbf{Z}} \mathrm{Trace}\, x_i y_{-i-1}$.

We introduce an algebra $\mathcal{O}(\widetilde{\mathfrak{gl}}_q)$ of functions on $\widetilde{\mathfrak{gl}}_q$ for which we can define a gradient and a Poisson bracket as in Example II.2.8, but which is large enough to contain functions of the type $X(\lambda) \mapsto \mathrm{Res}\, H(X^i(\lambda))$ (for $H \in \mathcal{O}(\mathfrak{gl}_q)$), which will be important later. We define on $\tilde{\mathfrak{g}}_{\leq n} = \lambda^n \mathfrak{gl}_q[[\lambda^{-1}]]$ an algebra of functions by

$$\mathcal{O}(\tilde{\mathfrak{g}}_{\leq n}) = \mathbf{C}\,[\xi^s]_{\substack{\xi \in \mathfrak{gl}_q^* \\ s \leq n}},$$

where ξ^s denotes, for $\xi \in \mathfrak{gl}_q^*$ and for $s \in \mathbf{Z}$, the linear map $\sum x_j \lambda^j \mapsto \xi(x_s)$. On $\widetilde{\mathfrak{gl}}_q$ we consider the following algebra of functions:

$$\mathcal{O}(\widetilde{\mathfrak{gl}}_q) = \left\{ F : \widetilde{\mathfrak{gl}}_q \to \mathbf{C} \mid \forall n \in \mathbf{Z} : F_{|\tilde{\mathfrak{g}}_{\leq n}} \in \mathcal{O}(\tilde{\mathfrak{g}}_{\leq n}) \right\}.$$

Thus, elements of $\mathcal{O}(\widetilde{\mathfrak{gl}}_q)$ restrict to polynomials on all subspaces $\tilde{\mathfrak{g}}_{\leq n}$. As in Example II.2.8 the gradient $\nabla F(X)$ of a function $F \in \mathcal{O}(\widetilde{\mathfrak{gl}}_q)$ at $X \in \widetilde{\mathfrak{gl}}_q$ is defined by

$$\langle \nabla F(X), Y \rangle = \frac{d}{dt}_{|t=0} F(X + tY) \qquad \forall Y \in \widetilde{\mathfrak{gl}}_q. \tag{3.1}$$

Proposition 3.1 *For any $X \in \widetilde{\mathfrak{gl}}_q$ and $F \in \mathcal{O}(\widetilde{\mathfrak{gl}}_q)$, $\nabla F(X)$ is well-defined by (3.1) and belongs to $\widetilde{\mathfrak{gl}}_q$. Moreover, for any $F, G \in \mathcal{O}(\widetilde{\mathfrak{gl}}_q)$ the Poisson bracket $\{F, G\}$, defined by*

$$\{F, G\}(X) = \langle X, [\nabla F(X), \nabla G(X)] \rangle$$

belongs to $\mathcal{O}(\widetilde{\mathfrak{gl}}_q)$, making $\mathcal{O}(\widetilde{\mathfrak{gl}}_q)$ into a Poisson algebra.

For $l \in \mathbf{Z}$ let R_l denote the endomorphism of $\widetilde{\mathfrak{gl}_q}$ defined by

$$R : \widetilde{\mathfrak{gl}_q} \to \widetilde{\mathfrak{gl}_q} : X \mapsto X_+ - X_-,$$
$$R_l : \widetilde{\mathfrak{gl}_q} \to \widetilde{\mathfrak{gl}_q} : X \mapsto R(\lambda^l X).$$

Since any linear combination of the endomorphisms R_l satisfies the modified classical Yang-Baxter equation (V.5.2) the brackets

$$\{F, G\}_l(X) = \frac{1}{2} \langle X, [R_l \nabla F(X), \nabla G(X)] + [\nabla F(X), R_l \nabla G(X)] \rangle , \tag{3.2}$$

form, for $l \in \mathbf{Z}$, a family of compatible Poisson brackets. We call them R-brackets to distinguish them from the above-defined canonical Lie-Poisson bracket $\{\cdot, \cdot\}$ on $\mathcal{O}(\widetilde{\mathfrak{gl}_q})$. Consider the Ad-invariant functions $K_i : x \mapsto \text{Trace} \frac{x^{i+1}}{i+1}$, $(i = 0, \ldots, q-1)$ on $\mathfrak{gl}_q$ and define, for any $j \in \mathbf{Z}$, the function

$$H_{ij} : \widetilde{\mathfrak{gl}_q} \to \mathbf{C} : X = X(\lambda) \mapsto \text{Res} \, \frac{K_i(X(\lambda))}{\lambda^{j+1}}. \tag{3.3}$$

Clearly $H_{ij} \in \mathcal{O}(\widetilde{\mathfrak{gl}_q})$ and $\nabla H_{ij}(X) = X^i \lambda^{-j-1}$. Therefore each H_{ij} is a Casimir for $\{\cdot, \cdot\}$ and Proposition V.5.1 implies that all functions H_{ij} are in involution with respect to all R-brackets $\{\cdot, \cdot\}_l$, which can also be deduced immediately from (3.2). As we have seen in Section V.5 the Hamiltonian vector fields that correspond to such functions can be written in Lax form; taking for example the 0-th R-bracket the function H_{ij} leads to the vector field $\mathcal{X}_{ij} = -\{\cdot, H_{ij}\}_0$, which takes the Lax form

$$\frac{dX}{dt_{ij}} = \frac{1}{2} [X, R \nabla H_{ij}(X)] . \tag{3.4}$$

Two alternative ways to write this are

$$\frac{dX}{dt_{ij}} = [X, (\nabla H_{ij}(X))_+] = -[X, (\nabla H_{ij}(X))_-]. \tag{3.5}$$

Notice that the equations (3.5) and (2.13) are formally the same but are defined on different spaces. The vector field (3.5) is in fact Hamiltonian with respect to all brackets $\{\cdot, \cdot\}_l$ since

$$\{\cdot, H_{ij}\}_0 = \{\cdot, H_{i,j+l}\}_l. \tag{3.6}$$

Therefore, $\mathcal{X}_{ij}$ can be written in Lax form with respect to all endomorphisms R_l,

$$\frac{dX}{dt_{ij}} = \frac{1}{2} [X, R_l \nabla H_{i,j+l}(X)] .$$

We now show that on $\widetilde{\mathfrak{gl}_q}$ all the R-brackets $\{\cdot, \cdot\}_l$, the Hamiltonians H_{ij} and the vector fields $\mathcal{X}_{ij}$ are linked by a vector field $\mathcal{V}$, which has the deformation property with respect to all these R-brackets. $\mathcal{V}$ is defined as the infinitesimal generator of the action of $\mathbf{C}$ on $\widetilde{\mathfrak{gl}_q}$ given by "shift in λ",

$$\left(s, \sum x_i \lambda^i \right) \mapsto \sum x_i (\lambda + s)^i ;$$

here we use for negative powers of λ the formal expansion

$$(\lambda + s)^{-1} = \sum_{i \geq 0} (-1)^i s^i \lambda^{-i-1},$$

which is actually convergent for small s, in particular it is the right definition if one wants to consider the fundamental vector field $\mathcal{V}$ of this action: the latter is easily computed as $\mathcal{V}_{X(\lambda)} = \frac{\partial}{\partial \lambda} X(\lambda)$, i.e., $\mathcal{V}\xi^i = (i+1)\xi^{i+1}$ for any $\xi \in \mathfrak{gl}_q^*$. The properties of $\mathcal{V}$ are given by the following proposition, whose proof is an easy consequence of the definitions.

Proposition 3.2   *Let $i, l \in \mathbf{Z}$.*

(1) *The Lie derivative of the l-th R-bracket is (up to a factor $-l$) the $(l-1)$-th R-bracket,*

$$\mathcal{L}_\mathcal{V}\{F, G\}_l - \{\mathcal{L}_\mathcal{V}F, G\}_l - \{F, \mathcal{L}_\mathcal{V}G\}_l = -l\{F, G\}_{l-1}, \qquad (3.7)$$

for any $F, G \in \mathcal{O}(\widetilde{\mathfrak{gl}_q})$;
(2) $\mathcal{L}_\mathcal{V}H_{ij} = (j+1)H_{i,j+1}$;
(3) $\mathcal{L}_\mathcal{V}\mathcal{X}_{ij} = [\mathcal{V}, \mathcal{X}_{ij}] = (j+1)\mathcal{X}_{i,j+1}$.

The conclusion is that we have for every $l, m \in \mathbf{Z}$ and for every H_i a bi-Hamiltonian hierarchy with respect to the R-brackets $\{\cdot, \cdot\}_l$ and $\{\cdot, \cdot\}_m$. A typical fragment of it is (for $m = 0$) depicted as follows (we omit the coefficients).

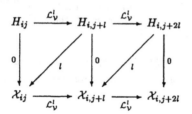

3.2. Reducing the R-brackets and the vector field $\mathcal{V}$

We will now apply Proposition II.2.27 to obtain a family of Poisson brackets and a vector field relating them on $\widetilde{\mathcal{M}}^{p,q}$. Before doing this we truncate $\widetilde{\mathfrak{gl}_q}$ to a finite-dimensional Poisson subspace with respect to some of the brackets.

Proposition 3.3   *Let $\mathcal{M}_S^{p,q}$ denote the affine subspace of $\widetilde{\mathfrak{gl}_q}$ defined by*

$$\mathcal{M}_S^{p,q} = \left\{ \sum_{i=0}^{[p/q]+1} M_i\lambda^i \in \widetilde{\mathfrak{gl}_q} \mid M_{[p/q]+1} = S \right\}.$$

$\mathcal{M}_S^{p,q}$ *is a Poisson subvariety of $\widetilde{\mathfrak{gl}_q}$ with respect to the Poisson structure $\sum_{l=0}^{[p/q]+1} c_l\{\cdot, \cdot\}_l$, where the complex numbers c_i are arbitrary. Moreover $\mathcal{V}$ is tangent to $\mathcal{M}_S^{p,q}$, hence it has the deformation property with respect to each of these restricted Poisson structures on $\mathcal{M}_S^{p,q}$.*

Proof

By Proposition II.2.18, $\mathcal{M}_S^{p,q}$ is a Poisson subvariety of $\widetilde{\mathfrak{gl}}_q$ if and only if its ideal $\mathcal{I}$ is a Poisson ideal of $\mathcal{O}(\widetilde{\mathfrak{gl}}_q)$. For $0 \leq i,j < q$ let $\xi_{ij} = \langle \cdot, e_{ij} \rangle \in \mathfrak{gl}_q^*$ and consider for $s \in \mathbf{Z}$ the linear function ξ_{ij}^s on $\widetilde{\mathfrak{gl}}_q$; it is convenient to write ξ_{ij}^s as follows

$$\xi_{ij}^s = \langle \cdot, e_{ij} \lambda^{s-1} \rangle.$$

The ideal $\mathcal{I}$ is generated by the functions ξ_{ij}^s where $s \notin \{0, \ldots, [p/q]+1\}$ and by the functions $\xi_{ij}^s - \xi_{ij}(S)$, where $s = [p/q] + 1$. Since $\nabla \xi_{ij}^s = e_{ij} \lambda^{-s-1}$ we find for any $l \in \mathbf{Z}$ the following formula for the R_l-bracket.

$$\{\xi_{ij}^s, \xi_{kl}^t\}_l = \epsilon_l^{st} \left(\delta_{kj} \xi_{il}^{s+t+1-l} - \delta_{il} \xi_{kj}^{s+t+1-l} \right), \tag{3.8}$$

where $\epsilon_l^{st} = 1$ if $s,t < l$ and $\epsilon_l^{st} = -1$ if $s,t \geq l$; otherwise $\epsilon_l^{st} = 0$. Substituting any of the above generators of $\mathcal{I}$ in (3.8), together with an arbitrary ξ_{kl}^t, one finds that the resulting linear function contains only terms ξ_{uv}^w where $w \notin \{0, \ldots, [p/q]+1\}$, showing that $\mathcal{I}$ is a Poisson ideal. The fact that $\mathcal{V}$ is tangent to $\mathcal{M}_S^{p,q}$ follows in a similar way from the explicit formula $\mathcal{V}\xi_{ij}^s = (s+1)\xi_{ij}^{s+1}$ for $\mathcal{V}$. ∎

The restriction of the above Poisson structures to $\mathcal{M}_S^{p,q}$ and the restriction of $\mathcal{V}$ to $\mathcal{M}_S^{p,q}$ will be denoted in the same way as the corresponding structures on $\widetilde{\mathfrak{gl}}_q$.

We are now precisely in the case of Proposition II.2.27: $\mathcal{M}_S^{p,q}$ is an affine Poisson variety (w.r.t. $\{\cdot, \cdot\}_l$, where $0 \leq l \leq [p/q]+1$) on which the stabilizer G_S of S acts; $\mathcal{M}_{RS}^{p,q}$ is a subvariety of $\mathcal{M}_S^{p,q}$ which is G_S-stable.

Proposition 3.4 *The triple $(\mathcal{M}_S^{p,q}, G_S, \mathcal{M}_{RS}^{p,q})$ is Poisson-reducible with respect to the Poisson bracket $\sum_{l=0}^{[p/q]+1} c_l \{\cdot, \cdot\}_l$, where the complex numbers c_i are arbitrary.*

Proof

We first show that the action of G_S on $\mathcal{M}_S^{p,q}$ is Poisson (where G_S is given the trivial Poisson structure). Actually the (diagonal) action of $GL(q)$ on $\widetilde{\mathfrak{gl}}_q$ is Poisson with respect to $\{\cdot, \cdot\}_l$ for any $l \in \mathbf{Z}$. To see this, take any $l \in \mathbf{Z}$ and $g \in GL(q)$. It is sufficient to show that $(\mathrm{Ad}_g)^* \{f_1, f_2\}_l = \{(\mathrm{Ad}_g)^* f_1, (\mathrm{Ad}_g)^* f_2\}_l$ for any $f_1, f_2 \in \mathcal{O}(\widetilde{\mathfrak{gl}}_q)$ that are linear. For f_i linear one has that $(\mathrm{Ad}_g)^* f_i$ is also linear hence $\nabla f_i(X)$ is independent of $X \in \widetilde{\mathfrak{gl}}_q$ and we can omit the argument X. Since

$$\frac{d}{dt}_{|t=0} f_1(\mathrm{Ad}_g(X+tY)) = f_1(\mathrm{Ad}_g Y) = \langle \nabla f_1, \mathrm{Ad}_g Y \rangle$$

we find that $\langle \nabla(\mathrm{Ad}_g)^* f_1, Y \rangle = \langle \mathrm{Ad}_{g^{-1}} \nabla f_1, Y \rangle$ giving $\nabla(\mathrm{Ad}_g)^* f_1 = \mathrm{Ad}_{g^{-1}} \nabla f_1$. Then

$$\begin{aligned}
\{(\mathrm{Ad}_g)^* f_1, (\mathrm{Ad}_g)^* f_2\}_l(X) &= \langle X, [\mathrm{Ad}_{g^{-1}} \nabla f_1, \mathrm{Ad}_{g^{-1}} \nabla f_2]_{R_l} \rangle \\
&= \langle \mathrm{Ad}_g X, [\nabla f_1, \nabla f_2]_{R_l} \rangle \\
&= \{f_1, f_2\}_l(\mathrm{Ad}_g X) \\
&= (\mathrm{Ad}_g)^* \{f_1, f_2\}_l(X).
\end{aligned}$$

Since $\tilde{\mathcal{M}}^{p,q}$ is G_S-invariant we are only left with the verification of condition (II.2.17). The ideal $\mathcal{I}$ of $\mathcal{M}_{RS}^{p,q}$ in $\mathcal{M}_S^{p,q}$ is generated by those elements of the form $\xi_{ij}^{[p/q]} - \xi_{ij}(R)$ for which $j - i \leq d - q$. For $l = 0, \ldots, [p/q]$ these elements are Casimirs of $\{\cdot, \cdot\}_l$. Indeed, if $X \in \mathcal{M}_S^{p,q}$ and $0 \leq m, n \leq q - 1$ and $0 \leq s \leq [p/q]$ then

$$\left\{ \xi_{ij}^{[p/q]}, \xi_{mn}^s \right\}_l (X) = \epsilon_l^{[p/q],s} \left(\delta_{mj} \xi_{in}^{[p/q]+s+1-l}(X) - \delta_{in} \xi_{mj}^{[p/q]+s+1-l}(X) \right),$$

which is zero: for $s \neq l$ this is seen at once, while for $s = l$ one computes

$$\begin{aligned}
\left\{ \xi_{ij}^{[p/q]}, \xi_{mn}^l \right\}_l (X) &= \delta_{in} \xi_{mj}^{[p/q]+1}(X) - \delta_{mj} \xi_{in}^{[p/q]+1}(X) \\
&= \langle \delta_{in} e_{mj} - \delta_{mj} e_{in}, S \rangle \\
&= \langle [e_{mn}, e_{ij}], S \rangle \\
&= \langle [e_{ij}, S], e_{mn} \rangle
\end{aligned}$$

and one finds again zero because $j - i + \deg S = j - i - d \leq -q$. We have shown (II.2.17) when $l \neq [p/q] + 1$. Suppose now that $l = [p/q] + 1$. Take $F \in \mathcal{O}(\mathcal{M}_S^{p,q}, \mathcal{M}_{RS}^{p,q})^{G_S}$ and notice that this implies that F satisfies the infinitesimal condition

$$\langle [\nabla F(X), X], \nu \rangle = 0, \qquad \forall X \in \tilde{\mathcal{M}}^{p,q}, \forall \nu \in \mathfrak{g}_S. \tag{3.9}$$

Verifying (II.2.17) amounts to showing that $\left\{ \xi_{ij}^{[p/q]}, F \right\}_{[p/q]+1} (X) = 0$ for any $X \in \tilde{\mathcal{M}}^{p,q}$ and for i, j such that $j - i \leq d - q$. But

$$\begin{aligned}
&\left\{ \xi_{ij}^{[p/q]}, F \right\}_{[p/q]+1} (X) \\
&= \frac{1}{2} \left\langle X, [e_{ij}, \nabla F(X)] + \left[\lambda^{-[p/q]-1} e_{ij}, R\left(\lambda^{[p/q]+1} \nabla F(X) \right) \right] \right\rangle \\
&= \frac{1}{2} \left\langle X, [e_{ij}, \nabla F(X)] + \left[\lambda^{-[p/q]-1} e_{ij}, \lambda^{[p/q]+1} \nabla F(X) - 2 \left(\lambda^{[p/q]+1} \nabla F(X) \right)_- \right] \right\rangle \\
&= \langle X, [e_{ij}, \nabla F(X)] \rangle - \left\langle X, \left[\lambda^{-[p/q]-1} e_{ij}, \left(\lambda^{[p/q]+1} \nabla F(X) \right)_- \right] \right\rangle.
\end{aligned}$$

Both terms vanish: the first one in view of (3.9) and the second one because it can be rewritten as

$$\left\langle S, \left[e_{ij}, \left(\lambda^{[p/q]+1} \nabla F(X) \right)_{-1} \right] \right\rangle = \left\langle [S, e_{ij}], \left(\lambda^{[p/q]+1} \nabla F(X) \right)_{-1} \right\rangle.$$

∎

In view of Proposition 2.5 we have shown that $\tilde{\mathcal{M}}^{p,q}$ carries a $([p/q] + 2)$-dimensional family of compatible Poisson structures. We will now show that the restriction of $\mathcal{V}$ to $\tilde{\mathcal{M}}^{p,q}$ has the deformation property with respect to these Poisson structures.

Proposition 3.5 *Let $\{\cdot, \cdot\} = \sum_{l=0}^{[p/q]+1} c_l \{\cdot, \cdot\}_l$ be any R-bracket on $\mathcal{M}_S^{p,q}$ and let $\{\cdot, \cdot\}_{\mathcal{V}} = \mathcal{L}_{\mathcal{V}} \{\cdot, \cdot\}$. Denote by $\mathcal{W}$ the restriction of $\mathcal{V}$ to $\tilde{\mathcal{M}}^{p,q}$. Then $\mathcal{L}_{\mathcal{W}} \{\cdot, \cdot\}' = \{\cdot, \cdot\}'_{\mathcal{V}}$, where $\{\cdot, \cdot\}'$ (resp. $\{\cdot, \cdot\}'_{\mathcal{V}}$) is the reduced bracket of $\{\cdot, \cdot\}$ (resp. $\{\cdot, \cdot\}_{\mathcal{V}}$) on $\tilde{\mathcal{M}}^{p,q}$. In particular $\mathcal{W}$ has the deformation property with respect to $\{\cdot, \cdot\}$.*

Proof

First notice that the triple $(\mathcal{M}_S^{p,q}, G_S, \mathcal{M}_{RS}^{p,q})$ is Poisson-reducible because

$$\{\cdot,\cdot\}_\mathcal{V} = - \sum_{l=1}^{[p/q]+1} c_l l \{\cdot,\cdot\}_{l-1},$$

an immediate consequence of (3.7). Let $f_1, f_2 \in \mathcal{O}\,(\mathcal{M}_{RS}^{p,q})^{G_S} \cong \tilde{\mathcal{M}}^{p,q}$ and let us denote the restriction of $\mathcal{V}$ to $\mathcal{M}_{RS}^{p,q}$ by $\mathcal{W}$ (notice that its further restricition to $\tilde{\mathcal{M}}^{p,q}$ is also denoted by $\mathcal{W}$). We need to show that

$$\{f_1, f_2\}_\mathcal{V}' = \mathcal{L}_\mathcal{W}\{f_1, f_2\}' - \{\mathcal{L}_\mathcal{W}f_1, f_2\}' - \{f_1, \mathcal{L}_\mathcal{W}f_2\}'.$$

To see that this formula makes sense one it suffices to prove that $\mathcal{L}_\mathcal{W}\,(\mathcal{M}_{RS}^{p,q})^{G_S} \subset (\mathcal{M}_{RS}^{p,q})^{G_S}$; to show this inclusion, use the fact that the actions of $\mathcal{L}_\mathcal{W}$ and G_S commute, a consequence of the fact that the flow of $\mathcal{L}_\mathcal{W}$ is given by $\phi_s : X(\lambda) \to X(\lambda+s)$, while G_S acts by simultaneous conjugation. If we let $F_1, F_2 \in \mathcal{O}\,(\mathcal{M}_S^{p,q})$ such that $f_1 = \rho(F_1)$ and $f_2 = \rho(F_2)$ then

$$\begin{aligned}
\{f_1, f_2\}_\mathcal{V}' &= \rho\,\{F_1, F_2\}_\mathcal{V} \\
&= \rho\mathcal{L}_\mathcal{V}\{F_1, F_2\} - \rho\{\mathcal{L}_\mathcal{V}F_1, F_2\} - \rho\{F_1, \mathcal{L}_\mathcal{V}F_2\} \\
&= \rho\mathcal{L}_\mathcal{V}\{F_1, F_2\} - \{\rho\mathcal{L}_\mathcal{V}F_1, f_2\} - \{f_1, \rho\mathcal{L}_\mathcal{V}F_2\}
\end{aligned}$$

so it suffices to show that $\rho\mathcal{L}_\mathcal{V}(F) = \mathcal{L}_\mathcal{W}\rho(F)$ for any $F \in \mathcal{O}\,(\mathcal{M}_S^{p,q}, \mathcal{M}_{RS}^{p,q})$. Let us denote by π^* the inclusion map $\mathcal{O}\,(\mathcal{M}_{RS}^{p,q})^{G_S} \to \mathcal{O}\,(\mathcal{M}_{RS}^{p,q})$ (which may be thought of as coming from the quotient map $\mathcal{M}_{RS}^{p,q} \to \mathcal{M}_{RS}^{p,q}/G_S$) and by $\imath : \mathcal{M}_{RS}^{p,q} \to \mathcal{M}_S^{p,q}$ the inclusion map. Then $\imath^*(F) = \pi^*\rho(F)$ for any $F \in \mathcal{O}\,(\mathcal{M}_S^{p,q}, \mathcal{M}_{RS}^{p,q})$. Therefore

$$\pi^*\mathcal{L}_\mathcal{W}\rho(F) = \mathcal{L}_\mathcal{W}\pi^*\rho(F) = \mathcal{L}_\mathcal{W}\imath^*(F) = \imath^*\mathcal{L}_\mathcal{V}F \overset{(i)}{=} \pi^*\rho\mathcal{L}_\mathcal{V}F$$

and the result follows from injectivity of π^*. In (i) we used the fact that

$$\mathcal{L}_\mathcal{V}\mathcal{O}\,(\mathcal{M}_S^{p,q}, \mathcal{M}_{RS}^{p,q})^{G_S} \subset \mathcal{O}\,(\mathcal{M}_S^{p,q}, \mathcal{M}_{RS}^{p,q})^{G_S},$$

an easy consequence of $\mathcal{L}_\mathcal{W}\,(\mathcal{M}_{RS}^{p,q})^{G_S} \subset (\mathcal{M}_{RS}^{p,q})^{G_S}$:

$$\begin{aligned}
\imath^*\mathcal{L}_\mathcal{V}\mathcal{O}\,(\mathcal{M}_S^{p,q}, \mathcal{M}_{RS}^{p,q})^{G_S} &= \mathcal{L}_\mathcal{W}\imath^*\mathcal{O}\,(\mathcal{M}_S^{p,q}, \mathcal{M}_{RS}^{p,q})^{G_S} = \mathcal{L}_\mathcal{W}\pi^*\mathcal{O}\,(\mathcal{M}_S^{p,q})^{G_S} \\
&= \pi^*\mathcal{L}_\mathcal{W}\mathcal{O}\,(\mathcal{M}_S^{p,q})^{G_S} \subset \mathcal{O}\,(\mathcal{M}_S^{p,q}, \mathcal{M}_{RS}^{p,q})^{G_S}.
\end{aligned}$$

∎

There is a final, innocent, reduction that can be done on these systems. Namely, since $\nabla H_{0j}(X) = \mathrm{Id}_q \lambda^{-j-1}$ the coefficients of $\mathrm{Trace}(X)$ are Casimirs and we can restrict phase space to traceless matrices. The multi-Hamiltonian structure, the commuting vector fields and the vector field $\mathcal{W}$ are just the restrictions of the original ones to this smaller phase space.

4. The odd and the even Mumford systems

4.1. The (odd) Mumford system

In the preceeding section we have described a family of commuting Hamiltonian vector fields on an affine subspace of $\widetilde{\mathfrak{gl}}_q$, which, as we said, can be taken as an affine subspace of $\widetilde{\mathfrak{sl}}_q$. For $q = 2$ and $p = 2g + 1$ we call the corresponding system the (genus g) *Mumford system*, because it was Mumford who first wrote down (in [Mum5]) explicitly these commuting vector fields and showed their algebraic complete integrability (the Hamiltonian structure is absent in [Mum5]). For reasons that will become clear in the next paragraph we will often refer to the Mumford system as the *odd Mumford system*.

Since we are considering special values for p and q we will simplify the notation, used earlier in this chapter; the new notation will also take into account the fact that we only consider traceless matrices from now on. We will denote the phase space of the genus g Mumford system, which consists of the traceless matrices in $\mathcal{M}^{2g+1,2}$ by $\mathcal{M}^g$. Explicitly, $\mathcal{M}^g$ consist of all matrices

$$A(\lambda) = \begin{pmatrix} v(\lambda) & u(\lambda) \\ w(\lambda) & -v(\lambda) \end{pmatrix},$$

where u and w are monic, $\deg u = g = \deg w - 1$ and $\deg v < g$. We have that $S = \begin{pmatrix} 0 & 0 \\ 1 & 0 \end{pmatrix}$ and $R = \begin{pmatrix} 0 & 1 \\ 0 & 0 \end{pmatrix}$ so that the group G_S consists of all matrices of the form $\begin{pmatrix} 1 & 0 \\ a & 1 \end{pmatrix}$ with Lie algebra $\mathfrak{g}_S = \mathfrak{g}_{-1}$. The space of all traceless matrices in $\mathcal{M}_{RS}^{2g+1,g}$ resp. $\mathcal{M}_S^{2g+1,g}$ will be denoted by $\mathcal{M}_{RS}^g$ resp. $\mathcal{M}_S^g$. A general element of either one of these spaces will be denoted by

$$\tilde{A}(\lambda) = \begin{pmatrix} \tilde{v}(\lambda) & \tilde{u}(\lambda) \\ \tilde{w}(\lambda) & -\tilde{v}(\lambda) \end{pmatrix},$$

for which $\tilde{w}$ is monic of degree $g+1$, while both $\tilde{u}$ and $\tilde{v}$ have degree at most g, with the extra condition that $\tilde{u}$ is monic when $\tilde{A}(\lambda) \in \mathcal{M}_{RS}^g$. Since G_S acts by conjugation the quotient map $\mathcal{M}_{RS}^g \to \mathcal{M}^g$ is explicitly given by

$$\begin{aligned} u_i &= \tilde{u}_i, \\ v_i &= \tilde{v}_i - \tilde{u}_i \tilde{v}_g, \qquad\qquad i = 0, 1, \ldots, g, \\ w_i &= \tilde{w}_i + 2\tilde{v}_i \tilde{v}_g - \tilde{u}_i \tilde{v}_g^2, \end{aligned} \qquad (4.1)$$

where $u_i, \tilde{u}_i, \ldots$ denotes the coefficient of λ^i in $u(\lambda), \tilde{u}(\lambda), \ldots$. Using (4.1) it is easy to compute the reduced brackets on $\mathcal{M}^g$, as given by (II.2.27). To do this, notice first that the Poisson brackets $\{\cdot, \cdot\}_l$ (for $0 \le l \le g + 1$) on $\mathcal{M}_S^g$ are given by

$$\begin{aligned} \{\tilde{u}_i, \tilde{v}_j\}_l &= \epsilon_l^{ij} \tilde{u}_{i+j+1-l}, \\ \{\tilde{v}_i, \tilde{w}_j\}_l &= \epsilon_l^{ij} \tilde{w}_{i+j+1-l}, \\ \{\tilde{w}_i, \tilde{u}_j\}_l &= 2\epsilon_l^{ij} \tilde{v}_{i+j+1-l}, \end{aligned}$$

(recall that $\epsilon_l^{st} = 1$ if $s, t < l$ and $\epsilon_l^{st} = -1$ if $s, t \ge l$; otherwise $\epsilon_l^{st} = 0$) and all other brackets (between linear functions) are zero. For example if $l \ne n$ then $\{v_i, w_j\}_l$ is found from

$$\{\tilde{v}_i - \tilde{u}_i \tilde{v}_g, \tilde{w}_j + 2\tilde{v}_j \tilde{v}_g - \tilde{u}_j \tilde{v}_g^2\}_l = \epsilon_l^{ij}(\tilde{w}_{i+j+1-l} - \tilde{u}_{i+j+1-l}\tilde{v}_g^2 + 2\tilde{v}_{i+j+1-l}\tilde{v}_g) + \tilde{u}_i \delta_j^l,$$

giving $\{v_i, w_j\}_l = \epsilon_l^{ij} w_{i+j+1-l} + u_i \delta_j^l$. In this way the reduced brackets $\{\cdot, \cdot\}_l$ are found to be given, for $l = 0, 1, \ldots, g$, by

$$
\begin{aligned}
\{u_i, v_j\}_l &= \epsilon_l^{ij} u_{i+j+1-l}, & \{u_i, u_j\}_l &= 0, \\
\{v_i, w_j\}_l &= \epsilon_l^{ij} w_{i+j+1-l} + u_i \delta_j^l, & \{v_i, v_j\}_l &= 0, \\
\{w_i, u_j\}_l &= 2\epsilon_l^{ij} v_{i+j+1-l}, & \{w_i, w_j\}_l &= 2\delta_i^l v_j - 2\delta_j^l v_i,
\end{aligned}
\tag{4.2}
$$

while the bracket $\{\cdot, \cdot\}_{g+1}$ is quadratic and is given by

$$
\begin{aligned}
\{u_i, v_j\}_{g+1} &= \epsilon_{g+1}^{ij} u_{i+j-g} - u_i u_j, & \{u_i, u_j\}_{g+1} &= 0, \\
\{v_i, w_j\}_{g+1} &= \epsilon_{g+1}^{ij} w_{i+j+-g} - u_i w_j, & \{v_i, v_j\}_{g+1} &= 0, \\
\{w_i, u_j\}_{g+1} &= 2\epsilon_{g+1}^{ij} v_{i+j-g} - 2u_j v_i, & \{w_i, w_j\}_{g+1} &= 2v_i w_j - 2v_j w_i.
\end{aligned}
\tag{4.3}
$$

For linear combinations of the Poisson structures $\{\cdot, \cdot\}_i, i = 0, \ldots, g$ a compact formula can be written. Namely, let $\varphi = \sum_{i=0}^{g} c_i \lambda^i$ be a polynomial of degree at most g and let[19] $\{\cdot, \cdot\}^\varphi = -\sum_{i=0}^{g} c_i \{\cdot, \cdot\}_i$. Then it follows from (4.2) that $\{\cdot, \cdot\}^\varphi$ can be written in terms of generating functions as follows:

$$
\begin{aligned}
\{u(x), u(y)\}^\varphi &= \{v(x), v(y)\}^\varphi = 0, \\
\{u(x), v(y)\}^\varphi &= \frac{u(x)\varphi(y) - u(y)\varphi(x)}{x - y}, \\
\{u(x), w(y)\}^\varphi &= -2\frac{v(x)\varphi(y) - v(y)\varphi(x)}{x - y}, \\
\{v(x), w(y)\}^\varphi &= \frac{w(x)\varphi(y) - w(y)\varphi(x)}{x - y} - u(x)\varphi(y), \\
\{w(x), w(y)\}^\varphi &= 2\left(v(x)\varphi(y) - v(y)\varphi(x)\right).
\end{aligned}
\tag{4.4}
$$

The Hamiltonians in involution are the coefficients of Trace $A(\lambda)^2/2$, i.e., the coefficients of $H(\lambda) = u(\lambda)w(\lambda) + v^2(\lambda)$. From (4.4) we can compute the Hamiltonian vector fields $\mathcal{X}_y = \{\cdot, H(y)\}^1$, where $y \in \mathbf{C}$ as well as the vector fields $\mathcal{X}_i = \{\cdot, H_i\}^1$, where H_i is the i-th coefficient of $H(\lambda)$ (notice that the function H_i, used here, coincides to the function H_{1i}, defined by (3.3)). For example

$$
\mathcal{X}_y(u(\lambda)) = \{u(\lambda), u(y)w(y) + v^2(y)\}^1 = 2\frac{u(\lambda)v(y) - u(y)v(\lambda)}{\lambda - y},
$$

leading to

$$
\frac{d}{dt_y} A(\lambda) = \left[A(\lambda), -\frac{A(y)}{\lambda - y} - \begin{pmatrix} 0 & 0 \\ u(y) & 0 \end{pmatrix} \right].
$$

Writing $H_i = \mathrm{Res}_{y=0} H(y)/y^{i+1}$ we find

$$
\frac{d}{dt_i} A(\lambda) = \left[A(\lambda), (A(\lambda)/\lambda^{i+1})_+ - \begin{pmatrix} 0 & 0 \\ u_i & 0 \end{pmatrix} \right].
\tag{4.5}
$$

---

[19] The minus sign has been put in to make these brackets $\{\cdot, \cdot\}^\varphi$ coincide (when $q = 2$) with the brackets $\{\cdot, \cdot\}^\varphi$ defined in Chapter III.

4.2. The even Mumford system

In the previous section we have constructed the (odd) Mumford system. Alternatively it can also be obtained from the hyperelliptic case considered in Paragraph III.2.4. We show how this can be done be constructing an integrable Hamiltonian system that is very similar to the Mumford system. It is an open problem to obtain this variant of the Mumford system (which was first constructed in [Van2]) from the algebra of pseudo-differential operators and/or from a reduction on the loop algebra $\widetilde{sl}_2$.

In the language of Paragraph III.2.4 we take F of the form $F(x, y) = y^2 - f(x)$, where $\deg f = 2g+2$, we take $\varphi = 1$ and choose $d = g$, the genus of the hyperelliptic curve $y^2 = f(x)$. Then the Lax equation (III.2.14) takes the form

$$\frac{d}{dt_i} A(\lambda) = \left[A(\lambda), [B_i(\lambda)]_+ \right],$$

where

$$A(\lambda) = \begin{pmatrix} v(\lambda) & u(\lambda) \\ w(\lambda) & -v(\lambda) \end{pmatrix}, \quad B_i(\lambda) = \frac{A(\lambda)}{u(\lambda)} \left[\frac{u(\lambda)}{\lambda^{i+1}} \right]_+ \quad \text{and} \quad w(\lambda) = -\left[\frac{F(\lambda, v(\lambda))}{u(\lambda)} \right]_+. \quad (4.6)$$

The highest flow $\dot{A}(\lambda) = X_{H_{g-1}} A(\lambda)$ is the most important for us and is simply given by

$$\dot{A}(\lambda) = [A(\lambda), B(\lambda)], \quad A(\lambda) = \begin{pmatrix} v(\lambda) & u(\lambda) \\ w(\lambda) & -v(\lambda) \end{pmatrix}, \quad B(\lambda) = \begin{pmatrix} 0 & 1 \\ b(\lambda) & 0 \end{pmatrix}, \quad (4.7)$$

where one computes $b(\lambda)$ from

$$b(\lambda) = \left[\frac{w(\lambda)}{u(\lambda)} \right]_+ \quad \text{and} \quad w(\lambda) = \left[\frac{f(\lambda) - v^2(\lambda)}{u(\lambda)} \right]_+. \quad (4.8)$$

We want these systems to be extended to a larger phase space in the sense of Amplification III.2.6. As deformation family M we take

$$M = \{ F_c(x, y) = y^2 - (x^{2g+2} + a_{2g+1} x^{2g+1} + \cdots + a_1 x + a_0) \},$$

which is an affine variety isomorphic to $\mathbf{C}^{2g+1}$. Notice that M contains an (odd) equation for every hyperelliptic curve of genus g. The $g+2$ coefficients $a_g, \ldots, a_{2g+1}$ are Casimirs for this larger system and the above Lax pair, supplemented with the equations coming from these Casimirs, describes the integrable Hamiltonian system on the larger phase space $\mathbf{C}^{g+2} \times \mathbf{C}^{2g} = \mathbf{C}^{3g+2}$ (which is equipped with coordinates $u_0, \ldots, u_{g-1}, v_0, \ldots, v_{g-1}, a_{2g+1}, \ldots, a_g$).

There is however a better way to look at this larger system. Notice that in the way in which we have presented it, the $g+1$ coefficients of

$$w(\lambda) = \lambda^{g+2} + w_{g+1} \lambda^{g+1} + w_g \lambda^g + \cdots + w_1 \lambda + w_0$$

are computed from (4.8); of course they will be polynomials in the u_i, v_i and a_i. One sees however that by inverting these relations the a_i can also be written as polynomials, but now in terms of u_i, v_i and w_i. The upshot is that we can look at this system as defined on $\mathbf{C}^{3g+2}$

with coordinates $\{u_0, \ldots, u_{g-1}, v_0, \ldots, v_{g-1}, w_0, w_1, \ldots, w_{g+1}\}$ and the Lax operator $A(\lambda)$ is now an arbitrary element

$$\begin{pmatrix} v(\lambda) & u(\lambda) \\ w(\lambda) & -v(\lambda) \end{pmatrix}$$

where

$$\deg u(\lambda) = g, \qquad u(\lambda) \text{ monic},$$
$$\deg v(\lambda) \leq g - 1,$$
$$\deg w(\lambda) = g + 2, \quad w(\lambda) \text{ monic}.$$

A family of Poisson brackets, similar to one for the odd Mumford system, is given, for any polynomial φ of degree at most g by

$$\{u(x), u(y)\}^{\varphi} = \{v(x), v(y)\}^{\varphi} = 0,$$
$$\{u(x), v(y)\}^{\varphi} = \frac{u(x)\varphi(y) - u(y)\varphi(x)}{x - y},$$
$$\{u(x), w(y)\}^{\varphi} = -2\frac{v(x)\varphi(y) - v(y)\varphi(x)}{x - y},$$
$$\{v(x), w(y)\}^{\varphi} = \frac{w(x)\varphi(y) - w(y)\varphi(x)}{x - y} - \alpha(x + y)u(x)\varphi(y),$$
$$\{w(x), w(y)\}^{\varphi} = 2\alpha(x + y)(v(x)\varphi(y) - v(y)\varphi(x)).$$

where $\alpha(u) = u + w_{g+1} - u_{g-1}$. From these formulas one obtains, as in the case of the odd Mumford system, easily Lax equations for the integrable vector fields $\mathcal{X}_y = \{\cdot, H(y)\}^1$, as well as the vector fields $\mathcal{X}_i = \{\cdot, H_i\}^1$, where H_i is the coefficient of λ^i in $H(\lambda) = u(\lambda)w(\lambda) + v^2(\lambda)$.

It is worthwhile to point out that when the odd Mumford system is constructed in this way then one arrives precisely at the Poisson structures $\{\cdot, \cdot\}^{\varphi}$, given in (4.4). This means that we have two different constructions for the multi-Hamiltonian structure of the odd Mumford system.

4.3. Algebraic complete integrability and Laurent solutions

Having shown integrability of the odd and the even Mumford systems in Chapter III we now comment on its algebraic complete integrability. It was shown by Mumford in [Mum5] that the general level set of the odd Mumford system is an affine part of a Jacobian (of the corresponding hyperelliptic curve) obtained by removing the theta divisor and that the flow of the integrable vector fields is linear on it, hence leading to algebraic complete integrability. In Section 5 we will modify our construction given in Chapter III in order to obtain similar systems which are a.c.i. Since this new construction coincides with the original one in the hyperelliptic case it will give a proof that the odd and even Mumford systems are a.c.i. For the odd case explicit solutions in terms of theta functions are given as follows (see [Mum5] for details). Let $D(t) = \mathrm{Ab}(\mathcal{D}(t))$ be an integral curve starting at $D = D(0)$ then $D(t) = At + D$, where A is a fixed vector which depends on the chosen vector field (going with t) and the corresponding polynomial $u(\lambda)$ is computed from its values at the Weierstrass points a_k of the curve using

$$U(a_k) = c_k \left(\frac{\vartheta[\delta + \eta_k](At + D)}{\vartheta[\delta](At + D)} \right)^2; \tag{4.9}$$

the vector η_k and δ are characteristics which are described explicitly in [Mum5]. In view of (IV.4.9) these define indeed meromorphic functions on the Jacobian.

4. The odd and the even Mumford systems

For both cases the smooth level sets of the momentum map appear as the first level of a stratification which we describe next (for more details and proofs see [Van4]).

Let Γ be a smooth (complete, irreducible) complex curve (say, a compact Riemann surface) of genus g which is hyperelliptic. The hyperelliptic involution σ extends linearly to $\mathrm{Div}(\Gamma)$ giving an involution $\mathcal{D} \mapsto \mathcal{D}^\sigma$. We introduce a decomposition of $\mathrm{Jac}(\Gamma)$ with respect to an arbitrary fixed point P on the (hyperelliptic) curve Γ. Let $\mathcal{I}_g$ denote the set

$$\mathcal{I}_g = \{(m,n) \in \mathbf{N} \times \mathbf{N} \mid 0 \leq m+n \leq g\}$$

which we order by $(m,n) \leq (m',n')$ if and only if $m \leq m'$ and $n \leq n'$. Then for $(m,n) \in \mathcal{I}_g$ we define a subset $\mathrm{Div}_{mn}(\Gamma,P)$ of $\mathrm{Div}(\Gamma)$ by

$$\mathrm{Div}_{mn}(\Gamma,P) = \left\{ \sum_{i=1}^{g-m-n} P_i + mP + nP^\sigma - gP \mid P_i \in \Gamma \setminus \{P,P^\sigma\} \text{ and } i \neq j \Rightarrow P_i \neq P_j^\sigma \right\};$$

the term gP is introduced here in order to make every element in $\mathrm{Div}_{mn}(\Gamma,P)$ of degree 0. If we introduce[20]

$$\mathrm{Div}_0(\Gamma,P) = \bigcup_{n=0}^{g} \bigcup_{m=0}^{g-n} \mathrm{Div}_{mn}(\Gamma,P)$$

then the Abel map $\mathrm{Ab} : \mathrm{Div}^0(\Gamma) \to \mathrm{Jac}(\Gamma) : \mathcal{D} \mapsto \mathrm{Ab}(\mathcal{D})$ restricts to a bijection $\mathrm{Ab} : \mathrm{Div}_0(\Gamma,P) \to \mathrm{Jac}(\Gamma)$. It is shown in [Van4] that the sets

$$J_{m,n}(\Gamma,P) \overset{\mathrm{def}}{=} \mathrm{Ab}(\mathrm{Div}_{mn}(\Gamma,P))$$

(or

$$J_m(\Gamma,P) \overset{\mathrm{def}}{=} \mathrm{Ab}\left(\mathrm{Div}_{m0}(\Gamma,P)\right)$$

in case $P = P^\sigma$) define a *stratification* of $\mathrm{Jac}(\Gamma)$, meaning that they are disjoint differentiable manifolds, whose boundary is a finite union of lower-dimensional sets $J_{s,t}(\Gamma,P)$ (resp. $J_s(\Gamma,P)$). In the case $P \neq P^\sigma$ the stratification is completely described by the following proposition.

Proposition 4.1 *If $P \neq P^\sigma$ then $\mathrm{Jac}(\Gamma)$ is stratified by the $(g-m-n)$-dimensional submanifolds $J_{m,n}(\Gamma,P)$, whose closure is given by the (finite) union*

$$\bar{J}_{m,n}(\Gamma,P) = \bigcup_{(k,l) \geq (m,n)} J_{k,l}(\Gamma,P). \tag{4.10}$$

Each stratum $J_{m,n}(\Gamma,P)$ has two boundary components which are translates of each other by

$$\vec{e} = \mathrm{Ab}(P^\sigma - P) = \left(\int_P^{P^\sigma} \omega_1, \ldots, \int_P^{P^\sigma} \omega_g \right) \pmod{\Lambda}.$$

---

[20] $\mathrm{Div}_0(\Gamma,P)$ is not to be confused with $\mathrm{Div}^0(\Gamma)$, the group of divisor of degree zero on Γ, of which it is a subset.

More generally, all $i+1$ strata of dimension $g-i$ are translates of each other by $n\vec{e}$ for some $n \in \{1,\dots,i\}$. The closures of the $(g-1)$-dimensional strata $J_{1,0}(\Gamma, P)$ and $J_{0,1}(\Gamma, P)$ are translates of the theta divisor and are tangent along their intersection $\bar{J}_{1,1}(\Gamma, P)$.

Thus the different strata fit together as dictated by the partial order $\leq$ on $\mathcal{I}_g$: if we represent the different spaces $\bar{J}_{m,n}(\Gamma, P)$ by $\bar{J}_{m,n}$, put those of equal dimension on the same horizontal line and depict inclusions by arrows, then the stratification is schematically represented by the following.

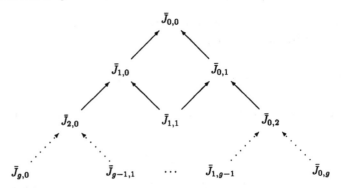

We also give the corresponding proposition for $P = P^\sigma$.

Proposition 4.2 *If $P = P^\sigma$ then $\mathrm{Jac}(\Gamma)$ is stratified by the $(g-m)$-dimensional submanifolds $J_m(\Gamma, P)$, whose closure is given by the (finite) union*

$$\bar{J}_m(\Gamma, P) = \bigcup_{k \geq m} \bar{J}_k(\Gamma, P)$$

and each stratum $\bar{J}_m(\Gamma, P)$ has just one boundary component.

In this case the stratification is simply depicted as

$$\bar{J}_g \to \bar{J}_{g-1} \to \bar{J}_{g-2} \to \cdots \to \bar{J}_1 \to \bar{J}_0$$

$\bar{J}_0 = \mathrm{Jac}(\Gamma)$, $\bar{J}_1$ is a translate of the theta divisor and $\bar{J}_g$ is the origin in $\mathrm{Jac}(\Gamma)$.

The relevance of these stratifications for the even and odd Mumford systems resides on the fact that each stratum corresponds to precisely one family of Laurent solutions for one special vector field of these systems (the most basic one). These Laurent solutions are given by the following two propositions (for the relation to the Sato Grassmannian, see [Van4]).

Proposition 4.3 *For the odd Mumford system there are $g+1$ families of Laurent solutions. The m-th family corresponds to the stratum $J_m(\Gamma, P)$ and the functions $u_1,\dots,u_g$ have the following Laurent expansions starting at points of the stratum $J_m(\Gamma, P)$*

$$
\begin{aligned}
u_{g-i} &= (-1)^i \frac{(2i-1)!!}{2^i i!} \frac{(m+i)!}{(m-i)!} \frac{1}{t^{2i}} + \mathcal{O}(t^{-2i+1}) \qquad (i=1,\dots,m), \\
u_{g-i} &= \mathcal{O}(t^{-2i+1}) \qquad\qquad\qquad\qquad\qquad\quad (i=m+1,\dots,g).
\end{aligned}
\tag{4.11}
$$

Proof

Equations (4.7) are written out in the case of the odd Mumford system (corresponding to $P = P^\sigma$) as

$$\dot{u}(x) = 2v(x),$$
$$\dot{v}(x) = -w(x) + (x - 2u_{g-1})u(x),$$
$$\dot{w}(x) = -2(x - 2u_{g-1})v(x),$$

or just as a third order equation,

$$\dddot{u}_i(x) = 4\left(\dot{u}_{i-1} - 2u_{g-1}\dot{u}_i - \dot{u}_{g-1}u_i\right) \qquad (i = 0, \ldots, g-1; \, u_{-1} = 0). \tag{4.12}$$

The ansatz[21]

$$u_i = \frac{1}{t^{2i}} \sum_{j=0}^{\infty} u_{ij} t^j$$

leads for the leading coefficients $a_i = u_{g-i,0}$ to the recursion relation

$$a_{i+1} = \frac{2i+1}{i+1}\left[\frac{i(i+1)}{2} + a_1\right] a_i. \tag{4.13}$$

To solve this recursion relation, notice that if $a_i = 0$ then $a_{i+1} = 0$; since $a_i = 0$ for at least one $i \leq g+1$, we find that

$$a_1 = -\frac{1}{2}m(m+1) \tag{4.14}$$

for some $m \in \{0, \ldots, g\}$ which leads by induction immediately to the formula

$$a_i = (-1)^i \frac{(2i-1)!!}{2^i i!} \frac{(m+i)!}{(m-i)!} \qquad (i = 1, \ldots, m),$$

and $a_{m+1} = \cdots = a_g = 0$, hence also to (4.11). The series for v_i and w_i follow immediately from it by differentiation, in particular they do not give rise to separate families of Laurent solutions. ∎

As a corollary of the proposition we see that the odd Mumford system induces a stratification on $\mathrm{Jac}(\Gamma)$ which coincides with the stratification by the subsets $J_m(\Gamma, P)$.

In the case of the even Mumford system we have the following result (for a proof see [Van4]).

Proposition 4.4 *For the even Mumford system there are $\frac{(g+1)(g+2)}{2}$ families of Laurent solutions, one for each element of the set $\mathcal{I}_g$. The (m,n)-th family corresponds to the stratum $J_{m,n}(\Gamma, P)$ and the functions $u_0, \ldots, u_{g-1}$ have the following Laurent expansion in t:*

$$u_{g-1} = \frac{m-n}{t} + \mathcal{O}(1),$$
$$u_{g-i} = \mathcal{O}(t^{-i}), \qquad (i = 2, \ldots, g). \tag{4.15}$$

In particular, the even Mumford system induces a stratification on $\mathrm{Jac}(\Gamma)$ which coincides with the stratification by the subsets $J_{m,n}(\Gamma, P)$.

---

[21] It can be shown that this gives all Laurent solutions, see [Van4].

5. The general case

We will now modify our construction of Chapter III to construct for a large class of curves an a.c.i. system whose general fiber of the momentum map is the affine part of the Jacobian of a deformation of this curve. This new construction will coincide with the previous one in one case (called the hyperelliptic case), considered in Section 4.

We start with a smooth plane curve $\Gamma_F \subset \mathbf{C}^2$, defined by an equation $F(x,y) = 0$. Let g denote the genus of the smooth completion $\bar{\Gamma}_F$ of Γ_F, which we assume to be non-zero. Each holomorphic differential ω on $\bar{\Gamma}_F$ can be written as

$$\omega = \frac{R(x,y)}{\frac{\partial F}{\partial y}(x,y)} dx,$$

for some polynomial $R(x,y)$, hence the choice of a basis of the space of holomorphic differentials leads to g polynomials $R_0(x,y), \ldots, R_{g-1}(x,y)$. Having fixed such a basis we define for any $c = (c_0, \ldots, c_{g-1}) \in \mathbf{C}^g$ a polynomial F_c with corresponding curve Γ_{F_c} by

$$F_c(x,y) = F(x,y) + \sum_{k=0}^{g-1} c_k R_k(x,y). \tag{5.1}$$

The following will in one of the statements below be assumed on the curve Γ_F:

Assumption For a general point $c \in \mathbf{C}^g$, a basis of the space of holomorphic differentials on Γ_{F_c} is given by

$$\frac{R_k(x,y)}{\frac{\partial F_c}{\partial y}(x,y)} dx \qquad (k = 0, \ldots, g-1).$$

This assumption, which is easily checked for any concrete curve at hand, is obviously valid for hyperelliptic, trigonal, say n-gonal curves. For curves with a bad singularity at infinity however, the condition may fail; an example of such a curve was kindly communicated to us by H. Knörrer and will be given below (Example 5.2).

We start by constructing the affine Poisson variety on which our a.c.i. system will live; the construction is closely related to the one in Paragraph III.2.2 to which we will refer several times. We suppose that Γ_F and the polynomials have been fixed (leaving aside the assumption at the moment). For simplicity we take the standard Poisson structure $\{y_i, x_j\} = \delta_{ij}$ on $\mathbf{C}^g$ (which corresponds to $\varphi = 1$ in Paragraph III.2.2); also we take d equal to the genus g of the curve Γ_F. Let $f, h \in \mathcal{O}(\mathbf{C}^g)$ be defined by

$$f = \prod_{i<j}(x_i - x_j)^2,$$

$$h = \det{}^2 (R_i(x_j, y_j))_{1 \le i,j \le n}$$

and let Δ' denote the union of their zero divisors, $\Delta' = (f) \cup (h) = (fh)$. Notice that (h) depends on F only (and not on the choice of R_k) so we will denote it by Δ_F. Thus $\Delta' =$

$\Delta \cup \Delta_F$. Proposition II.2.35 leads to an affine Poisson variety $(M_1, \{\cdot, \cdot\})$ and a Poisson morphism $M_1 \to \mathbf{C}^g \setminus \Delta'$. Explicitly M_1 is given by

$$M_1 = \left\{ (x_0, x_0', (x_1, y_1), \ldots, (x_g, y_g)) \mid x_0 \prod_{i<j}(x_i - x_j)^2 = x_0' \det^2\left(R_i(x_j, y_j)\right) = 1 \right\}.$$

In view of the squares in the definition of f and h the symmetric group S_g defines a Poisson action on M_1 leading to a quotient $(M_2, \{\cdot, \cdot\}_2)$,

$$M_2 = \{(t, t', u(\lambda), v(\lambda)) \mid t\,\mathrm{disc}\,u(\lambda) = t' R(u, v) = 1\}.$$

Here we have computed[22] $\det^2(R_i(x_j, y_j))$ in terms of the u_i and v_i and called the resulting polynomial $R(u, v)$. Finally we define

$$M_F^{2g} = \{(t', u(\lambda), v(\lambda)) \mid t' R(u, v) = 1\},$$

which is the affine variety we were aiming at. This variety is determined up to isomorphism by F (it does not depend on the choice of basis R_i so we use a subscript F instead of R). As in Paragraph III.2.2 we get a commutative diagram,

$$
\begin{array}{ccc}
M_1 & \longrightarrow & M_2 \\
\downarrow{\scriptstyle p_1} & & \downarrow{\scriptstyle p_2} \\
(\mathbf{C}^2)^g \setminus \Delta' & \underset{S}{\longrightarrow} & M_F^{2g}
\end{array}
$$

in which S is (a restriction of) the map (III.2.3) and all maps turn out to be Poisson maps: as in Paragraph III.2.2 the Poisson structure $\{\cdot, \cdot\}_2$ on M_2 descends to a Poisson structure on M_F^{2g}, also denoted by $\{\cdot, \cdot\}_2$. Apart from the verification that the brackets of the u_i and the v_i do not depend on t, in the present case one also has to verify that the same is true for the brackets of these with t'. However, by Proposition II.2.35 the latter are given by

$$\{u_i, t'\}_1 = -t'^2 \{u_i, R(u, v)\}_1,$$
$$\{v_i, t'\}_1 = -t'^2 \{v_i, R(u, v)\}_1,$$

so they are independent of t. For hyperelliptic curves $f = h$ hence $\Delta' = \Delta = \Delta_F$ and the present construction reduces to the construction which was given in Paragraph III.2.2.

We now adapt our construction given in Section III.2.3. The natural map $\hat{H}_{F,d}$ is now replaced by the map[23]

$$\hat{H}_F : (\mathbf{C}^2)^{2g} \setminus \Delta' \to \mathbf{C}^g$$

---

[22] In order to effectively compute $R(u, v)$ for a concrete example one first replaces y_i by $v(x_i)$ to obtain a symmetric polynomial in $x_1, \ldots, x_g$. This polynomial is easily rewritten in terms of the elementary symmetric functions $u_1, \ldots, u_g$.

[23] We do not add the dimension as a subscript here, because it is implicit in F, namely $d = g$ is the genus of Γ_F.

whose components $\hat{H}_0, \ldots, \hat{H}_{g-1}$ are defined by requiring that the polynomial

$$\sum_{i=0}^{g-1} \hat{H}_i(x_1, y_1), \ldots, (x_g, y_g))R_i(x, y) \tag{5.2}$$

has for $(x, y) = (x_j, y_j)$ the value $F(x_j, y_j)$, where $j = 1, \ldots, g$.. Solving for the $\hat{H}_i$ involves only the determinant of the matrix with elements $R_i(x_j, y_j)$, hence we arrive at a regular morphism $H_F : M \to \mathbf{C}^g$ which makes the following diagram commutative:

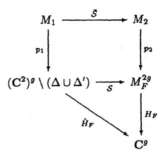

Although the components of H_F depend on the choice of R_i, the algebra $\mathcal{A}_F$ generated by these is clearly independent of it. It leads to an integrable Hamiltonian system (which is an a.c.i. system if the assumption is satified) by the following proposition (for a different proof of integrability see [Van5]).

Proposition 5.1 *Let $F(x, y) \in \mathbf{C}[x, y]$ be such that $F(x, y) = 0$ defines a non-singular curve $\Gamma_F \subset \mathbf{C}^2$ of positive genus g. Then $\left(M_F^{2g}, \{\cdot, \cdot\}_2, \mathcal{A}_F\right)$ defines an integrable Hamiltonian system whose level set over 0 is isomorphic to an affine part of the Jacobian of Γ_F and the flow of the restriction to this level of all integrable vector fields is linearized by this isomorphism.*

When the above assumption about Γ_F is satisfied then this integrable Hamiltonian system is an a.c.i. system whose general level set over $c \in \mathbf{C}^g$ is isomorphic to an affine part of the Jacobian of the deformation Γ_{F_c} defined by (5.1) of the curve Γ_F.

Proof
    Since S is a Poisson map it suffices to show that the components of $\hat{H}_F$ are in involution. Clearly $\{F(x_i, y_i), F(x_j, y_j)\}_h = 0$ for all i, j. From the definition of the components $\hat{H}_k$ of $\hat{H}_F$ we have

$$\{\hat{H}_0 R_0(x_i, y_i) + \cdots + \hat{H}_{g-1}R_{g-1}(x_i, y_i), \hat{H}_0 R_0(x_j, y_j) + \cdots + \hat{H}_{g-1}R_{g-1}(x_j, y_j)\}_1 = 0. \tag{5.3}$$

Since the second component depends only on (x_j, y_j) (although each individual H_i depends on all $(x_1, \ldots, y_g)$) it is in involution with all coefficients of $R_k(x_i, y_i)$ and (5.3) can be rewritten as

$$\sum_{k,l=0}^{g-1} \{\hat{H}_k, \hat{H}_l\} R_k(x_i, y_i) R_l(x_j, y_j) = 0.$$

5. The general case

Since the polynomials R_k are independent it follows that $\{\hat{H}_k, \hat{H}_l\} = 0$ for all k and l. Thus the algebra $\mathcal{A}_F$ generated by $H_1, \ldots, H_g$ is involutive. It has dimension g by the same argument as used in the proof of Proposition III.2.3, hence $\mathcal{A}_F = \mathbf{C}[H_1, \ldots, H_g]$. As for completeness of $\mathcal{A}$, notice that choosing a closed point in the spectrum of $\mathcal{A}_F$ consists in fixing the values of the H_i, hence the level sets (in M_F^{2g}) are given by

$$
F(\lambda, v(\lambda)) - \sum_{i=0}^{g-1} c_i R_i(\lambda, v(\lambda)) = 0 \mod u(\lambda),
$$

where the c_i are these fixed values. Denoting the polynomial on the left hand side by F_c we see that they correspond to the level sets $\mathcal{F}_{F_c,g}$ described by Proposition III.3.3 and Lemma III.3.4. More precisely they are obtained from these by removing a divisor. In particular the general level set is still irreducible and every level set has dimension g. This completes the proof that $\mathcal{A}_F$ is an integrable algebra, hence that $\left(M_F^{2g}, \{\cdot, \cdot\}_2, \mathcal{A}_F\right)$ defines an integrable Hamiltonian system.

In order to say more about the general level sets we restrict ourselves from now on to those curves Γ_F for which the assumption (announced at the beginning of this paragraph) is valid. If not then our argument is still valid for the fiber over 0 (i.e., the one which corresponds to $H_0 = \cdots = H_{g-1} = 0$).

The main observation to be made here is that the restriction $\Delta_{F,c}$ of the divisor Δ_F of h to a general fiber c of the momentum map (which is an (affine part of) a g-fold symmetric product of the curve Γ_{F_c}) is precisely the divisor which is blown down by the Abel-Jacobi map in order to construct the Jacobi variety of Γ_{F_c} from the g-fold symmetric product of Γ_{F_c}. To see this, notice that the points of $\Delta_{F,c}$ correspond exactly to those divisors of degree g on Γ_{F_c} for which the matrix $R_i(x_j, y_j)$ becomes singular, meaning that the space of holomorphic differentials with zeros at these points has positive dimension, hence (by Riemann-Roch) the dimension of the linear system defined by this divisor has positive dimension and this whole linear system is collapsed to a point by the Abel-Jacobi map. Thus, removing the zero divisor of (h) from $\mathbf{C}^{2g}$, which led precisely to our phase space M_F^{2g}, has the effect of removing from each fiber the divisor which is blown down by the Abel-Jacobi map.

Recall however that we had already removed a divisor (with several irreducible components) from the symmetric product which included at least one component which maps to a translate of the theta divisor (in the notation of Paragraph III.3.4 the latter components are the $\mathcal{E}_{F,d}(\infty_k)$). Clearly each of the divisors in Δ'_c is linear equivalent to an effective divisor which contains ∞_k, hence is collapsed to a point in the image of $\mathcal{E}_{F,d}(\infty_k)$ (under the Abel-Jacobi map); since this is true for all k the image of Δ'_c is contained in the intersection of the translates of the theta divisor which were all missing already in our original affine part of the symmetric product. It follows that our level set in M_F^{2g} is now isomorphic to an affine part of the Jacobian of Γ_{F_c}. If the assumption on the curve Γ_F is not satisfied then we still have that the fiber over 0 is isomorphic to an affine part of the Jacobian of Γ_F (this property may still hold for some other fibers).

We have checked now (under the assumption) that the general fiber is an affine part of an Abelian variety; we may take as a base N the space of all curves Γ_{F_c} which are non-singular and construct $\mathcal{T}$ as the universal Jacobian over it (replacing every point by the corresponding Jacobian), giving us the necessary ingredients for the proof that we have an a.c.i. system.

Only linearity of the flow on the Jacobians remains to be checked. By construction we have at a general point

$$F(x_j, y_j) = \sum_{i=0}^{g-1} \hat{H}_i R_i(x_j, y_j),$$

where $\hat{H}_i$ are the components of $\hat{H}_F$. Taking the bracket $\{\cdot, \cdot\}_g$ with x_k and recalling that $\{y_i, x_j\}_g = \delta_{ij}$, we have

$$\frac{\partial F}{\partial y}(x_j, y_j)\delta_{jk} = -\sum_{i=0}^{g-1} R_i(x_j, y_j) X_{\hat{H}_i} x_k + \sum_{i=0}^{g-1} \hat{H}_i \frac{\partial R_i}{\partial y}(x_j, y_j)\delta_{jk}.$$

Restricted to the invariant manifolds $\hat{H}_i = c_i$ we have

$$\sum_{i=0}^{g-1} R_i(x_j, y_j) X_{\hat{H}_i} x_k = -\frac{\partial F_c}{\partial y}(x_j, y_j)\delta_{jk},$$

$$\sum_{i=0}^{g-1} \frac{R_i(x_j, y_j)}{\frac{\partial F_c}{\partial y}(x_j, y_j)} X_{\hat{H}_i} x_k = -\delta_{jk},$$

which is easily rewritten as

$$\sum_{i=1}^{g} \frac{R_j(x_i, y_i)}{\frac{\partial F_c}{\partial y}(x_i, y_i)} X_{\hat{H}_k} x_i = -\delta_{jk}.$$

Since Γ_F satisfies the assumption, we have on the left exactly a basis of the space of holomorphic differentials on Γ_{F_c}, so we find that the vector fields $X_{\hat{H}_i}$ (hence also the vector fields X_{H_i}) linearize under the Abel-Jacobi map, i.e., their flow is linear on the Jacobian of Γ_{F_c}. This shows that we have an a.c.i. system. Without the assumption we still have linearity of the flow on the Jacobian of Γ_F (the invariant manifold over 0).

Let us also show that these a.c.i. systems satisfy the Adler-van Moerbeke condition for algebraic complete integrability (which we did not impose in our definition of an a.c.i. system). That is, we show that along each of the components of the divisor which is missing from $\mathrm{Jac}(\bar{\Gamma}_{F_c})$ at least one of the functions u_i, v_i, t' has a pole. Notice that u_i and v_i (when restricted to Γ_{F_c}) are given as symmetric functions on Γ_{F_c} hence giving indeed meromorphic functions on $\mathrm{Jac}(\bar{\Gamma}_{F_c})$. In the notation of III.3.4 the missing components are the images under the Abel-Jacobi map of the divisors $\bar{\mathcal{D}}_{F,g}$ and $\bar{\mathcal{E}}_{F,g}$. As for the former, the functions v_i obviously all have a pole along it, for the latter, on a component $\bar{\mathcal{E}}_{F,g}(\infty_k)$ where $x(\infty_k)$ is infinite, all u_i have a pole, and if $y(\infty_k)$ is infinite, then at least v_{g-1} has a pole along it. ∎

Example 5.2   Here is Knörrer's example which shows that the above assumption is not valid for all curves. Notice that the assumption implies that the genus of the curves Γ_{F_c} is constant and equal to the genus of Γ_F (for generic c); in the present example the genus of the general curves will be higher than the genus of Γ_F providing the counterexample. Take

$$F(x, y) = x^8 y^2 + y^{10} - 1$$

and $R_1(x, y) = x^6 y$. It is a curve of degree 10 whose closure $x^8 y^2 + y^{10} - z^{10} = 0$ has $(1 : 0 : 0)$ as its only singular point. It is checked that this point is a cusp whose tangent has a contact of order 10 to the curve. It follows that the genus of the smooth completion of the curve equals $\frac{(10-1)(10-2)}{2} - 5 = 31$. If however one adds a multiple of R_1 to the curve, which gives

$$F_c(x, y) = x^8 y^2 + y^{10} + cx^6 y - 1,$$

then for general $c \in \mathbf{C}$ it still has only one singularity at infinity but it is now a cusp whose tangent has a contact of order 6 to the curve, giving 33 as its genus. Thus for general c the genus of Γ_{F_c} is bigger than the genus of Γ_F; if other holomorphic differentials are taken into account then it can of course only get worse. Thus Γ_F does not satisfy the assumption.

Chapter VII

Two-dimensional a.c.i. systems and applications

1. Introduction

The last chapter is devoted to the study of several two-dimensional integrable Hamiltonian system which are of interest, especially from the point of view of algebraic geometry.

We specialize the odd and even Mumford systems to the case of genus 2 in Section 2. We give explicit equations for these systems, which will be useful when studying the other examples, which will all be shown to be related to the genus 2 Mumford systems. An application of the (genus two) even Mumford system was worked out in collaboration with José Bertin (see [BV2]) and is given in Section 3. We consider an arbitrary curve of genus two which has an automorphism of order three. This automorphism can be extended to its Jacobian and leads to a singular quotient, similar to the classical Kummer surface. We exhibit a 9_4 configuration on this quotient and show that it is a complete intersection of a quadric and cubic hypersurface in $\mathbf{P}^4$. By using the even Mumford system we explicitly compute equations for the quadric and cubic hypersurface, thereby giving a projective realization of these generalized Kummer surfaces.

We study in Section 4 an integrable quartic potential on the plane, which was discovered by Garnier as a very special case of a large family of integrable Hamiltonian systems which he derived from the Schlesinger equations (see [Gar]). In Paragraph 4.1 we will show in two different ways that the general fiber of its momentum map is an affine part of an Abelian surface of type $(1,4)$. One method uses some of their specific geometry, as given in the beautiful paper [BLS], the other one is based on a morphism to the odd Mumford system. It follows that the Garnier potential is an a.c.i. system of type $(1,4)$ Moreover the morphism to the odd Mumford system leads to a Lax representation of the Garnier potential. In Paragraph 4.5 we consider the limiting case in which the potential is a central potential.

Then the general fiber of the momentum map is not an affine part of an Abelian variety but is a $\mathbf{C}^*$ bundle over an elliptic curve.

To the line bundle which defines a $(1,4)$-polarization there is associated a rational map to $\mathbf{P}^3$ (we assume here that the Abelian surface is not a product of elliptic curves). The image of this map was shown in [BLS] to be an octic and its coefficients parametrize a $24:1$ cover of the moduli space $\mathcal{A}_{(1,4)}$ of $(1,4)$-polarized Abelian surfaces. The octic is an unramified cover of the Kummer surface of a Jacobian which is canonically associated to the $(1,4)$-polarized Abelian surface. One way to characterize this Jacobian is that it is the unique Jacobian J for which the map 2_J (multiplication by 2) can be factorized via the Abelian surface. We will give in Paragraph 4.3 some other characterizations of this Jacobian. It should also be remarked that the morphism to the odd Mumford system precisely maps the general level set, which is an affine part of an Abelian surface of type $(1,4)$ to its canonical Jacobian, in particular we can use this morphism to explicitly describe the natural map between $\mathcal{A}_{(1,4)}$ and the moduli space of Riemann surfaces of genus two. The result is surprisingly simple; looked at in another way it shows how to write down explicitly for a curve (given by its equation) the equation of the Kummer surface of its Jacobian in the classical Rosenhain form.

We also get an explicit description of the moduli space $\mathcal{A}_{(1,4)}$. It is obtained as a quotient of the parameter space which is given by the coefficients of the octic. The group which is acting on it has order 24 and is isomorphic to $\mathbf{Z}/4\mathbf{Z} \times S_3$; the action was suggested by an automorphism of the system. The final result is that $\mathcal{A}_{(1,4)}$ is birational to a cone $\mathcal{M}^3$ in weighted projective space $\mathbf{P}^{(1,2,2,3,4)}$ for which we give explicit equations.

Another a.c.i. system that is naturally related to Abelian surfaces of type $(1,4)$ is the geodesic flow on $SO(4)$ corresponding to some special metric, as was first pointed out in [BV4]. In this case, studied in Section 5, the general fiber of the momentum map is an affine part of a hyperelliptic Jacobian, but there is a group of translations acting on this fiber, so that there is naturally associated to this fiber an Abelian surface of type $(1,4)$. The resulting map from the base space to the cone $\mathcal{M}^3$ in $\mathbf{P}^{(1,2,2,3,4)}$ will be explicitly computed. See [BV4] for its relation to the moduli problem for a.c.i. systems.

We look in Section 6 at a hierarchy of integrable potentials V_n on the plane. Their integrability is easily checked but revealing the geometry of the general level sets of the momentum map requires more work. For the first non-trivial member V_3 we define a morphism to the odd Mumford system and deduce from it that V_3 is algebraic completely integrable and we identify the general fiber of the momentum map as an Abelian surface of polarization type $(1,2)$ (this result was previously established by using Painlevé analysis by Adler and van Moerbeke, see [AM9]). For V_4 there is a similar morphism to the even Mumford system and we show that its general fiber is a $2:1$ cover of a hyperelliptic Jacobian, which is ramified along two touching translates of the theta divisor; this result could also be obtained by using a morphism to the Bechlivanidis-van Moerbeke system, which is discussed in Paragraph 2.3. For V_n we use a morphism to show that the level sets are $2:1$ unramified covers of a certain two-dimensional stratum in the Jacobian of a hyperelliptic curve of genus $\lceil \frac{n+1}{2} \rceil$. We also exhibit Lax equations (with a spectral parameter) for the hierarchy.

In the last section we reconsider the periodic three body Toda lattice. We use again a morphism to give a new proof of its algebraic complete integrability. The order three automorphism which comes from a cyclic permutation of the three particles leads to an a.c.i. system of type $(1,3)$ and we give a concrete realization of this quotient system.

176

2. The genus two Mumford systems

2.1. The genus two odd Mumford system

For future use we now give explicit formulas for the genus 2 Mumford systems. Before doing this, let us restrict the odd and even Mumford systems to a Casimir which is common to all the Poisson structures $\{\cdot,\cdot\}^\varphi$, where $\deg \varphi \leq 2$, without restricting the class of (hyperelliptic) curves that appear as spectral curves for these systems. Namely, we consider for the odd (resp. even) Mumford system the hyperplane given by $H_{2g} = 0$ (resp. $H_{2g+1} = 0$). Since this corresponds to fixing a level of a Casimir all formulas are easily transcribed: for the odd (resp. even) case, simply substitute $w_g \mapsto u_{g-1}$ (resp. $w_{g+1} \mapsto u_{g-1}$) in all formulas. We start with the odd case. Then the Lax operator is given by

$$A(\lambda) = \begin{pmatrix} v_1\lambda + v_0 & \lambda^2 + u_1\lambda + u_0 \\ \lambda^3 - u_1\lambda^2 + w_1\lambda + w_0 & -v_1\lambda - v_0 \end{pmatrix} \tag{2.1}$$

and the Lax equations for the Hamiltonian vector fields $\mathcal{X}_1$ and $\mathcal{X}_2$ are given by $X_{H_i}A(\lambda) = [A(\lambda), B_i(\lambda)]$, where

$$B_1(\lambda) = \begin{pmatrix} 0 & 1 \\ \lambda - 2u_1 & 0 \end{pmatrix}, \qquad B_0(\lambda) = \begin{pmatrix} v_1 & \lambda + u_1 \\ \lambda^2 - u_1\lambda + w_1 - u_0 & -v_1 \end{pmatrix}.$$

If we denote derivation in the direction of the first vector field by a dot and in the direction of the zeroth vector field by a prime, these are written out as

$$\begin{aligned}
\dot{u}(\lambda) &= 2v(\lambda), & u'(\lambda) &= 2v(\lambda)(\lambda + u_1) - 2v_1u(\lambda), \\
\dot{v}(\lambda) &= (\lambda - 2u_1)u(\lambda) - w(\lambda), & v'(\lambda) &= u(\lambda)(\lambda^2 - u_1\lambda + w_1 - u_0) - (\lambda + u_1)w(\lambda), \\
\dot{w}(\lambda) &= -2v(\lambda)(\lambda - 2u_1), & w'(\lambda) &= 2w(\lambda)v_1 - 2v(\lambda)(\lambda^2 - u_1\lambda + w_1 - u_0).
\end{aligned}$$

In order to write down explicitly the compatible Poisson structures it is useful to have these even further written out, namely in the following form:

$$\begin{aligned}
\dot{u}_1 &= 2v_1, & u'_1 &= 2v_0, \\
\dot{u}_0 &= 2v_0, & u'_0 &= 2(u_1v_0 - u_0v_1), \\
\dot{v}_1 &= u_0 - 2u_1^2 - w_1, & v'_1 &= -2u_1u_0 - w_0, \\
\dot{v}_0 &= -2u_1u_0 - w_0, & v'_0 &= -u_1w_0 + u_0(w_1 - u_0), \\
\dot{w}_1 &= 4u_1v_1 - 2v_0, & w'_1 &= 2(u_1v_0 + u_0v_1), \\
\dot{w}_0 &= 4u_1v_0, & w'_0 &= 2(u_0v_0 + v_1w_0 - v_0w_1).
\end{aligned}$$

Recall also that the corresponding integrable algebra $\mathcal{A}$ is a polynomial algebra generated by the coefficients of $v^2(\lambda) + u(\lambda)w(\lambda)$, i.e., by

$$\begin{aligned}
H_3 &= u_0 + w_1 - u_1^2, \\
H_2 &= -u_1u_0 + u_1w_1 + w_0 + v_1^2, \\
H_1 &= u_1w_0 + u_0w_1 + 2v_1v_0, \\
H_0 &= u_0w_0 + v_0^2,
\end{aligned} \tag{2.2}$$

(recall that we have put $H_4 = 0$). We now explicit for this system the formulas for the three compatible Poisson structures $\{\cdot,\cdot\}^\varphi$, where $\varphi = 1, \lambda, \lambda^2$. Their Poisson matrices are easily computed from (VI.4.4) in terms of the system of generators $u_1, u_0, v_1, v_0, w_1, w_0$ (in that order). Notice that since S_4 is not a Casimir of $\{\cdot,\cdot\}^{\lambda^3}$ the latter Poisson structure does not restrict to the level $S_4 = 0$. The Poisson structure $\{\cdot,\cdot\}^1$ is given by

$$\begin{pmatrix} 0 & 0 & 0 & 1 & 0 & 0 \\ 0 & 0 & 1 & u_1 & 0 & -2v_1 \\ 0 & -1 & 0 & 0 & 1 & -2u_1 \\ -1 & -u_1 & 0 & 0 & -u_1 & w_1 - u_0 \\ 0 & 0 & -1 & u_1 & 0 & 2v_1 \\ 0 & 2v_1 & 2u_1 & u_0 - w_1 & -2v_1 & 0 \end{pmatrix}. \tag{2.3}$$

It is easy to compute that H_2 and H_3 are Casimirs and that H_1 and H_0 give the first and zeroth vector fields. Notice also that this structure is affine (i.e., it is a modified Lie-Poisson structure, see Example II.2.14). The Poisson matrix of the (affine) structure $\{\cdot,\cdot\}^\lambda$ is given by

$$\begin{pmatrix} 0 & 0 & 1 & 0 & 0 & 0 \\ 0 & 0 & 0 & -u_0 & 0 & 2v_0 \\ -1 & 0 & 0 & 0 & -2u_1 & 0 \\ 0 & u_0 & 0 & 0 & -u_0 & -w_0 \\ 0 & 0 & 2u_1 & u_0 & 0 & -2v_0 \\ 0 & -2v_0 & 0 & w_0 & 2v_0 & 0 \end{pmatrix}. \tag{2.4}$$

In this case H_0 and H_3 are the Casimirs and H_1 and H_2 give the two vector fields. Finally the Poisson matrix of $\{\cdot,\cdot\}^{\lambda^2}$ is given by

$$\begin{pmatrix} 0 & 0 & -u_1 & -u_0 & 2v_1 & 2v_0 \\ 0 & 0 & -u_0 & 0 & 2v_0 & 0 \\ u_1 & u_0 & 0 & 0 & -w_1 & -w_0 \\ u_0 & 0 & 0 & 0 & -w_0 & 0 \\ -2v_1 & -2v_0 & w_1 & w_0 & 0 & 0 \\ -2v_0 & 0 & w_0 & 0 & 0 & 0 \end{pmatrix}. \tag{2.5}$$

In this case H_0 and H_1 are Casimirs and H_2 and H_3 give the two vector fields. Notice that this structure is even linear. The Hamiltonian vector fields which correspond to the polynomials $H_0, \ldots, H_3$ using these structures are summarized in the following table.

$\{\cdot, H\}^\varphi$	H_3	H_2	H_1	H_0
$\{\cdot,\cdot\}^1$	0	0	$\mathcal{X}_1$	$\mathcal{X}_0$
$\{\cdot,\cdot\}^\lambda$	0	$\mathcal{X}_1$	$\mathcal{X}_0$	0
$\{\cdot,\cdot\}^{\lambda^2}$	$\mathcal{X}_1$	$\mathcal{X}_0$	0	0

Table 3

2.2. The genus two even Mumford system

We now give the corresponding formulas for the genus 2 even Mumford system. The Poisson structures $\{\cdot,\cdot\}^\varphi$, which we computed from Proposition III.2.1. These structures are neither linear nor affine. The Lax operator is given by

$$A(\lambda) = \begin{pmatrix} v_1\lambda + v_0 & \lambda^2 + u_1\lambda + u_0 \\ \lambda^4 - u_1\lambda^3 + w_2\lambda^2 + w_1\lambda + w_0 & -v_1\lambda - v_0 \end{pmatrix}$$

and the Lax equations for the Hamiltonian vector fields $\mathcal{X}_1$ and $\mathcal{X}_0$ are given by $X_{H_i}A(\lambda) = [A(\lambda), B_i(\lambda)]$, where

$$B_1(\lambda) = \begin{pmatrix} 0 & 1 \\ \lambda^2 - 2u_1\lambda + 2u_1^2 - u_0 + w_2 & 0 \end{pmatrix},$$

and

$$B_0(\lambda) = \begin{pmatrix} v_1 & \lambda + u_1 \\ \lambda^3 - u_1\lambda^2 + (w_2 - u_0)\lambda + 2u_1u_0 + w_1 & -v_1 \end{pmatrix}.$$

They are written out as vector fields on $\mathbf{C}^7$ as

$$\dot{u}_1 = 2v_1, \qquad\qquad u_1' = 2v_0,$$
$$\dot{u}_0 = 2v_0, \qquad\qquad u_0' = 2(u_1v_0 - u_0v_1),$$
$$\dot{v}_1 = 2u_1^3 + u_1w_2 - 3u_1u_0 - w_1, \qquad v_1' = 2u_1^2u_0 - u_0^2 + u_0w_2 - w_0,$$
$$\dot{v}_0 = 2u_1^2u_0 - u_0^2 + u_0w_2 - w_0, \qquad v_0' = u_0w_1 + u_1(2u_0^2 - w_0),$$
$$\dot{w}_2 = 2(2u_1v_1 - v_0), \qquad w_2' = 2(u_1v_0 + u_0v_1),$$
$$\dot{w}_1 = 2(-2u_1^2v_1 + 2u_1v_0 - v_1w_2 + v_1u_0), \qquad w_1' = -2(2u_1u_0v_1 + v_0(w_2 - u_0)),$$
$$\dot{w}_0 = -2v_0(2u_1^2 - u_0 + w_2), \qquad w_0' = 2(v_1w_0 - v_0w_1 - 2u_1u_0v_0).$$

The corresponding integrable algebra $\mathcal{A}$ is now the polynomial algebra generated by

$$\begin{aligned}
H_4 &= w_2 - u_1^2 + u_0, \\
H_3 &= w_1 - u_1u_0 + u_1w_2, \\
H_2 &= u_0w_2 + u_1w_1 + w_0 + v_1^2, \\
H_1 &= u_1w_0 + u_0w_1 + 2v_1v_0, \\
H_0 &= u_0w_0 + v_0^2.
\end{aligned} \tag{2.6}$$

We also write down the three compatible Poisson structures $\{\cdot,\cdot\}^\varphi$ for the genus 2 even Mumford system (for $\varphi = 1, \lambda, \lambda^2$); their Poisson matrices are now written in terms of the system of generators $u_1, u_0, v_1, v_0, w_2, w_1, w_0$ (in that order). Here is the Poisson matrix of $\{\cdot,\cdot\}^1$

$$\begin{pmatrix}
0 & 0 & 0 & 1 & 0 & 0 & 0 \\
0 & 0 & 1 & u_1 & 0 & 0 & -2v_1 \\
0 & -1 & 0 & 0 & 1 & -2u_1 & w_2 - u_0 + 2u_1^2 \\
-1 & -u_1 & 0 & 0 & -u_1 & w_2 - u_0 & w_1 + 2u_1u_0 \\
0 & 0 & -1 & u_1 & 0 & 0 & 2v_1 \\
0 & 0 & 2u_1 & u_0 - w_2 & 0 & 0 & -4u_1v_1 \\
0 & 2v_1 & u_0 - 2u_1^2 - w_2 & -2u_1u_0 - w_1 & -2v_1 & 4u_1v_1 & 0
\end{pmatrix}. \tag{2.7}$$

It has H_2, H_3 and H_4 as Casimirs and H_1 and H_0 give the first and zeroth vector fields. The Poisson matrix of $\{\cdot,\cdot\}^\lambda$ is given by

$$
\begin{pmatrix}
0 & 0 & 1 & 0 & 0 & 0 & 0 \\
0 & 0 & 0 & -u_0 & 0 & 0 & 2v_0 \\
-1 & 0 & 0 & 0 & -2u_1 & w_2 - u_0 + 2u_1^2 & 0 \\
0 & u_0 & 0 & 0 & -u_0 & 2u_1 u_0 & -w_0 \\
0 & 0 & 2u_1 & u_0 & 0 & 0 & -2v_0 \\
0 & 0 & u_0 - w_2 - 2u_1^2 & -2u_1 u_0 & 0 & 0 & 4u_1 v_0 \\
0 & -2v_0 & 0 & w_0 & 2v_0 & -4u_1 v_0 & 0
\end{pmatrix}. \tag{2.8}
$$

This one has H_0, H_3 and H_4 as Casimirs and H_1 and H_2 give the two vector fields. A third structure is given by

$$
\begin{pmatrix}
0 & 0 & -u_1 & -u_0 & 0 & 2v_1 & 2v_0 \\
0 & 0 & -u_0 & 0 & 0 & 2v_0 & 0 \\
u_1 & u_0 & 0 & 0 & 2u_1^2 - u_0 & -w_1 & -w_0 \\
u_0 & 0 & 0 & 0 & 2u_1 u_0 & -w_0 & 0 \\
0 & 0 & u_0 - 2u_1^2 & -2u_1 u_0 & 0 & -2v_0 + 4u_1 v_1 & 4u_1 v_0 \\
-2v_1 & -2v_0 & w_1 & w_0 & 2v_0 - 4u_1 v_1 & 0 & 0 \\
-2v_0 & 0 & w_0 & 0 & -4u_1 v_0 & 0 & 0
\end{pmatrix}. \tag{2.9}
$$

In this case H_0, H_1 and H_4 are Casimirs and H_2 and H_3 give the two vector fields. Finally, a fourth structure, with H_2, H_3 and H_4 as Casimirs, H_1 and H_2 giving the two commuting vector fields, is given by the following (in order to make this anti-symmetric fit on this page we only print its upper triangular part)

$$
\begin{pmatrix}
0 & 0 & u_1^2 - u_0 & u_0 u_1 & 2v_1 & 2(v_0 - u_1 v_1) & -2u_1 v_0 \\
 & 0 & u_0 u_1 & u_0^2 & 2v_0 & -2u_0 v_1 & -2u_0 v_0 \\
 & & 0 & 0 & u_1 w_2 - w_1 & u_1 w_1 - w_0 & u_1 w_0 \\
 & & & 0 & u_0 w_2 - w_0 & u_0 w_1 & u_0 w_0 \\
 & & & & 0 & 2v_1 w_2 & 2v_0 w_2 \\
 & & & & & 0 & 2(v_0 w_1 - v_1 w_0) \\
 & & & & & & 0
\end{pmatrix}. \tag{2.10}
$$

As above we summarize the Hamiltonian vector fields which correspond to $H_0, \ldots, H_4$ upon using these structures in a table.

$\{\cdot, H_i\}^\varphi$	H_4	H_3	H_2	H_1	H_0
$\{\cdot,\cdot\}^1$	0	0	0	$\mathcal{X}_1$	$\mathcal{X}_0$
$\{\cdot,\cdot\}^\lambda$	0	0	$\mathcal{X}_1$	$\mathcal{X}_0$	0
$\{\cdot,\cdot\}^{\lambda^2}$	0	$\mathcal{X}_1$	$\mathcal{X}_0$	0	0
$\{\cdot,\cdot\}^{\lambda^3}$	$\mathcal{X}_1$	$\mathcal{X}_0$	0	0	0

Table 4

It can be checked by direct computation that these brackets also verify (VI.3.7), but a conceptual proof (given for the odd Mumford systems through the loop algebra $\mathfrak{gl}_q$) is missing.

2.3. The Bechlivanidis-van Moerbeke system

In this paragraph we wish to consider an integrable Hamiltonian system on $\mathbf{C}^7$ which was constructed by Bechlivanidis and van Moerbeke (see [BM]) in order to understand the geometry of the Goryachev-Chaplygin top. This system is sometimes called *the seven-dimensional system*, but since the dimension of an integrable Hamiltonian system was defined as the dimension of the general fiber of the momentum map, which is two in this case, we prefer not to call it this way; we prefer to call it the *Bechlivanidis-van Moerbeke system*. It was constructed as a pair of commuting vector fields together with five independent functions whose derivatives in the direction of these vector fields was zero; this was done without making reference to any Poisson structure for which the equations of motion were Hamiltonian. Several such Poisson structures (which are compatible) were constructed for it by us in [Van1]. Not only will these be reproduced here but we will show that the Bechlivanidis-van Moerbeke system is isomorphic to the genus 2 even Mumford system, thereby proving its algebraic complete integrability (a proof of this has been given nowhere) and "explaining" where the bi-Hamiltonian structure of this system comes from. There is for this system a third (compatible) structure which is found in a natural way; it is closely related to the Poisson structure of the Toda lattice discussed in the last section of this chapter; of course this structure can also be transported to the even Mumford system, thereby leading to another compatible structure for that system.

We pick coordinates $s_1, \ldots, s_7$ on $\mathbf{C}^7$ and consider the following algebra of functions,

$$\mathcal{A} = \mathbf{C}[S_1, S_2, S_3, S_4, S_5],$$

where

$$
\begin{aligned}
S_1 &= s_1 - 4s_2^2 - 8s_4, \\
S_2 &= s_1 s_2 + 4s_6, \\
S_3 &= s_4^2 - s_5^2 + s_3, \\
S_4 &= s_4 s_6 + s_5 s_7 + s_2 s_3, \\
S_5 &= -s_6^2 + s_7^2 - s_1 s_3.
\end{aligned}
\tag{2.11}
$$

Also consider the Poisson matrix $(\{s_i, s_j\})_{1 \le i, j \le 8}$, given by

$$
\begin{pmatrix}
0 & 0 & -16s_5 & 0 & -8 & 0 & 16s_2 \\
0 & 0 & 0 & 0 & 0 & 0 & 4 \\
16s_5 & 0 & 0 & 2s_5 & 2s_4 & -4s_2 s_5 & 4s_2 s_4 \\
0 & 0 & -2s_5 & 0 & -1 & 0 & -2s_2 \\
8 & 0 & -2s_4 & 1 & 0 & -2s_2 & 0 \\
0 & 0 & 4s_2 s_5 & 0 & 2s_2 & 0 & -4s_2^2 - s_1 \\
-16s_2 & -4 & -4s_2 s_4 & 2s_2 & 0 & 4s_2^2 + s_1 & 0
\end{pmatrix}.
\tag{2.12}
$$

Although we do not suggest to the reader to do this (because it is long), it can be checked by direct computation that this matrix defines indeed a Poisson structure on $\mathbf{C}^7$; we will check it in another way. What is however checked at once is that S_1, S_2 and S_3 are Casimirs for this structure and that S_4 and S_5 are in involution; they lead to the vector fields (up to a constant)

$$\dot{s}_1 = -8s_7, \qquad\qquad s_1' = 8(s_1s_5 + 2s_2s_7),$$
$$\dot{s}_2 = 4s_5, \qquad\qquad s_2' = 4s_7,$$
$$\dot{s}_3 = 2(s_4s_7 + s_5s_6), \qquad\qquad s_3' = 4(s_2s_5s_6 + s_2s_4s_7 - 2s_3s_5),$$
$$\dot{s}_4 = -4s_2s_5 - s_7, \qquad\qquad s_4' = s_1s_5 - 2s_5s_7,$$
$$\dot{s}_5 = s_6 - 4s_2s_4, \qquad\qquad s_5' = s_1s_4 + 2s_2s_6 - 4s_3,$$
$$\dot{s}_6 = -s_1s_5 + 2s_2s_7, \qquad\qquad s_6' = -s_1s_7 - 2s_1s_2s_5 - 4s_2^2s_7,$$
$$\dot{s}_7 = s_1s_4 + 2s_2s_6 - 4s_3. \qquad\qquad s_7' = 8s_2s_3 + 4s_1s_2s_4 - 4s_2^2s_6 - s_1s_6.$$

Before writing down other Poisson structures, let us prove that the Bechlivanidis-van Moerbeke system is isomorphic to the even Mumford system by giving explicitly the isomorphism; this map was constructed in [Van2] as an illustration of the algorithm (recalled in Section V.4) which was developed there. The isomorphism is given by the map

$$\phi : \mathbf{C}^7 \to \mathbf{C}^7 : (s_1, \ldots, s_7) \mapsto (u(\lambda), v(\lambda), w(\lambda))$$

where

$$u(\lambda) = \lambda^2 + 2s_2\lambda + s_1,$$
$$v(\lambda) = 8s_5\lambda - 8s_7,$$
$$w(\lambda) = \lambda^4 - 2s_2\lambda^3 + (s_1 - 16s_4 - 4s_2^2)\lambda^2 - 4(s_2(s_1 - 8s_4 - 2s_2^2) + 4s_6)\lambda$$
$$+ 4(s_1s_2^2 + 8s_2s_6 - 16s_3).$$

Of course this map is regular, but it is even biregular: s_1, s_2, s_5 and s_7 are of course regular in terms of the coefficients of $u(\lambda)$ and $v(\lambda)$; given this s_4 is regular in terms of these and the coefficient of λ^2 in $w(\lambda)$; the same follows for s_6 and s_3 upon using respectively the coefficients in λ and λ^0 in $w(\lambda)$. Next it is easy to check that

$$\phi^*(w_2 - u_1^2 + u_0) = 2(s_1 - 4s_2^2 - 8s_4),$$
$$\phi^*(w_1 - u_1u_0 + u_1w_2) = -4(s_1s_2 + 4s_6),$$
$$\phi^*(u_1w_1 + u_0w_2 + w_0 + v_1^2) = -64(s_4^2 - s_5^2 + s_3) + (s_1 - 4s_2^2 - 8s_4)^2,$$
$$\phi^*(u_1w_0 + u_0w_1 + 2v_1v_0) = -128(s_4s_6 + s_5s_7 + s_2s_3) - 4(s_1 - 4s_2^2 - 8s_4)(s_1s_2 + 4s_6),$$
$$\phi^*(u_0w_0 + v_0^2) = 64(s_7^2 - s_6^2 - s_1s_3) + 4(s_1s_2 + 4s_6)^2.$$

$$(2.13)$$

Thus $\phi^*\mathcal{A}' = \mathcal{A}$, where $\mathcal{A}'$ denotes here the integrable algebra of the (genus 2) even Mumford system. Since ϕ is a biregular map the fact that $\mathcal{A}'$ is integrable for some Poisson structure, $\phi^*\mathcal{A}$ is integrable for the corresponding Poisson structure; moreover since $\mathcal{A}'$ is a.c.i. the same holds true for $\mathcal{A}$. The Poisson structure which corresponds to (2.9) is given (up to a constant) by (2.12) hence (2.12) defines a Poisson structure and ϕ is a Poisson map with respect to the Poisson structures (2.9) and (2.12). Since ϕ is biregular we can also transport the other

2. The genus two Mumford systems

Poisson structures. If we transport the Poisson structure given by (2.8) then we get (up to a constant) the following Poisson matrix:

$$\begin{pmatrix}
0 & 0 & -16s_7 & 0 & 0 & 0 & -8s_1 \\
0 & 0 & 0 & 0 & -4 & 0 & 0 \\
16s_7 & 0 & 0 & 2s_7 & -2s_6 & -4s_2s_7 & 8s_3 - 4s_2s_6 \\
0 & 0 & -2s_7 & 0 & 4s_2 & 0 & -s_1 \\
0 & 4 & 2s_6 & -4s_2 & 0 & -s_1 & 0 \\
0 & 0 & 4s_2s_7 & 0 & s_1 & 0 & 2s_1s_2 \\
8s_1 & 0 & 4s_2s_6 - 8s_3 & s_1 & 0 & -2s_1s_2 & 0
\end{pmatrix}.$$

In this case S_1, S_2 and S_5 are Casimirs and S_3 and S_4 generate (up to a constant) the vector fields X_1 and X_2; this can be checked directly or by using (2.13). One might think now that the structure (2.7) will give a Poisson bracket for which S_1, S_4 and S_5 are Casimirs but S_2 and S_3 generate the two vector fields, but this is not the case. To see this, let us rewrite (2.13) as follows:

$$\phi^* H_4 = 2S_1,$$
$$\phi^* H_3 = -4S_2,$$
$$\phi^* H_2 = S_1^2 - 64S_3,$$
$$\phi^* H_1 = -128S_4 - 4S_1S_2,$$
$$\phi^* H_0 = 64S_5 + 4S_2^2.$$

Since ϕ is a Poisson isomorphism it maps Casimirs to Casimirs. Thus, if H_2, H_3 and H_4 are Casimirs then S_1, S_2 and S_3 are Casimirs for the corresponding structure. Similarly, if H_0, H_3 and H_4 are Casimirs then S_1, S_2 and S_5 are Casimirs for the corresponding structure. However, if H_0, H_1 and H_4 are Casimirs then S_1, $32S_4 + S_1S_2$ and $16S_5 + S_2^2$ are Casimirs, so that (up to a multiple) S_2 and S_3 generate X_1 but S_4 and S_5 give respectively S_1X_2 and S_2X_2. It follows even that there is no linear combination of these three Poisson structures which has S_1, S_4 and S_5 as Casimirs, while S_2 and S_3 give the vector fields X_1 and X_2. However one finds by trial and error easily that the following structure does this job:

$$\begin{pmatrix}
0 & 0 & -4(s_5s_6 + s_4s_7) & 2s_7 & -2s_6 & 2s_1s_5 + 4s_2s_7 & 4s_2s_6 - 2s_1s_4 \\
0 & 0 & 0 & -s_5 & -s_4 & s_7 & s_6 \\
4(s_5s_6 + s_4s_7) & 0 & 0 & 0 & 0 & -2s_3s_5 & 2s_3s_4 \\
-2s_7 & s_5 & 0 & 0 & 0 & 0 & -s_3 \\
2s_6 & s_4 & 0 & 0 & 0 & -s_3 & 0 \\
-4s_2s_7 - 2s_1s_5 & -s_7 & 2s_3s_5 & 0 & s_3 & 0 & -2s_2s_3 \\
2s_1s_4 - 4s_2s_6 & -s_6 & -2s_3s_4 & s_3 & 0 & 2s_2s_3 & 0
\end{pmatrix}.$$

(2.14)

This one is important because it will appear later when studying the Toda lattice. It is compatible with the previous ones in view of Theorem 14 in [Van2] which says that if in a two-dimensional integrable Hamiltonian system all integrable vector fields are Hamiltonian with respect to two different Poisson brackets, then these brackets are compatible. Of course it leads also to a fifth Hamiltonian structure for the genus 2 even Mumford system. For a generalization of this special structure to all even Mumford systems, see [FV].

Using the isomorphism everything can be transported from the genus 2 even Mumford system to the Bechlivanidis-van Moerbeke system. Since we have Lax equations for the even Mumford system we also have them for this system. Similarly we obtain all possible Laurent solutions from those of the even Mumford system, and since all coefficients of the map are real we can use our knowledge about the topology of the real level sets of the even Mumford system to describe the topology of these level sets for the Bechlivanidis-van Moerbeke system. Finally, in the next section we will use the even Mumford system to study a certain problem in algebraic geometry; clearly we could use the isomorphic Bechlivanidis-van Moerbeke system instead.

3. Application: generalized Kummer surfaces

We now give an application of integrable Hamiltonian systems to algebraic geometry, which was worked out in collaboration with José Bertin (see [BV2]). For other applications which use the same technique, see [PV2].

3.1. Genus two curves with an automorphism of order three

We consider a curve Γ of genus two, equipped with an automorphism of order three, denoted by τ. By the Riemann-Hurwitz formula the quotient Γ/τ has genus zero and τ has four fixed points. Since Γ has genus two it is also hyperelliptic; as before the hyperelliptic involution will be denoted by σ. We have the following diagram

τ necessarily maps Weierstrass points to Weierstrass points, hence the commutator $[\tau, \sigma]$ fixes all these points and we see that $\tau\sigma = \sigma\tau$ since the only automorphisms which fix all Weierstrass points are σ and identity. It follows on the one hand that τ induces on $\mathbf{P}^1$ a fractional linear transformation $\tilde{\tau}$ of order three, and on the other hand that the four fixed points of τ consist of two σ-orbits. We may therefore suppose that $\tilde{\tau}$ is given by $\tilde{\tau}(x) = \epsilon x$, $\epsilon = \exp(\frac{2\pi i}{3})$, by choosing a coordinate x on $\mathbf{P}^1$ such that these two orbits correspond to $x = 0$ and $x = \infty$. The images of the Weierstrass points form two orbits of three points under $\tilde{\tau}$, which correspond to the roots of the equation $x^3 = \lambda^3$ and $x^3 = \lambda^{-3}$, possibly after a rescaling of x. Obviously $\lambda \neq 0$; since both orbits are different, $\lambda^3 \neq \lambda^{-3}$, i.e., $\lambda^6 \neq 1$. This shows that Γ has an equation

$$\begin{aligned} y^2 &= (x^3 - \lambda^3)(x^3 - \lambda^{-3}) \\ &= x^6 + 2\kappa x^3 + 1, \end{aligned} \tag{3.1}$$

with $\kappa \neq \pm 1$.

Clearly, every equation of the form (3.1) with $\kappa \neq \pm 1$, defines a smooth curve of genus two with an automorphism $(x, y) \mapsto (\epsilon x, y)$ of order three; also, if κ in (3.1) is replaced by $-\kappa$ then an isomorphic curve is obtained. Conversely, let there be given two isomorphic curves Γ and Γ' with respective automorphisms τ and τ' of order three. We may suppose that the isomorphism $\phi : \Gamma \to \Gamma'$ respects the automorphism, i.e., $\phi\tau = \tau'\phi$. We claim that if Γ_1 and Γ_2 are written as above as

$$\Gamma_i : y^2 = x^6 + 2\kappa_i x^3 + 1,$$

then $\kappa_1^2 = \kappa_2^2$. To see this, notice that ϕ obviously commutes with σ hence there is an induced linear transformation $\tilde{\phi}$ which satisfies $\tilde{\phi}(\epsilon x) = \epsilon\tilde{\phi}(x)$, for all $x \in \mathbf{P}^1$. Thus $\tilde{\phi}(x) = \rho x$ and $\phi(x, y) = (\rho x, y)$, iving $\rho^6 = 1$. It follows that $\kappa^2 \neq 1$ can be taken as modular parameter.

The automorphism group of Γ contains a subgroup which is isomorphic to $S_3 \times \mathbf{Z}/2\mathbf{Z}$, as is seen immediately from (3.1); it actually coincides with this group, unless $\kappa = 0$ (in which case the group of automorphisms jumps to $D_6 \times \mathbf{Z}/2\mathbf{Z}$). Namely, there is, apart from the hyperelliptic involution σ, an action of S_3 by means of which the Weierstrass points belonging to one τ-orbit can be at random permuted. For future use we choose an element μ of order two in this symmetry group S_3 corresponding to a transposition in S_3, say $\mu(x, y) = (x^{-1}, yx^{-3})$ and notice that it commutes with σ but not with τ. Its fixed points are the two points in $\pi_\sigma^{-1}\{1\}$, hence Γ/μ is an elliptic curve.

We will find it convenient to denote the fixed points of τ, which are mapped by π_σ to 0 (resp. ∞) by o_1 and o_2 (resp. ∞_1 and ∞_2). Then $o_1^\sigma = o_2$, $\infty_1^\sigma = \infty_2$ and we may suppose $\mu(o_1) = \infty_1$ giving also $\mu(o_2) = \infty_2$. In the same way we denote the Weierstrass points corresponding to the $x^3 = \lambda^3$-orbit by λ_i, $\tau(\lambda_i) = \lambda_{i+1}$ (indices are taken modulo 3) and the ones corresponding to the $x^3 = \lambda^{-3}$-orbit by $\bar{\lambda}_i$, $\mu(\lambda_i) = \bar{\lambda}_i$. Then the action of $S_3 \times \mathbf{Z}/2\mathbf{Z}$ on these points is contained in the following table.

	order	o_1	o_2	∞_1	∞_2	λ_i	$\bar{\lambda}_i$
τ	3	o_1	o_2	∞_1	∞_2	λ_{i+1}	$\bar{\lambda}_{i-1}$
σ	2	o_2	o_1	∞_2	∞_1	λ_i	$\bar{\lambda}_i$
μ	2	∞_1	∞_2	o_1	o_2	$\bar{\lambda}_i$	λ_i

Table 5

3.2. The 9_4 configuration on the Jacobian of Γ

Let $J(\Gamma)$ denote the Jacobian of Γ and for a divisor $\mathcal{D}$ of degree 0, let $\mathrm{Ab}(\mathcal{D})$ denote the corresponding point in $J(\Gamma)$ as before. Recall that for any fixed $Q_1, Q_2 \in \Gamma$, every element $\omega \in J(\Gamma)$ can be written as $\omega = \mathrm{Ab}(P_1 + P_2 - Q_1 - Q_2)$; moreover this representation is unique if and only if $P_1 \neq P_2^\sigma$, all $P + P^\sigma$ and $Q + Q^\sigma$ $(P, Q \in \Gamma)$ being linearly equivalent, $P + P^\sigma \sim_l Q + Q^\sigma$. In the present case of curves (3.1) which have an automorphism τ of order three, the cover π_τ associated to τ provides in addition (using the notations of the previous section for the fixed points of τ) the following linear equivalences

$$3o_1 \sim_l 3o_2 \sim_l 3\infty_1 \sim_l 3\infty_2. \tag{3.2}$$

The automorphism τ extends in a natural way to an automorphism on $J(\Gamma)$, also denoted by τ. It is given and well-defined for $\omega = \mathrm{Ab}(P_1 + P_2 - Q_1 - Q_2)$ as follows: $\tau(\omega) = \mathrm{Ab}(\tau(P_1) + \tau(P_2) - \tau(Q_1) - \tau(Q_2))$.

Proposition 3.1 *The automorphism τ has nine fixed points and nine invariant theta curves on $J(\Gamma)$.*

Proof

The principal polarisation on $J(\Gamma)$ is invariant under $\mathrm{Aut}(\Gamma)$, hence the isomorphism $J(\Gamma) \to \hat{J}(\Gamma)$ from $J(\Gamma)$ to its dual $\hat{J}(\Gamma)$ is $\mathrm{Aut}(\Gamma)$-invariant and the second statement follows from the first one.

3. Application: generalized Kummer surfaces

We count the number of fixed points of τ in two different ways. At first we use the holomorphic Lefschetz fixed point formula

$$\sum_p (-1)^p \operatorname{Trace} f^*|_{H^{p,0}(M)} = \sum_{f(p_\alpha)=p_\alpha} \frac{1}{\det(I - B_\alpha)}, \tag{3.3}$$

for a holomorphic map $f : M \to M$, where B_α is the linear part of f at the fixed point p_α. We apply it for $f = \tau$ and $M = J(\Gamma)$; in this case $H^{p,0}(J(\Gamma))$ may be identified with the p-th skew-symmetric power of the cotangent bundle at any point of $J(\Gamma)$. For the left-hand side in (3.3), the basis of $H^{p,0}(J(\Gamma))$ may thus be taken in a point $\mathrm{Ab}(P_1 + P_2 - Q_1 - Q_2)$ as $\{\Omega_1, \Omega_2\} = \{\omega_1(P_1) + \omega_1(P_2), \omega_2(P_1) + \omega_2(P_2)\}$, where $\omega_i = x^{i-1} dx/y$ and $\Omega_1 \wedge \Omega_2$ generates $H^{2,0}(J(\Gamma))$. Since $\tau^* \Omega_i = \epsilon^i \Omega_i$, $(i = 1, 2)$, the left hand side in (3.3) gives

$$\sum_{p=0}^{2} (-1)^p \operatorname{Trace} \tau^*|_{H^{p,0}(J(\Gamma))} = 1 - \operatorname{Trace} \begin{pmatrix} \epsilon & 0 \\ 0 & \epsilon^2 \end{pmatrix} + 1 = 3.$$

As for the right hand side, obviously all B_α are equal, in fact

$$B_\alpha = \begin{pmatrix} \epsilon & 0 \\ 0 & \epsilon^2 \end{pmatrix} \tag{3.4}$$

when local coordinates dual to Ω_1 and Ω_2 are picked around the point P_α. Therefore

$$\det(I - B_\alpha) = (1 - \epsilon)(1 - \epsilon^2) = 3,$$

and the number of fixed points of τ is indeed nine.

A second way to determine the number of fixed points of τ is by writing down an explicit list: if we write every point $\omega \in J(\Gamma)$ as $\omega = \mathrm{Ab}(P_1 + P_2 - 2\infty_1)$ then $\tau\omega = \omega$ if and only if $\tau(P_1) + \tau(P_2) \sim_l P_1 + P_2$, i.e., $P_1 = P_2^\sigma$ or P_1 and P_2 are both fixed points for τ. Using (3.2) we arrive at the following list

$$\{O, \; 0_1 - 0_2, \; 0_2 - 0_1, \; \infty_1 - \infty_2, \; \infty_2 - \infty_1, \; 0_1 - \infty_1, \; 0_2 - \infty_2, \; 0_1 - \infty_2, \; 0_2 - \infty_1\}. \tag{3.5}$$

The nine invariant curves are then given by the nine translates over these points of the image of Γ in $J(\Gamma)$ by the map $x \mapsto \mathrm{Ab}(x - \infty_1)$. Since this curve obviously contains exactly the four fixed points

$$\{O, \; \infty_2 - \infty_1, \; 0_1 - \infty_1, \; 0_2 - \infty_1\},$$

each of the nine invariant curves will contain exactly four fixed points. Dually, every fixed point belongs to four invariant curves since the origin O belongs to the four curves

$$\{x \mapsto \mathrm{Ab}(x - \infty_i), \; x \mapsto \mathrm{Ab}(x - 0_i), \; i = 1, 2\}.$$

∎

Notice that the fixed points form a group F (isomorphic to $\mathbf{Z}/3\mathbf{Z} \oplus \mathbf{Z}/3\mathbf{Z}$) which is a subgroup of $J_3(\Gamma)$, the three-torsion subgroup of $J(\Gamma)$. On $J_3(\Gamma)$ there is a non-degenerated alternating form $(\cdot, \cdot)$ induced by the Riemann form corresponding to the principal polarisation. The subgroup $F \subset J_3(\Gamma)$ has the following property.

Proposition 3.2 *The group F of fixed points of τ on $J(\Gamma)$ is a totally isotropic subgroup of $J_3(\Gamma)$ with respect to the Riemann form $(\cdot, \cdot)$.*

Proof

τ is a symplectic automorphism of $J_3(\Gamma) \cong (\mathbf{Z}/3\mathbf{Z})^4$, which satisfies $1 + \tau + \tau^2 = 0$; also $\dim \ker(\tau - 1) = 2$. It follows that F consists exactly of the elements of the form $\tau(x) - x$ where $x \in J_3(\Gamma)$. Finally, if $y \in F$, then obviously $(y, \tau(x) - x) = 0$. ∎

Apart from the Riemann form, which coincides on $J_3(\Gamma)$ with Weil's pairing e_3 (see [LB]) a function can be defined on F with values in the group of cubic roots of unity. It is analogous to Mumford's quadratic form (theta characteristic) on the two-torsion subgroup $J_2(\Gamma)$ of $J(\Gamma)$ and can be defined in complete generality. It measures the obstruction for a line bundle $\mathcal{L}$ to descend to the quotient $J(\Gamma)/\tau$. One can define it as follows. Choose a linearisation of $\mathcal{L}$ with respect to the cyclic group $\mathbf{Z}/3\mathbf{Z}$ generated by τ, i.e., an isomorphism $\phi : \mathcal{L} \tilde{\to} \tau^*(\mathcal{L})$ with $\phi(0) = \mathrm{Id}_{\mathcal{L}_0}$. When x is a fixed point of τ, then ϕ induces an isomorphism of $\mathcal{L}_x$ which is multiplication by a root of unity $e(x)$, and $e : x \mapsto e(x)$ is the desired function. It depends on the choice of $\mathcal{L}$ itself and not only on the polarisation. If Θ is the (theta) divisor which corresponds to $\mathcal{L}$, i.e., $\mathcal{L} = [\Theta]$, then the corresponding $e = e_\Theta$ may be computed as follows. Let $f = 0$ be a local defining function for Θ in x. Since the divisor Θ is non-singular, the leading part h of f is linear and we have $\tau^*(h) = e(x)h$. Since the singular points are of type A_2, as is seen from (3.4), there exist local coordinates $\{u, v\}$ at x such that $\tau^*(u) = \epsilon u$ and $\tau^*(v) = \epsilon^2 v$. Therefore we have either $h = u$ and $e(x) = \epsilon$, or $h = v$ and $e(x) = \epsilon^2$. Also if $x \notin \Theta$ then $e(x) = 1$. It follows that e_Θ is explicitly given for all $x \in F$ by $e_\Theta(x) = \tau_{*|T_x\Theta}$, or equivalently

$$e_\Theta(x)v = \tau_* v \text{ for all } v \in T_x\Theta. \tag{3.6}$$

The automorphisms μ and σ act on F as well as on the set of invariant theta curves. It is desirable to have a "totally symmetric" theta curve, i.e., invariant by τ, σ and μ. The main observation of this paragraph, from which the 9_4-configuration is a consequence, is the following.

Proposition 3.3 *There is a unique totally symmetric theta curve among the nine invariant theta curves. The function e_Θ associated to this curve Θ is a quadratic form on F; it is given in a suitable basis of F and upon identification of the group of cubic roots of 1 with $\mathbf{F}_3$ by*

$$e_\Theta(r, s) = r^2 - s^2 \pmod{3}.$$

Proof

The existence of the curve is clear: since the polarisation is invariant by the group $\mathrm{Aut}(\Gamma)$, we may find an invariant line bundle which gives this polarisation, hence also an invariant divisor. It is unique since if there are two $\mathrm{Aut}(\Gamma)$-invariant curves, then their (two) intersection points must be invariant under $\mathrm{Aut}(\Gamma)$ which is impossible (see Table 5). It is easy to identify Θ: it is given by the image of $P \mapsto \mathrm{Ab}(P + \infty_1 - 2\infty_2)$. To see this, notice that this image can be written as

$$P \mapsto \mathrm{Ab}(P + S_1 + S_1^\sigma - 3S_2),$$

independent of the choice of $S_1, S_2 \in \{o_1, o_2, \infty_1, \infty_2\}$. From this representation it is also clear that Θ contains the four points $\mathrm{Ab}(S_1^\sigma - S_1)$, $S_1 \in \{o_1, o_2, \infty_1, \infty_2\}$.

188

3. Application: generalized Kummer surfaces

Let us determine e_Θ in terms of the basis $\{\zeta_1, \zeta_2\}$ where $\zeta_1 = \mathrm{Ab}(\infty_2 - \infty_1)$ and $\zeta_2 = \mathrm{Ab}(o_2 - o_1)$. Since $\tau = \sigma\tau\sigma$ and $\sigma^2 = 1$ it follows using the chain rule that if $\tau(x) = x$ and $v \in T_{x^\sigma}\Theta$ then

$$e_\Theta(x^\sigma)v = \tau_* v = \sigma_* \tau_* \sigma_* v = e_\Theta(x)\sigma_* \sigma_* v = e_\Theta(x)v,$$

hence $e_\Theta(x^\sigma) = e_\Theta(x)$. In the same way it follows from $\tau = \mu\tau^{-1}\mu$ that $e_\Theta(\mu(x)) = e_\Theta(x)^{-1}$. Therefore, if we identify the group of cubic roots of unity with $\mathbf{F}_3$ by $e_\Theta(\zeta_1) = 1$ then e_Θ is given by

$$e_\Theta(r\zeta_1 + s\zeta_2) = r^2 - s^2 \pmod 3. \tag{3.7}$$

∎

The 9_4 configuration is now described as follows. if ω and ω' are two fixed points, then

$$\omega \in \Theta + \omega' \iff e_\Theta(\omega - \omega') \neq 0.$$

It follows that every invariant theta curve passes through four fixed points and that every fixed point belongs to four invariant theta curves. Moreover we have seen that the function e_Θ determines the direction of the tangent to Θ in the fixed points of τ. Therefore, if $\omega, \omega' \in F$ then $\Theta + \omega$ and $\Theta + \omega'$ are tangent in a common point $x \in F$ if and only if

$$e_{\Theta+\omega}(x) = e_{\Theta+\omega'}(x).$$

Since $e_{\Theta+\omega}(x) = e_\Theta(x - \omega)$, this condition is rewritten as

$$e_\Theta(x - \omega) = e_\Theta(x - \omega')$$

which is satisfied for $\omega' = 2x - \omega$ (only). We conclude that the four invariant curves running through one fixed point come in two pairs: since any two theta curves always intersect in two points (which may coincide), the curves of one pair are tangent in their unique intersection point and the curves of opposite pairs intersect in two different points (see Figure 7, which also contains the dual picture, equally present in the 9_4 configuration).

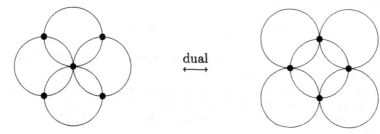

Figure 7

There is also a neat way to display the incidence relations of the points and the curves of the 9_4 configuration by an incidence diagram analogous to the one for the 16_6 configuration on the Kummer surface, as we explained in Paragraph IV.4.4. Let us define $W_{rs} = r\zeta_1 + s\zeta_2$ and $\Gamma_{rs} = \Theta + W_{rs}$, where $\zeta_1 = \text{Ab}(\infty_2 - \infty_1)$ and $\zeta_2 = \text{Ab}(o_2 - o_1)$ as in the proof of Proposition 3.3; also Θ is the totally symmetric theta curve given by Proposition 3.3. It follows from (3.7) that one can use for the 9_4 configuration the following incidence diagrams.

$$
\begin{array}{ccc}
W_{00} & W_{01} & W_{02} \\
W_{10} & W_{11} & W_{12} \\
W_{20} & W_{21} & W_{22}
\end{array}
\qquad
\begin{array}{ccc}
\Gamma_{00} & \Gamma_{01} & \Gamma_{02} \\
\Gamma_{10} & \Gamma_{11} & \Gamma_{12} \\
\Gamma_{20} & \Gamma_{21} & \Gamma_{22}
\end{array}
$$

3.3. A projective embedding of the generalised Kummer surface

In this section we will compute explicit equations for the quotient $S = J(\Gamma)/\tau$ as an algebraic surface in $\mathbf{P}^4$. Since τ has nine fixed points, S has nine singular points and we have seen that they are of type A_2. The minimal resolution of these singularities of S leads to a K-3 surface (a generalized Kummer surface), which we will denote by X (see [Bea1]). Let $\pi : J(\Gamma) \to S$ be the quotient and denote by Θ the unique divisor given by Proposition 3.3.

Proposition 3.4 *Let M be the divisor on S for which $\pi^*(M) = [3\Theta]$. Then M is very ample, leading to an embedding of S as the complete intersection of a quadric and a cubic threefold in $\mathbf{P}^4$.*

Proof

Using the quadratic form e_Θ we see that $\mathcal{L}^{\otimes 3} = [3\Theta]$ descends to a line bundle M on S, i.e., $\pi^*(M) = \mathcal{L}^{\otimes 3}$. Let us denote by N the line bundle on X which is the pull-back of M by the canonical map from X to S. Then using $\mathcal{L} \cdot \mathcal{L} = 2$ we find

$$18 = \mathcal{L}^{\otimes 3} \cdot \mathcal{L}^{\otimes 3} = (\deg \pi)M \cdot M = 3\,M \cdot M,$$

so that $M \cdot M = 6$, which is also the self-intersection of N. Therefore, we find by the Riemann-Roch Theorem (for K-3 surfaces),

$$\chi(N) = \chi(\mathcal{O}_X) + \frac{N \cdot N}{2} = 2 + 3 = 5.$$

It follows moreover from Serre duality and Kodaira vanishing that $\dim H^i(X, \mathcal{O}(N)) = 0$ for $i > 0$, so that $\dim H^0(X, \mathcal{O}(N)) = \chi(N) = 5$.

The morphism ϕ_N corresponding to N can be factorized via the blow-up $p : X \to S$ and is shown to provide an injective morphism $\phi : S \to \mathbf{P}^4$. More precisely, it can be seen by analyzing theta curves on J that ϕ_N is one to one away from the exceptional curves. If we consider now the surjective map

$$\text{Sym}\, H^0(X, N) \to \oplus_{t \geq 0} H^0\left(X, N^{\otimes t}\right),$$

whose kernel leads to the defining equations for the image of S in $\mathbf{P}^4$, we see by a dimension count as above that the kernel contains a quadratic as well as an (independent) cubic form. Since the degree of N equals six, we see that the image is the complete intersection of a quadric and a cubic hypersurface in $\mathbf{P}^4$. ∎

3. Application: generalized Kummer surfaces

Consider now the genus 2 even Mumford system; our system of generators (2.6) of its integrable algebra have the property of being weight homogeneous when the variables $u_0, \ldots, w_0$ are assigned the weights given by the weight table below. We call a Poisson bracket *weight homogeneous of degree i* when the bracket of any two weight homogeneous elements is weight homogeneous and

$$\{\lambda^k x_k, \lambda^l x_l\} = \lambda^{i+s}\{x_k, x_l\}$$

for all weight homogeneous elements x_k and x_l (of weight k resp. l), where s denotes the weight of $\{x_k, x_l\}$. Then the Poisson brackets $\{\cdot, \cdot\}^1$, $\{\cdot, \cdot\}^\lambda$ and $\{\cdot, \cdot\}^{\lambda^2}$ have weights 2, 3 and 4 respectively. Everything is contained in the following table.

weight	1	2	3	4	5	6
	u_1	u_0	v_0	w_2		
		v_1	w_1			
		w_0				
		H_4	H_3	H_2	H_1	H_0
		$\{\cdot, \cdot\}^{\lambda^2}$	$\{\cdot, \cdot\}^\lambda$	$\{\cdot, \cdot\}^1$		

It follows from the table that the genus 2 even Mumford system has an automorphism defined by

$$(u_1, u_0, v_1, v_0, w_1, w_2) \mapsto (\epsilon u_1, \epsilon^2 u_0, \epsilon^2 v_1, v_0, \epsilon^2 w_0, w_1, \epsilon w_2) \tag{3.8}$$

upon taking the second structure as Poisson structure. Having a (finite) automorphism allows us to construct a quotient system (which is also a.c.i. in view of Proposition V.2.5). Here however we are only interested in the level sets which are invariant for this automorphism. Again by the weight table these are the subsets of $\mathbf{C}^7$ given by

$$
\begin{aligned}
w_2 - u_1^2 + u_0 &= 0, \\
w_1 - u_1 u_0 + u_1 w_2 &= \kappa, \\
u_0 w_2 + u_1 w_1 + w_0 + v_1^2 &= 0, \\
u_1 w_0 + u_0 w_1 + 2 v_1 v_0 &= 0, \\
u_0 w_0 + v_0^2 &= \mu \, (= 1),
\end{aligned}
\tag{3.9}
$$

for κ and μ arbitrary. This is an affine part of the Jacobian of the curve

$$y^2 = x^6 + 2\kappa x^3 + \mu,$$

whenever this curve is smooth, i.e., $\mu \neq 0$. These are presicely the curves of genus two with an automorphism of order 3; since we have normalized the equation of our curve as $y^2 = x^6 + 2\kappa x^3 + 1$ we will focus in the sequel only at the level sets for which $\mu = 1$.

From our description of the level sets as affine parts of the symmetric square of the curve it is also clear that the automorphism on these level sets comes from the order three automorphisms on the curves. Explicitly our map $\mathcal{S}$ (defined in general by (III.2.3)) is given by

$$
((x_1, y_1), (x_2, y_2)) \mapsto
\begin{cases}
u(\lambda) = \lambda^2 - (x_1 + x_2)\lambda + x_1 x_2, \\
v(\lambda) = \dfrac{y_1 - y_2}{x_1 - x_2}\lambda + \dfrac{x_1 y_2 - x_2 y_1}{x_1 - x_2}
\end{cases}
\tag{3.10}
$$

and $w(\lambda)$ is defined by the fundamental relation $u(\lambda)w(\lambda) + v^2(\lambda) = f(\lambda)$, where $f(x)$ is the right hand side of our equation $y^2 = x^6 + 2\kappa x^3 + 1$ for the curve Γ. Since the action on the curve is given $(x, y) \mapsto (\epsilon x, y)$ we see from (3.10) that it corresponds indeed with (3.9), hence our automorphism extends the automorphism on the special Jacobians.

Now that we have explicit equations for an affine part of the Jacobian of Γ_κ (i.e., (3.9) with $\mu = 1$), we want to construct the regular functions on this affine part which extend to meromorphic functions on the Jacobian of Γ_κ having a pole of order ≤ 3 along a fixed component of the divisor to be adjoined at infinity — recall that two translates of the theta divisor need to be adjoined to each affine part in order to complete them into Jacobians. As we explained in Section V.3 this is done by using the Laurent solution (the principal balances) e.g. to the vector field $\mathcal{X}_1$ of which we computed the first terms in Paragraph VI.4.2. From these first terms one finds easily (i.e., by solving linear equations) a few extra terms, to wit

$$u_0 = \frac{1}{t}(2a - 2a^2 t + dt^2 + (2ac - ad + 2a^2 b)t^3$$
$$+ (2ae + 2ab^2 + 2bc - bd - 2a^3 b + a^2 d - 2f)t^4 + \cdots),$$
$$u_1 = \pm\frac{1}{t}(1 + at + bt^2 + ct^3 + et^4 + ft^5 + \cdots).$$

Here $a, \ldots, f$ denote the free parameters; the Laurent solutions for v_1 and v_0 follow from it by differentiation and those for $w(\lambda)$ upon using the invariants. The $\pm$ distinguishes between the principal balances which correspond to the two different divisors. In order to have the functions which descend to the quotient we look at those invariant polynomials (invariant with respect to the action of τ) which have no poles upon substituting the Laurent solutions when picking the $-$ sign and have a pole of order at most 3 when picking the $+$ sign. Notice that invariance just means here being weight homogeneous of weight $0 \bmod 3$.

We arrive at the following list of functions.

$$z_0 = 1,$$
$$z_1 = u_1 u_0 - v_0,$$
$$z_2 = 2u_1(u_0 + v_1 - u_1^2),$$
$$z_3 = 2u_0 v_1^2 + 2v_0^2 + 2u_0 v_1(2u_0 - u_1^2) + 2u_1 v_0(u_1^2 - v_1 - 3u_0) + 2u_0^3,$$
$$z_4 = 2v_1^3 - 2(u_1^2 + 4u_0)v_1^2 + 10v_0(u_1 v_1 - v_0) + 2v_1(7u_0 u_1^2 - u_1^4 - 11u_0^2)$$
$$+ 2v_0(2\kappa + 15u_1 u_0 - 5u_1^3) + 2(u_1^2 - u_0)^3 - 10u_0^3 - 4\kappa u_1 u_0.$$

$$(3.11)$$

To find the image of $J(\Gamma)$ in $\mathbf{P}^4$ it suffices to eliminate the variables u_i, v_i and w_i from (3.9) and (3.11). In fact, from the first three equations of (3.9) the variables w_i are eliminated linearly and the other equations reduce to

$$2\kappa(u_0 - u_1^2) + 3u_1 u_0^2 - u_1 v_1^2 - 4u_0 u_1^3 + 2v_1 v_0 + u_1^5 = 0,$$
$$-2\kappa u_1 u_0 + u_0 u_1^4 - u_0 v_1^2 - 3u_1^2 u_0^2 + u_0^3 + v_0^2 = 1,$$

$$(3.12)$$

so it suffices to eliminate u_1, u_0, v_1 and v_0 from (3.12) and (3.11) (we have already eliminated the w_i-variables in (3.11)). In the latter z_1 and z_2 are solved linearly for v_1 and v_0,

$$v_0 = u_1 u_0 - z_1,$$
$$v_1 = u_1^2 - u_0 + \frac{z_2}{2u_1},$$

$$(3.13)$$

and the new equation for z_3, obtained by substituting (3.13) in (3.11) is then solved linearly for u_0 as

$$u_0 = \frac{2u_1^2}{z_2^2}\left(z_3 - z_1 z_2 - 2z_1^2\right). \tag{3.14}$$

After substitution of (3.13) and (3.14) in the last equation of (3.11) and in the equations of (3.12), we are left with three linear equations in u_1^3, which reflects the fact that $J(\Gamma)$ will be a $3:1$ cover of its image in $\mathbf{P}^4$. If we eliminate u_1^3 we arrive at the following two equations:

$$8z_1^3 - 24\kappa z_1^2 - 4\left(2\kappa z_2 + 6z_3 + z_4\right)z_1 + 4\kappa z_3 - 2\kappa z_2^2 - 3z_2 z_3 - z_2 z_4 = 0,$$

$$8z_1^4 - 16\kappa z_1^3 - 4\left(2 + 2\kappa z_2 + 6z_3 + z_4\right)z_1^2 + \left(8\kappa + 4\kappa z_3 - 2\kappa z_2^2 - 4z_2 - 3z_2 z_3 - z_2 z_4\right)z_1$$
$$+ 2\kappa z_2 z_3 + 14z_3 + 2z_4 - 2z_2^2 + z_3 z_4 + 5z_3^2 = 0.$$

Using the first equation, the second equation can be replaced by

$$8\left(1 - 3\kappa^2\right)z_1^2 + 4\left(-2\kappa + \left(1 - 2\kappa^2\right)z_2 - 6\kappa z_3 - \kappa z_4\right)z_1 + 2\left(1 - \kappa^2\right)z_2^2$$
$$- 5z_3^2 - \kappa z_2\left(5z_3 + z_4\right) + 2z_3\left(2\kappa^2 - 7\right) - z_3 z_4 - 2z_4 = 0,$$

or equivalently by ${}^t ZAZ = 0$, where

$$A = \begin{pmatrix} 0 & -8\kappa & 0 & 2(2\kappa^2 - 7) & -2 \\ -8\kappa & 16(1 - 3\kappa^2) & 4(1 - 2\kappa^2) & -24\kappa & -4\kappa \\ 0 & 4(1 - 2\kappa^2) & 4(1 - \kappa^2) & -5\kappa & -\kappa \\ 2(2\kappa^2 - 7) & -24\kappa & -5\kappa & -10 & -1 \\ -2 & -4\kappa & -\kappa & -1 & 0 \end{pmatrix}.$$

Although at this point these equations for the quadric and cubic hypersurfaces which define S as a subset of $\mathbf{P}^4$ (which we will identify in the sequel with S) may not seem very attractive, we will see that natural coordinates can be picked for $\mathbf{P}^4$ in which these equations take a very symmetric form. Indeed, the projective basis of $\mathbf{P}^4$ that we used is rather arbitrary: for example, the coordinates of the nine fixed points for τ do not possess special coordinates in terms of the present basis. One observes that if the five fixed points for τ which do not lie on Θ are taken as base points for $\mathbf{P}^4$ then the coordinates of the four fixed points on Θ take a simple form and are independent of κ. It is read off from the incidence diagram that

$$\{W_{00}, W_{11}, W_{12}, W_{21}, W_{22}\}$$

are the five points which do not lie on Θ. To find their coordinates, use a local parameter t and take $x = t$,

$$y = \pm\left(1 + \kappa t^3\right) + \mathcal{O}\left(t^6\right),$$

picking either sign around o_1 or o_2, and in the same way, $x = t^{-1}$ and

$$y = \pm\left(t^{-3} + \kappa + \frac{1 - 2\kappa^2}{2}t^3\right) + \mathcal{O}\left(t^4\right)$$

for ∞_1 and ∞_2. Then a careful computation yields the following coordinates:

$$
\begin{aligned}
W_{00} &: (0:0:0:0:1), \\
W_{11}, W_{12} &: (0:0:1:\pm 1:\mp 3-2\kappa), \\
W_{21}, W_{22} &: (1:\pm 1:\mp 2-2\kappa:\mp 2\kappa:4\kappa^2\pm 14\kappa+4).
\end{aligned}
$$

We take the points

$$
\{W_{00}, W_{12}, W_{21}, W_{12}, W_{11}\}
$$

as base points for $\mathbf{P}^4$ (in that order), i.e., $O = (1:0:0:0:0)$, etc., with associated coordinates $y_0, \ldots, y_4$. Then the four fixed points on Θ have as coordinates

$$
\begin{aligned}
W_{10} &= (1:1:1:0:0), & W_{20} &= (1:0:0:1:1), \\
W_{01} &= (1:1:0:1:0), & W_{02} &= (1:0:1:0:1),
\end{aligned}
$$

we see that they lie on the (2-dimensional!) plane

$$
y_0 = y_2 + y_3 = y_1 + y_4,
$$

and it is easy to see that in fact Θ is contained in this plane. The translations $t_{W_{10}}$ and $t_{W_{01}}$ correspond to projective transformations of the surface and take in terms of these coordinates the simple form

$$
t_{W_{10}} = \begin{pmatrix} -1 & 1 & 1 & 0 & 0 \\ -1 & 0 & 1 & 1 & 0 \\ -1 & 1 & 0 & 0 & 1 \\ 0 & 0 & 1 & 0 & 0 \\ 0 & 1 & 0 & 0 & 0 \end{pmatrix} \quad \text{and} \quad t_{W_{01}} = \begin{pmatrix} -1 & 1 & 0 & 1 & 0 \\ -1 & 0 & 1 & 1 & 0 \\ 0 & 0 & 0 & 1 & 0 \\ -1 & 1 & 0 & 0 & 1 \\ 0 & 1 & 0 & 0 & 0 \end{pmatrix}
$$

from which the equations for the planes to which the other invariant curves belong, are obtained at once. This configuration of nine points in $\mathbf{P}^4$ is characterized by the fact that there exist nine planes with the property that each of these planes contains four of the nine points and every point belongs to four of the planes. Thus we have recovered in a direct way a configuration that has been studied in the work of Segre and Castelnuovo on nets of cubic hypersurfaces in $\mathbf{P}^4$ (see [Cas] and [Seg]).

The equations of the quadric and cubic hypersurfaces Q and C take in terms of the new coordinates the following symmetric form:

$$
\begin{aligned}
Q : & \; c(y_1+y_4)(y_2+y_3+y_4-y_0) + \bar{c}(y_2+y_3)(y_1+y_3+y_4-y_0) = cy_4^2 + \bar{c}y_3^2, \\
C : & \; \bar{c}^3 y_2^2(y_1+y_3+y_4-y_0) + \bar{c}^2 c y_2 \left((y_1+y_4)(y_2-y_0) + y_0 y_3 + y_1 y_4\right) - \\
& \; c^3 y_1^2(y_2+y_3+y_4-y_0) - \bar{c}c^2 y_1 \left((y_2+y_3)(y_1-y_0) + y_0 y_4 + y_2 y_3\right) = 0,
\end{aligned}
$$

where $c = 1 + \kappa$ and $\bar{c} = 1 - \kappa$. The cubic equation can be simplified in a significant way by adding to it the equation for Q multiplied with $c^2 y_1 - \bar{c}^2 y_2$. The result is

$$
c^2 y_1 y_4(y_2+y_3-y_0) - \bar{c}^2 y_2 y_3(y_1+y_4-y_0) = 0.
$$

3. Application: generalized Kummer surfaces

If we define

$$x_1 = y_0 - y_2 - y_3, \qquad x_4 = y_1 + y_4 - y_0,$$
$$x_2 = -y_1, \qquad\qquad x_5 = y_2,$$
$$x_3 = -y_4, \qquad\qquad x_6 = y_3,$$

then S is given as an algebraic variety in $\mathbf{P}^5$ by

$$
\begin{aligned}
\mathcal{C} &: c^2 x_1 x_2 x_3 + \bar{c}^2 x_4 x_5 x_6 = 0, \\
\mathcal{Q} &: c(x_1 x_2 + x_2 x_3 + x_1 x_3) + \bar{c}(x_4 x_5 + x_5 x_6 + x_4 x_6) = 0, \\
\mathcal{H} &: x_1 + x_2 + x_3 + x_4 + x_5 + x_6 = 0,
\end{aligned}
\qquad (3.15)
$$

and the 9_4 configuration is presented in the form used by Segre and Castelnuovo. In fact, the coordinates of the singular points W_{rs} have a 1 on position $r + s \bmod 3$, a -1 on position $3 + (r - s \bmod 3)$ and zeros elsewhere; the nine planes they belong to are given by $\mathcal{H} \cap (x_i = x_{j+3} = 0)$ for $i, j = 1, \ldots, 3$. Moreover the theta curves are mapped to the nine conics C_{ij}, $(1 \le i, j \le 3)$, given by $\mathcal{H}_{ij} = 3C_{ij}$. For example, C_{16} is given as

$$
\begin{aligned}
c x_2 x_3 + \bar{c} x_4 x_5 &= 0, \\
x_2 + x_3 + x_4 + x_5 &= 0.
\end{aligned}
$$

Notice that if one changes the sign of κ in the equations (3.15) then an isomorphic surface is obtained (interchange $c \leftrightarrow \bar{c}$ and $x_i \leftrightarrow x_{i+3}$ for $i = 1, \ldots, 3$), in agreement with the fact that κ^2 is the modular parameter.

4. The Garnier potential

4.1. The Garnier potential and its integrability

It is shown in [CC] that for any $\lambda = (\lambda_1, \ldots, \lambda_n)$, the potential

$$V_\lambda = \left(\sum_{i=1}^n q_i^2 \right)^2 + \sum_{i=1}^n \lambda_i q_i^2, \tag{4.1}$$

defines an integrable Hamiltonian system on $\mathbf{R}^{2n} = \{(q_1, \ldots, q_n, p_1, \ldots, p_n) \mid q_i, p_i \in \mathbf{R}\}$, equipped with the standard symplectic structure $\omega = \sum dq_i \wedge dp_i$, when the Hamiltonian is taken as the total energy

$$H_\lambda = T + V_\lambda, \quad T = \frac{1}{2} \sum_{i=1}^n p_i^2,$$

(T is the *kinetic energy*). It was pointed out to me by A. Perelomov that the integrability of these potentials was already known to Garnier; therefore we will call V_λ the Garnier potential. We study here the case $n = 2$ (two degrees of freedom) writing

$$V_{\alpha\beta} = (q_1^2 + q_2^2)^2 + \alpha q_1^2 + \beta q_2^2.$$

It would be interesting to study also the higher-dimensional potentials from the point of view of algebraic geometry.

Since we are only interested here in the complex geometry we will from now consider the potential as being defined on the affine Poisson variety $\mathbf{C}^4$ with the standard Poisson structure. However it is sometimes useful to extend it by the parameters α and β, i.e., we consider $\mathbf{C}^6$ as the phase space with coordinates q_i, p_i, α and β and with $\{q_i, p_i\} = 1$ and all other brackets between these coordinates are zero (see Proposition II.3.24). The Poisson structure on both $\mathbf{C}^4$ and $\mathbf{C}^6$ will be denoted by $\{\cdot, \cdot\}$.

For $\alpha \neq \beta$ we define $\mathcal{A}_{\alpha\beta} = \mathbf{C}[F, G]$ where F and G are defined by

$$F = (q_1 p_2 - q_2 p_1)^2 + (\beta - \alpha)(p_1^2 + 2q_1^4 + 2q_1^2 q_2^2 + 2\alpha q_1^2),$$
$$G = (q_1 p_2 - q_2 p_1)^2 + (\alpha - \beta)(p_2^2 + 2q_2^4 + 2q_1^2 q_2^2 + 2\beta q_2^2).$$

Notice that the Hamiltonian

$$H = \frac{1}{2}(p_1^2 + p_2^2) + (q_1^2 + q_2^2)^2 + \alpha q_1^2 + \beta q_2^2,$$

which corresponds to the potential $V_{\alpha\beta}$ belongs to $\mathcal{A}_{\alpha\beta}$ since

$$F - G = 2(\beta - \alpha)H.$$

We also define

$$\mathcal{A} = \frac{\mathbf{C}[F, G, H, \alpha, \beta]}{\mathrm{idl}(F - G + 2(\alpha - \beta)H)}.$$

It is easy to check that $\{F, G\} = 0$. This can for example be done by using the explicit form of the vector field X_H which is given by

$$\begin{aligned}
\dot{q}_1 &= p_1, & \dot{p}_1 &= -2q_1(2q_1^2 + 2q_2^2 + \alpha), \\
\dot{q}_2 &= p_2, & \dot{p}_2 &= -2q_2(2q_1^2 + 2q_2^2 + \beta).
\end{aligned} \tag{4.2}$$

When this vector field is taken on $\mathbf{C}^6$ then of course $\dot{\alpha} = \dot{\beta} = 0$ are added. It follows that both $\mathcal{A}_{\alpha\beta}$ and $\mathcal{A}$ are involutive and clearly they have maximal dimension. All level sets of the momentum map associated to $\mathcal{A}_{\alpha\beta}$ are two-dimensional and the general fiber is irreducible hence $\mathcal{A}_{\alpha\beta}$ is complete. It follows that $\mathcal{A}$ is also complete. Thus we have verified that $\left(\mathbf{C}^4, \{\cdot, \cdot\}, \mathcal{A}_{\alpha\beta}\right)$ and $(\mathbf{C}^6, \{\cdot, \cdot\}, \mathcal{A})$ are integrable Hamiltonian systems (for all $\alpha \neq \beta$).

We will be interested in the level sets of the momentum map over closed points. Apart from Paragraph 4.5 we will only be interested in those for which $\alpha \neq \beta$. We will denote these by $\mathcal{F}_{(\alpha,\beta,f,g)}$ or $\mathcal{F}_{fg}$ when $\alpha \neq \beta$ have been fixed. Explicitly $\mathcal{F}_{fg}$ is given as a subset of $\mathbf{C}^4$ by the following equations:

$$\begin{aligned}
(q_1 p_2 - q_2 p_1)^2 + (\beta - \alpha)(p_1^2 + 2q_1^4 + 2q_1^2 q_2^2 + 2\alpha q_1^2) &= f, \\
(q_1 p_2 - q_2 p_1)^2 + (\alpha - \beta)(p_2^2 + 2q_2^4 + 2q_1^2 q_2^2 + 2\beta q_2^2) &= g.
\end{aligned}$$

In order to simplify some of the formulas in the sequel we let, for given f and g, the constant h be determined by $f - g = 2(\beta - \alpha)h$.

Clearly this system has a lot of automorphisms. For future reference we list them in the table below. We also add two quasi-automorphisms which are of interest (i.e., $\jmath_1$ and $\jmath_2$). κ_2 is an automorphism of order three, ϵ being a primitive cubic root of unity and $\jmath_2$ is not of finite order, λ being an arbitrary non-zero complex number.

	q_1	q_2	p_1	p_2	α	β	F	G	H
$\imath_1$	$-q_1$	q_2	$-p_1$	p_2	α	β	F	G	H
$\imath_2$	q_1	$-q_2$	p_1	$-p_2$	α	β	F	G	H
$\jmath_1$	q_1	q_2	$-p_1$	$-p_2$	α	β	F	G	H
$\jmath_2$	λq_1	λq_2	$\lambda^2 p_1$	$\lambda^2 p_2$	$\lambda^2 \alpha$	$\lambda^2 \beta$	$\lambda^6 F$	$\lambda^6 G$	$\lambda^4 H$
κ_1	q_2	q_1	p_2	p_1	β	α	G	F	H
κ_2	$\epsilon^2 q_1$	$\epsilon^2 q_2$	ϵp_1	ϵp_2	$\epsilon \alpha$	$\epsilon \beta$	f	g	$\epsilon^2 h$

Table 6

The involutions $\imath_1$, $\imath_2$ and $\jmath$ restrict to all level surfaces $\mathcal{F}_{fg}$ and their restriction will be denoted by the same letter. Notice that the automorphisms in the table lead to many different quotients and they will all play a special role in this text; they are not all the automorphism but they are the ones which are seen "at once". In view of the automorphism $\jmath_2$ it is natural to consider (α, β, f, g) as belonging to the weighted projective space[24] $\mathbf{P}^{(1,1,3,3)}$.

---

[24] A quick introduction to weighted projective spaces is given in an appendix to [AM7].

Notice that if $\alpha = \beta$ then $F(= G)$ is just the square of the momentum

$$q = q_1 p_2 - q_2 p_1, \qquad (4.3)$$

which is obviously in involution with the energy corresponding to a central potential. What is remarkable however is that if $\alpha \neq \beta$ then the equations defining $\mathcal{F}_{fg}$ can be rewritten (birationally) in terms of q_1, q_2 and the momentum q, giving precisely the equations (IV.5.2) of the octic $\mathcal{O}$ with

$$
\begin{aligned}
\lambda_0^2 &= 4(\alpha - \beta)^2 (\alpha + \beta) - 2(f + g), & y_0 &= 1, \\
\lambda_1^2 &= g, & y_1 &= q_1 \sqrt[4]{2(\alpha - \beta)/f}, \\
\lambda_2^2 &= 2(\alpha - \beta)^3, & y_2 &= q/\sqrt[4]{fg}, \\
\lambda_3^2 &= f, & y_3 &= q_2 \sqrt[4]{2(\alpha - \beta)/g}.
\end{aligned}
\qquad (4.4)
$$

It follows that for general f, g the surface $\mathcal{F}_{fg}$ is birationally equivalent to the affine part $\mathcal{O}_0 = \mathcal{O} \cap \{y_0 \neq 0\}$ of the octic $\mathcal{O}$ which is itself birationally equivalent to an Abelian surface of type $(1, 4)$. We show in the following proposition that $\mathcal{F}_{fg}$ actually *is* (isomorphic to) an affine part of an Abelian surface of type $(1, 4)$.

Proposition 4.1 *Fixing any $\alpha \neq \beta \in \mathbf{C}$, the affine surface $\mathcal{F}_{fg} \subset \mathbf{C}^4$ defined by*

$$
\begin{aligned}
(q_1 p_2 - q_2 p_1)^2 + (\beta - \alpha)(p_1^2 + 2q_1^4 + 2q_1^2 q_2^2 + 2\alpha q_1^2) &= f, \\
(q_1 p_2 - q_2 p_1)^2 + (\alpha - \beta)(p_2^2 + 2q_2^4 + 2q_1^2 q_2^2 + 2\beta q_2^2) &= g,
\end{aligned}
$$

is for general[25] $f, g \in \mathbf{C}$ isomorphic to an affine part of an Abelian surface T_{fg}^2, of type $(1, 4)$, obtained by removing a smooth curve $\mathcal{D}_{fg}$ of genus 5,

$$\mathcal{F}_{fg} = T_{fg}^2 \setminus \mathcal{D}_{fg},$$

and the vector field X_H extends to a linear vector field on T_{fg}^2.

Proof
(i) Let G be the group generated by the involutions $\imath_1, \imath_2$, and $\jmath$. Our first aim is to show that $\mathcal{F}_{fg}/G$ is (isomorphic to) an affine part of a Kummer surface. Since f and g are general, we may suppose that $(\lambda_0 : \lambda_1 : \lambda_2 : \lambda_3)$ given by (4.4) do not belong to S. For these λ_i, let Q be the quadric (Kummer surface)

$$
\begin{aligned}
&\lambda_0^2 z_0 z_1 z_2 z_3 + \lambda_1^2 (z_0^2 z_1^2 + z_2^2 z_3^2) + \lambda_2^2 (z_0^2 z_2^2 + z_1^2 z_3^2) + \lambda_3^2 (z_0^2 z_3^2 + z_1^2 z_2^2) + \\
&2\lambda_1 \lambda_2 (z_0 z_1 + z_2 z_3)(z_1 z_3 - z_0 z_2) + 2\lambda_1 \lambda_3 (z_0 z_3 - z_1 z_2)(z_0 z_1 - z_2 z_3) + \\
&2\lambda_2 \lambda_3 (z_1 z_2 + z_0 z_3)(z_1 z_3 + z_0 z_2) = 0,
\end{aligned}
\qquad (4.5)
$$

which is obtained from (IV.5.2) by setting $z_i = y_i^2$, i.e., there is an unramified $8 : 1$ cover $\mathcal{O} \to Q$; this map restricts to a map $\bar{p}_0 : \mathcal{O}_0 \to Q_0$, where $Q_0 = Q \cap \{z_0 \neq 0\}$. Also the rational

---

[25] Precise conditions will be given later (Proposition 4.6).

map $\phi : \mathcal{F}_{fg} \to \mathcal{O}_0$ given by (4.3) and (4.4) induces a birational map $\tilde{\phi} : \mathcal{F}_{fg}/G \to Q_0$, giving rise to a commutative diagram

$$
\begin{array}{ccc}
\mathcal{F}_{fg} & \xrightarrow{\ \phi\ } & \mathcal{O}_0 \\
{\scriptstyle \pi}\downarrow & & \downarrow{\scriptstyle \bar{p}_0} \\
\mathcal{F}_{fg}/G & \xrightarrow[\ \tilde{\phi}\ ]{} & Q_0
\end{array}
\qquad (4.6)
$$

Since Q_0 is normal, it suffices to show that $\tilde{\phi}$ is bijective. Obviously $\tilde{\phi}$ is surjective: if $(x_1, x_2, x_3) \in Q_0$, let (y_1, y_2, y_3) be such that $y_i^2 = x_i$ and let q_1, q_2, q be determined from (4.4). Then these satisfy the condition under which p_1, p_2 exist such that $(q_1, q_2, p_1, p_2) \in \mathcal{F}_{fg}$ and $q = q_1 p_2 - q_2 p_1$. Then $\tilde{\phi}(q_1, q_2, p_1, p_2) = (x_1, x_2, x_3)$. On the other hand, if $(\tilde{\phi} \circ \pi)(q_1, q_2, p_1, p_2) = (\tilde{\phi} \circ \pi)(q_1', q_2', p_1', p_2')$ then $q_1 = \epsilon_1 q_1', q_2 = \epsilon_2 q_2', q = \epsilon q'$, (where $q' = q_1' p_2' - q_2' p_1'$) for $\epsilon_1, \epsilon_2, \epsilon \in \{-1, 1\}$. Then one sees that

$$
(q_1, q_2, p_1, p_2) = \imath_1^{\epsilon_1} \imath_2^{\epsilon_2} \imath^{\epsilon} (q_1', q_2', p_1', p_2'),
$$

where $\imath_k^{\epsilon_k}$ means $\imath_k$ in case $\epsilon_k = -1$ and identity for $\epsilon_k = 1$. It follows that $\pi(q_1, q_2, p_1, p_2) = \pi(q_1', q_2', p_1', p_2')$, and $\tilde{\phi}$ is injective. This shows that $\tilde{\phi}$ is an isomorphism, hence $\mathcal{F}_{fg}/G$ is isomorphic to the (affine) Kummer surface defined by Q_0.

(ii) We proceed to show that $\mathcal{F}_{fg}$ is isomorphic to an affine part of an Abelian surface, more precisely to the normalization $\mathcal{A}$ of $\mathcal{O}_0$ (the octic is singular along the coordinate planes). This normalization can be obtained via the birational map $\phi_{\mathcal{L}} : \mathcal{T}^2 \to \mathcal{O}$. In particular, by restriction of (IV.5.4) to an affine piece we get a commutative diagram

$$
\begin{array}{ccc}
\mathcal{A} & \xrightarrow{\ \phi_{\mathcal{L}}\ } & \mathcal{O}_0 \\
{\scriptstyle p_0}\downarrow & & \downarrow{\scriptstyle \bar{p}_0} \\
K_0 & \xrightarrow[\ \phi_{\mathcal{N}^2}\ ]{} & Q_0
\end{array}
\qquad (4.7)
$$

where $\phi_{\mathcal{N}^2}$ is an isomorphism. If we combine both diagrams (4.6) and (4.7) we get

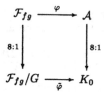

$$
\begin{array}{ccc}
\mathcal{F}_{fg} & \xrightarrow{\ \varphi\ } & \mathcal{A} \\
{\scriptstyle 8:1}\downarrow & & \downarrow{\scriptstyle 8:1} \\
\mathcal{F}_{fg}/G & \xrightarrow[\ \tilde{\varphi}\ ]{} & K_0
\end{array}
$$

with φ the birational map $\phi_{\mathcal{L}}^{-1}\phi$ and $\tilde{\varphi}$ the isomorphism $\phi_{\mathcal{N}^2}^{-1}\tilde{\phi}$. Now the two covers $\mathcal{F}_{fg} \to \mathcal{F}_{fg}/G$ and $\mathcal{A} \to K_0$ are only ramified in discrete points; the same holds true if $\mathcal{A}$ and $\mathcal{F}_{fg}$ are replaced by their closures: the closure of $\mathcal{A}$ is just $\mathcal{T}^2$ and the closure of $\mathcal{F}_{fg}$ is obtained from the explicit embedding which will be given in 4.4.1. By Zariski's main Theorem the normality of $\mathcal{T}^2$ implies that the lifting φ of $\bar{\varphi}$ must also be an isomorphism and we get

$$\mathcal{F}_{fg} = \mathcal{T}_{fg}^2 \setminus \mathcal{D}_{fg}$$

for some divisor $\mathcal{D}_{fg}$ on a $(1,4)$-polarized Abelian surface $\mathcal{T}_{fg}^2$. It is seen that $\mathcal{D}_{fg}$ is a $4 : 1$ unramified cover of a translate of the Riemann theta divisor of the canonical Jacobian, hence $\mathcal{D}_{fg}$ is smooth and has genus 5; an equation for $\mathcal{D}_{fg}$ will be given in Paragraph 4.4.

(iii) Finally we show that X_H extends to a linear vector field on $\mathcal{T}_{fg}^2$. Letting $\theta_0 = 1$, $\theta_1 = q_1^2$ and $\theta_3 = q^2$, we have shown that an equation for the Kummer surface of the canonical Jacobian associated to $\mathcal{F}_{fg}$ is a quartic in these variables. From (4.4) and (IV.5.2) the leading term in θ_3^2 is given by $((\alpha + \beta)\theta_0 + \theta_1 + \theta_2)^2 - 4(\alpha\beta\theta_0 + \beta\theta_1 + \alpha\theta_2)$, or, in terms of the original variables,

$$(q_1^2 + q_2^2 + \alpha + \beta)^2 - 4(\alpha\beta + \alpha q_2^2 + \beta q_1^2). \tag{4.8}$$

We let x_1 and x_2 be the roots of the polynomial

$$x^2 + \left(q_1^2 + q_2^2 + \alpha + \beta\right) x + \alpha\beta + \alpha q_2^2 + \beta q_1^2,$$

as suggested by the algorithm recalled in Section V.4 ("suggested" because we did not prove yet that the system is a.c.i.). Explicitly, let

$$\begin{aligned} x_1 + x_2 &= -(q_1^2 + q_2^2 + \alpha + \beta), & \dot{x}_1 + \dot{x}_2 &= -2(q_1 p_1 + q_2 p_2), \\ x_1 x_2 &= \alpha\beta + \alpha q_2^2 + \beta q_1^2, & x_1 \dot{x}_2 + \dot{x}_1 x_2 &= 2(\beta q_1 p_1 + \alpha q_2 p_2), \end{aligned} \tag{4.9}$$

then it is not hard to rewrite the equations $F = f$, $G = g$, defining $\mathcal{F}_{fg}$, in terms of $x_1, x_2, \dot{x}_1, \dot{x}_2$. This gives

$$\dot{x}_i^2 = \frac{8(x_i + \alpha)(x_i + \beta)\left(x_i^3 + (\alpha + \beta)x_i^2 + (\alpha\beta - h)x_i + (\beta f - \alpha g)/2(\alpha - \beta)\right)}{(x_1 - x_2)^2}$$

so that

$$\begin{aligned} \frac{dx_1}{\sqrt{f(x_1)}} + \frac{dx_2}{\sqrt{f(x_2)}} &= 0, \\ \frac{x_1 dx_1}{\sqrt{f(x_1)}} + \frac{x_2 dx_2}{\sqrt{f(x_2)}} &= 2\sqrt{2}dt, \end{aligned} \tag{4.10}$$

where

$$f(x) = (x + \alpha)(x + \beta)\left(x^3 + (\alpha + \beta)x^2 + (\alpha\beta - h)x + \frac{\beta f - \alpha g}{2(\alpha - \beta)}\right).$$

Integrating (4.10) we see that X_H is a linear vector field on $\mathcal{F}_{fg}$, which obviously extends to a linear vector field on $\mathcal{T}_{fg}^2$. From this expression and (VI.4.9) explicit solutions for q_k ($k = 1, 2$) are given by

$$q_k \sim \frac{\theta[\delta + \eta_k](At + D)}{\theta[\delta](At + D)},$$

where η_1 and η_2 are halfperiods which correspond to $-\alpha$ and $-\beta$.

Notice that as a by-product we find an equation

$$y^2 = (x + \alpha)(x + \beta)\left(x^3 + (\alpha + \beta)x^2 + (\alpha\beta - h)x + \frac{\beta f - \alpha g}{2(\alpha - \beta)}\right). \tag{4.11}$$

for the curve whose Jacobian is the canonical Jacobian associated to $\mathcal{T}_{fg}^2$. ∎

4. The Garnier potential

Alternatively, the linearizing variables (4.9) suggest a morphism ϕ from the potential $V_{\alpha\beta}$ on $\mathbf{C}^4$ to the genus 2 odd Mumford system, namely one defines $\phi(q_1, q_2, p_1, p_2) = (u(\lambda), v(\lambda), w(\lambda))$ where

$$u(\lambda) = \lambda^2 + (q_1^2 + q_2^2 + \alpha + \beta)\lambda + \alpha\beta + \alpha q_2^2 + \beta q_1^2,$$

$$v(\lambda) = \frac{1}{\sqrt{2}}\left[(q_1 p_1 + q_2 p_2)\lambda + (\beta q_1 p_1 + \alpha q_2 p_2)\right],$$

$$w(\lambda) = \lambda^3 + (\alpha + \beta - q_1^2 - q_2^2)\lambda^2 - \left(\frac{p_1^2 + p_2^2}{2} + (\alpha + \beta)\left(q_1^2 + q_2^2\right) - \alpha\beta\right)\lambda \qquad (4.12)$$

$$- \alpha\beta\left(\frac{p_1^2}{2\alpha} + \frac{p_2^2}{2\beta} + q_1^2 + q_2^2\right).$$

The genus 2 odd Mumford system which is considered here is the one on $\mathbf{C}^7$ and the Poisson structure is a linear combination of the Poisson structures we have considered before[26], namely it corresponds to $\varphi = -x^2 + (\alpha+\beta)x - \alpha\beta$. This morphism ϕ is neither injective nor surjective, however the latter is easily cured since the image is a level set of the Casimirs of the odd Mumford system which carries an induced a.c.i. system. As for the injectivity of ϕ, clearly it can be factorized through the quotient system obtained by dividing $\mathbf{C}^4$ out by the group (of order four) generated by $\imath_1$ and $\imath_2$. It follows from Example II.2.31 that the regular functions on this quotient are generated by

$$Q_i = q_i^2, \; P_i = p_i^2, \; R_i = P_i Q_i, \quad (i = 1, 2)$$

with only two relations: $R_i^2 = Q_i P_i$, $(i = 1, 2)$. Then the quotient Poisson structure is given by

$$\begin{pmatrix} T_1 & 0 \\ 0 & T_2 \end{pmatrix}, \quad \text{where} \quad T_i = \begin{pmatrix} 0 & 4R_i & 2Q_i \\ -4R_i & 0 & -2P_i \\ -2Q_i & 2P_i & 0 \end{pmatrix}$$

and the map $\bar{\phi}$ via which ϕ factors is

$$u(\lambda) = \lambda^2 + (Q_1 + Q_2 + \alpha + \beta)\lambda + \alpha\beta + \alpha Q_2 + \beta Q_1,$$

$$v(\lambda) = \frac{1}{\sqrt{2}}\left[(R_1 + R_2)\lambda + (\beta R_1 + \alpha R_2)\right],$$

$$w(\lambda) = \lambda^3 + (\alpha + \beta - R_1 - R_2)\lambda^2 - \left(\frac{P_1 + P_2}{2} + (\alpha + \beta)(Q_1 + Q_2) - \alpha\beta\right)\lambda$$

$$- \alpha\beta\left(\frac{P_1}{2\alpha} + \frac{P_2}{2\beta} + Q_1 + Q_2\right).$$

Since everything in $\bar{\phi}$ is linear, this map is a biregular map. Thus the quotient system is isomorphic to a trivial subsystem of the genus 2 odd Mumford system and is a.c.i. We may conclude as in Paragraph 6.1 that (for $\alpha \neq \beta$) $(\mathbf{C}^4, \{\cdot, \cdot\}, A_{\alpha\beta})$ is an a.c.i. system of type $(1, 4)$. The same is of course true for $(\mathbf{C}^6, \{\cdot, \cdot\}, A)$. As a by-product we get also a Lax representation for the potentials $V_{\alpha\beta}$. For the vector field X_H for example it is given by

$$\frac{d}{dt}\begin{pmatrix} v(\lambda) & u(\lambda) \\ w(\lambda) & -v(\lambda) \end{pmatrix} = \sqrt{2}\left[\begin{pmatrix} v(\lambda) & u(\lambda) \\ w(\lambda) & -v(\lambda) \end{pmatrix}, \begin{pmatrix} 0 & 1 \\ \lambda - 2(q_1^2 + q_2^2) & 0 \end{pmatrix}\right],$$

where $u(\lambda)$, $v(\lambda)$ and $w(\lambda)$ are given by (4.12). $\blacksquare$

---

[26] These were given in 2.3, 2.4 and 2.5 on $\mathbf{C}^6$ and are easily rewritten on $\mathbf{C}^7$.

4.2. Some moduli spaces of Abelian surfaces of type (1,4)

In this paragraph we describe a map ψ from the moduli space $\mathcal{A}_{(1,4)}$ of polarized Abelian surfaces of type $(1,4)$ into an algebraic cone $\mathcal{M}^3$ in some weighted projective space. To be precise we recall from Section IV.5 that $(1,4)$-polarized Abelian surfaces which are products of elliptic curves (with the product polarization) are excluded from $\mathcal{A}_{(1,4)}$. The map will be bijective on the dense subset $\tilde{\mathcal{A}}_{(1,4)}$ which is the moduli space of polarized Abelian surfaces $(\mathcal{T}^2, \mathcal{L})$ for which the rational map $\phi_{\mathcal{L}} : \mathcal{T}^2 \to \mathbf{P}^3$ is birational. Thus we construct a projective model for the moduli space $\mathcal{A}_{(1,4)}$. The main idea in this construction is to see how the Galois group of the cover $\mathcal{A}^0_{(1,4)} \to \mathcal{A}_{(1,4)}$ acts on $\mathcal{P}$ and define $\mathcal{M}^3$ to be the quotient. This quotient will be easy to calculate since it is a quotient of (a Zariski open subset of) $\mathbf{P}^3$ by a group which acts linearly. The fact that this action is so simple is surprising and was suggested to us on the one hand by the formulas (4.4) which show that the sign of the λ_i does not matter and on the other hand by the automorphism κ_1 which shows that the Abelian surface which corresponds to α, β, f, g is isomorphic to the one which corresponds to β, α, g, f which indicates that the modular parameters λ_1 and λ_3 can be permuted (upon adding the proper i's or signs for λ_0 and λ_2). A posteriori we can forget about the integrable Hamiltonian system and proceed as follows.

Recall from Paragraph IV.5 that $\mathcal{A}^0_{(1,4)}$ maps onto

$$\mathcal{P} = \frac{\mathbf{P}^3 \setminus S}{\lambda_0 \sim -\lambda_0} \bigcup \text{(three rational curves in } S \text{, each missing eight points)},$$

bijectively on the first component (which is dense); the three rational curves are thought of as lying in $\mathbf{P}^3/(\lambda_0 \sim -\lambda_0)$ at the boundary of this component. $\mathcal{A}^0_{(1,4)}$ is a $24:1$ (ramified) cover of $\mathcal{A}_{(1,4)}$: let σ and τ be elements of order 4 such that $K(\mathcal{L}) = \langle \sigma \rangle \oplus \langle \tau \rangle$, and define

$$K_1 = \{0, \sigma, 2\sigma, 3\sigma\}, \qquad\qquad K_4 = \{0, \sigma + 2\tau, 2\sigma, 3\sigma + 2\tau\},$$
$$K_2 = \{0, \tau, 2\tau, 3\tau\}, \qquad\qquad K_5 = \{0, 2\sigma + \tau, 2\tau, 2\sigma + 3\tau\},$$
$$K_3 = \{0, \sigma + \tau, 2\sigma + 2\tau, 3\sigma + 3\tau\}, \qquad K_6 = \{0, \sigma + 3\tau, 2\sigma + 2\tau, 3\sigma + \tau\}.$$

These are the only cyclic subgroups of order 4 of $K(\mathcal{L})$. It is easy to see that taking all possible isomorphisms $K(\mathcal{L}) \cong \mathbf{Z}/4\mathbf{Z} \oplus \mathbf{Z}/4\mathbf{Z}$ we find exactly the 24 decompositions

$$K(\mathcal{L}) = K_i \oplus K_j, \ (1 \le i, j \le 6, \ |i - j| \ne 0, 3).$$

We describe the cover

$$\mathcal{A}^0_{(1,4)} \xrightarrow{24:1} \mathcal{A}_{(1,4)}$$

and we construct a $24 : 1$ cover $\mathcal{P} \to \mathcal{M}^3$ and a map

$\psi : \mathcal{A}_{(1,4)} \to \mathcal{M}^3$, where $\mathcal{M}^3$ is an algebraic variety (lying in weighted projective space $\mathbf{P}^{(1,2,2,3,4)}$), such that there results a commutative diagram

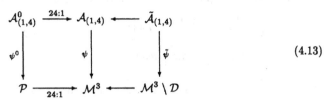

$$(4.13)$$

202

in which the restriction $\tilde{\psi}$ of ψ to $\tilde{\mathcal{A}}_{(1,4)}$ is a bijection ($\mathcal{D}$ is a divisor on $\mathcal{M}^3$ which will be determined explicitly).

The group $G = GL(2, \mathbf{Z}/4\mathbf{Z})$ acts transitively on (ordered!) bases as follows: if σ, τ are such that $K(\mathcal{L}) = \langle \sigma \rangle \oplus \langle \tau \rangle$ and $\begin{pmatrix} a & b \\ c & d \end{pmatrix} \in G$ then

$$\begin{pmatrix} a & b \\ c & d \end{pmatrix} \cdot (\sigma, \tau) = (a\sigma + b\tau, c\sigma + d\tau),$$

giving a new decomposition $K(\mathcal{L}) = \langle a\sigma + b\tau \rangle \oplus \langle c\sigma + d\tau \rangle$. We denote by H the normal subgroup of G which consists of those elements of G which are congruent to the identity matrix, modulo 2. Then H acts on the set of decompositions of $K(\mathcal{L})$, thus H acts on $\mathcal{A}^0_{(1,4)}$; to determine the corresponding action on the isomorphic space $\mathcal{P}$, it is sufficient to take any element of H, act to obtain a new basis and determine the new coordinates $(y_0 : y_1 : y_2 : y_3)$ according to (IV.5.1). Substituting these in (IV.5.2) the new parameters $(\pm\lambda_0 : \lambda_1 : \lambda_2 : \lambda_3)$ are found immediately. The result is contained in the following table (since diagonal matrices act trivially, only one representative of each coset modulo diagonal matrices is shown):

basis	$K(\mathcal{L})$	coo. for $\mathbf{P}^3$	moduli in $\mathcal{P}$
(σ, τ)	$K_1 \oplus K_2$	$(y_0 : y_1 : y_2 : y_3)$	$(\pm\lambda_0 : \lambda_1 : \lambda_2 : \lambda_3)$
$(\sigma + 2\tau, \tau)$	$K_4 \oplus K_2$	$(y_0 : y_1 : iy_2 : iy_3)$	$(\pm\lambda_0 : -\lambda_1 : \lambda_2 : \lambda_3)$
$(\sigma, 2\sigma + \tau)$	$K_1 \oplus K_5$	$(y_0 : iy_1 : y_2 : iy_3)$	$(\pm\lambda_0 : \lambda_1 : -\lambda_2 : \lambda_3)$
$(\sigma + 2\tau, 2\sigma + \tau)$	$K_4 \oplus K_5$	$(y_0 : iy_1 : iy_2 : -y_3)$	$(\pm\lambda_0 : \lambda_1 : \lambda_2 : -\lambda_3)$

Table 7

The upshot of the table is that all $(\pm\lambda_0 : \pm\lambda_1 : \pm\lambda_2 : \pm\lambda_3)$ correspond to the same Abelian surface. The quotient space is given by

$$\mathcal{P}' = \frac{\mathcal{P}}{(\pm\lambda_0 : \lambda_1 : \lambda_2 : \lambda_3) \sim (\pm\lambda_0 : \pm\lambda_1 : \pm\lambda_2 : \pm\lambda_3)} \tag{4.14}$$
$$\cong \left(\mathbf{P}^3 \setminus S'\right) \bigcup (\text{three rational curves in } S', \text{ each missing three points}),$$

upon defining $\mu_i = \lambda_i^2$ as coordinates for the quotient $\mathbf{P}^3$, from which in particular equations for the three rational curves as well as for the three points are immediately obtained (the fact that there are three missing points instead of two is due to ramification of the quotient map at two of the three points). The divisors S and S' will be calculated later. We will also interpret this "intermediate" moduli space $\mathcal{P}'$.

Notice that G/H is isomorphic to the permutation group S_3, so we have an action of S_3 on $\mathcal{P}'$ (which extends to all of $\mathbf{P}^3$ since it is linear). Choosing six representatives for G/H we find as above the following table:

basis	$K(\mathcal{L})$	coo. for $\mathbf{P}^3$	moduli in $\mathcal{P}'$
(σ, τ)	$K_1 \oplus K_2$	$(y_0 : y_1 : y_2 : y_3)$	$(\mu_0 : \mu_1 : \mu_2 : \mu_3)$
$(\tau, 3\sigma)$	$K_2 \oplus K_1$	$(y_0 : y_2 : y_1 : iy_3)$	$(-\mu_0 : \mu_2 : \mu_1 : \mu_3)$
$(\sigma, \sigma+\tau)$	$K_1 \oplus K_3$	$(\sqrt{i}y_2 : y_1 : \sqrt{i}y_0 : y_3)$	$(\mu_0 : \mu_3 : -\mu_2 : \mu_1)$
$(\sigma+\tau, \tau)$	$K_3 \oplus K_2$	$(y_1 : y_0 : \sqrt{i}y_2 : \sqrt{i}y_3)$	$(\mu_0 : -\mu_1 : \mu_3 : \mu_2)$
$(3\tau, \sigma+\tau)$	$K_2 \oplus K_3$	$(\sqrt{i}y_1 : iy_2 : \sqrt{i}y_0 : y_3)$	$(\mu_0 : -\mu_3 : \mu_1 : -\mu_2)$
$(\sigma+\tau, 3\sigma)$	$K_3 \oplus K_1$	$(\sqrt{i}y_2 : \sqrt{i}y_0 : -y_1 : -iy_3)$	$(\mu_0 : \mu_2 : -\mu_3 : -\mu_1)$

Table 8

The Tables 7 and 8 together show how to reconstruct explicitly the decomposition of $K(\mathcal{L})$ from the equation of the octic. More important, it allows us to construct the quotient space $\mathcal{M}^3$ as is shown in the following proposition.

Proposition 4.2   *There is a bijective map* $\tilde{\psi} : \tilde{\mathcal{A}}_{(1,4)} \to \mathcal{M}^3 \setminus \mathcal{D}$, *where* $\mathcal{M}^3$ *is the cone defined by*

$$f_4^2 = f_1(4f_2^3 - 27f_3^2)$$

in weighted projective space $\mathbf{P}^{(1,2,2,3,4)}$ *(with coordinates* $(f_0 : \cdots : f_4)$*) and* $\mathcal{D} = \mathcal{D}_1 + \mathcal{D}_2$ *is the divisor whose two irreducible components are cut off from* $\mathcal{M}^3$ *by the hypersurfaces*

$$\mathcal{D}_1 : f_4 = f_1(f_1 - 3f_2), \qquad\qquad (4.15)$$
$$\mathcal{D}_2 : 512f_4 = -16\left(16f_2^3 + 72f_1f_2 - 27f_1^2 - 48f_0f_3\right) + 3f_0^2\left(f_0^2 + 24f_1 - 32f_2\right).$$

In particular the moduli space $\tilde{\mathcal{A}}_{(1,4)}$ *has the structure of an affine variety. The map* $\tilde{\psi}$ *extends in a natural way to a map*

$$\psi : \mathcal{A}_{(1,4)} \to \mathcal{M}^3,$$

the image of the (two-dimensional) boundary $\mathcal{A}_{(1,4)} \setminus \tilde{\mathcal{A}}_{(1,4)}$ *being* $C \setminus \{P, Q\}$, *where* C *is the rational curve (inside* $\mathcal{D}$*) given by*

$$C : 3f_0^2 = 4(4f_2 - f_1),$$

and $P, Q \in C$ *are given by* $P = (4 : 0 : 3 : 2 : 0)$, *and* $Q = (2 : 1 : 1 : 0 : -2)$. *Moreover, apart from its vertex* $(1 : 0 : 0 : 0 : 0)$, *all points in the cone* $\mathcal{M}^3$ *correspond to some level surface* $\mathcal{F}_{(\alpha, \beta, f, g)}$ *for some* α, β, f *and* g, *with* $\alpha \neq \beta$.

Proof
    First we describe the quotient of $\mathbf{P}^3$ by the action of S_3, and show that it is (isomorphic to) the algebraic variety $\mathcal{M}^3$ given by an equation $f_4^2 = f_1(4f_2^3 - 27f_3^2)$ in weighted projective space $\mathbf{P}^{(1,2,2,3,4)}$. To do this we use the (induced) action of S_3 on $\mathbf{C}^3$ which is given in terms of affine coordinates $x_i = \mu_i/\mu_0$ for $\mathbf{C}^3$ by

$$(1,2) \cdot (x_1, x_2, x_3) = (-x_2, -x_1, -x_3),$$
$$(1,2,3) \cdot (x_1, x_2, x_3) = (-x_3, x_1, -x_2).$$

4. The Garnier potential

Since the action is orthogonal, it must be reducible, having an invariant line and an invariant plane orthogonal to it. Indeed let

$$u_1 = x_1 + x_2 - x_3,$$
$$u_2 = x_1 - x_2, \tag{4.16}$$
$$u_3 = x_1 + x_3,$$

then u_1 is anti-invariant for $(1,2)$ and is invariant for $(1,2,3)$; u_2 and u_3 are chosen orthogonal to u_1. Then invariants

$$f_2 = u_2^2 - u_2 u_3 + u_3^2,$$
$$f_3 = u_2 u_3 (u_2 - u_3),$$

for the action of S_3 are found. Also there is

$$\Delta = u_2^2 (2u_2 - 3u_3) + u_3^2 (2u_3 - 3u_2)$$

which is $(1,2)$-anti-invariant and $(1,2,3)$-invariant, giving a new invariant $f_4 = u_1 \Delta$. Since f_2 and f_3 generate the invariants depending on u_2, u_3 the invariant Δ^2 is expressible in terms of f_2 and f_3,

$$\Delta^2 = 4f_2^3 - 27 f_3^2,$$

i.e., Δ^2 is nothing else than the discriminant of the cubic polynomial $x^3 - f_2 x + f_3$. It follows that

$$f_4^2 = f_1 (4f_2^3 - 27 f_3^2), \tag{4.17}$$

where $f_1 = u_1^2$. Notice that (f_1, f_2, f_3, f_4) have degree $(2, 2, 3, 4)$ so that the quotient of $\mathbf{P}^3$ by the action of S_3 is given by (4.17) viewed as an equation in weighted projective space $\mathbf{P}^{(1,2,2,3,4)}$ with respect to coordinates $(f_0 : f_1 : f_2 : f_3 : f_4)$. In conclusion we have established the cover $\mathcal{P} \to \mathcal{M}^3$ and there is an induced map $\psi : \mathcal{A}_{(1,4)} \to \mathcal{M}^3$ which makes

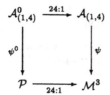

into a commutative diagram (since the actions on $\mathcal{A}_{(1,4)}^0$ are the same by construction).

The reducible divisor $\mathcal{D}$ is easily computed once explicit equations for S (or S') are known. Since we know of no easy direct way to determine S, we postpone the computation of S to Paragraph 4.4, where the potentials will be used to compute S in a straightforward way; we will show there that S' breaks up in four irreducible pieces $\mu_1 = 0$, $\mu_2 = 0$, $\mu_3 = 0$ and $\mathrm{disc}(P_3^\mu(x)) = 0$ where P_3 is the polynomial

$$P_3 = 4\mu_2 x^3 - (\mu_0 + 2\mu_1 + 6\mu_2 + 2\mu_3)x^2 + (\mu_0 - 2\mu_1 + 2\mu_2 - 6\mu_3)x - 4\mu_3,$$

and $\mathrm{disc}(P_3^\mu(x)) = 0$ denotes its discriminant (in x). Granted this, we take $\mu_1 = 0$, let $x_1 = 0$ and eliminate x_2 and x_3 from f_1, f_2 and f_4. Then the relation

$$f_4 = f_1 (f_1 - 3f_2)$$

is found at once; obviously the same equation is found for $\mu_2 = 0$, $\mu_3 = 0$. The computation for disc$(P_3^\mu(x)) = 0$ is longer but also straightforward. Namely, by a simple translation in x the monic polynomial $P_3^\mu(x)/(4\mu_2)$ can be written as $x^3 - ax + b$, with discriminant $4a^3 - 27b^2$. When this discriminant (depending on μ_i) is written in terms of u_i using the inverse of (4.16), the equation (4.15) for $\mathcal{D}_2$ is read off immediately.

As for the curve to be added to $\tilde\psi(\tilde{\mathcal{A}}_{(1,4)})$ to obtain $\psi\left(\mathcal{A}_{(1,4)}\right)$, notice that the action of S_3 identifies the three rational curves in (4.14), leading to a single curve. To compute its equation (as a subvariety of $\mathcal{D}_1$) in terms of the coordinates f_i, let, according to (IV.5.5), $\mu_1 = 0$ and $\mu_0 = 2(\mu_2 + \mu_3)$. Then in terms of μ_0 and μ_2 we get

$$f_0 = \mu_0,$$
$$f_1 = (2\mu_2 - \mu_0/2)^2,$$
$$f_2 = \mu_2^2 - \frac{\mu_0\mu_2}{2} + \frac{\mu_0^2}{4},$$

leading to

$$3f_0^2 = 4(4f_2 - f_1),$$

by elimination of μ_0 and μ_2. As for the two special points P and Q on this curve, it is easy to check that picking $\mu_1 = 0$, $\mu_2 = \mu_3$ and $\mu_0 = 2(\mu_2 + \mu_3)$ leads to the point $(4 : 0 : 3 : 2 : 0)$ and alternatively taking $\mu_1 = \mu_2 = 0$, $\mu_0 = 2\mu_3$ leads to the point $(2 : 1 : 1 : 0 : -2)$. This gives explicit equations for all these spaces and proves the announced result in (4.13).

Finally, let $(f_0 : \cdots : f_4) \in \mathcal{M}^3$ be any point different from the vertex $(1 : 0 : 0 : 0 : 0)$ of this cone. Then $\mu_2 \neq 0$ for at least one of the six points $(\mu_0 : \mu_1 : \mu_2 : \mu_3)$ lying over this point. Define α, β, f, g by

$$\alpha = \mu_0 + 2\mu_1 + 2\mu_2 + 2\mu_3,$$
$$\beta = \mu_0 + 2\mu_1 - 2\mu_2 + 2\mu_3,$$
$$f = 128\mu_2^2\mu_3, \tag{4.18}$$
$$g = 128\mu_2^2\mu_1,$$

then $\alpha \neq \beta$ and α, β, f and g satisfy (4.4). This shows that, apart from the vertex, all points in the cone $\mathcal{M}^3$ correspond to some level surface $\mathcal{F}_{(\alpha,\beta,f,g)}$ for some $\alpha \neq \beta$, f and g. This concludes the proof of the proposition. ∎

4.3. The precise relation with the canonical Jacobian

In this paragraph we want to show that a $(1,4)$-polarized Abelian surface $\mathcal{T}^2 \in \tilde{\mathcal{A}}_{(1,4)}$ is intimately related to its canonical Jacobian $J(\mathcal{T}^2)$ (introduced in Paragraph IV.5), hence also to some curve of genus two, denoted $\Gamma(\mathcal{T}^2)$. In fact there is more: at the level of the Jacobian, let $J(\mathcal{T}^2)$ be represented as $\mathbf{C}^2/\Lambda$, then $\mathcal{T}^2$ induces a non-degenerate decomposition of the lattice Λ and at the level of the curve, $\mathcal{T}^2$ induces a decomposition of the set of Weierstrass points of $\Gamma(\mathcal{T}^2)$ which in turn corresponds to an incidence diagram for the 16_6 configuration on its Kummer surface; moreover, the Abelian surface can be reconstructed from either of these data (Proposition 4.3).

4. The Garnier potential

Recall that the canonical Jacobian of a $(1,4)$-polarized Abelian surface $\mathcal{T}^2 = (T^2, \mathcal{L}) \in \tilde{\mathcal{A}}_{(1,4)}$ is defined as the (irreducible principally polarized) Abelian surface $J(\mathcal{T}^2) = T^2/K$, where K is the (unique) subgroup of two-torsion elements of $K(\mathcal{L})$. We have seen that such an Abelian surface is the Jacobian of a smooth curve Γ of genus two, i.e., it is given as $\mathbf{C}^2/\Lambda$, where Λ is the *period lattice*

$$\Lambda = \left\{ \oint_\gamma \vec{\omega} \mid \gamma \in H_1(\Gamma, \mathbf{Z}) \right\}$$

consisting of all periods of $\vec{\omega} = {}^t(\omega_1, \omega_2)$, the ω_i being (independent) holomorphic differentials on Γ. The Abelian group $H_1(\Gamma, \mathbf{Z})$ has an (alternating) intersection form ${}^\#(\cdot)$ and $H_1(\Gamma, \mathbf{Z})$ can be decomposed into non-degenerate planes (in many different ways),

$$H_1(\Gamma, \mathbf{Z}) = H_1 \oplus H_2, \qquad {}^\#(\cdot)_{H_1} \text{ and } {}^\#(\cdot)_{H_2} \text{ non-degenerate.}$$

Such a decomposition leads to a decomposition $\Lambda = \Lambda_1 \oplus \Lambda_2$ upon defining

$$\Lambda_i = \left\{ \oint_\gamma \vec{\omega} \mid \gamma \in H_i \right\} ; \tag{4.19}$$

both $H_1(\Gamma, \mathbf{Z}) = H_1 \oplus H_2$ and $\Lambda = \Lambda_1 \oplus \Lambda_2$ will be called *non-degenerate decompositions*. They are called in addition *simple* if each H_i is generated by cycles which come from simple closed curves (Jordan curves) in $\mathbf{P}^1$ under some (hence any) double cover $\pi : \Gamma \to \mathbf{P}^1$.

The relevance of simple, non-degenerate decompositions and incidence diagrams (recalled in Paragraph IV.4.4) for $(1,4)$-polarized Abelian surfaces is seen from the following proposition.

Proposition 4.3 *There is a natural correspondence between the following four (isomorphism classes of) data:*

(1) *a $(1,4)$-polarized Abelian surface $\mathcal{T}^2 \in \tilde{\mathcal{A}}_{(1,4)}$,*

(2) *a Jacobi surface $J = \mathbf{C}^2/\Lambda$ + a simple, non-degenerate decomposition $\Lambda = \Lambda_1 \oplus \Lambda_2$ of Λ,*

(3) *a smooth genus two curve Γ + a decomposition $\mathcal{W} = \mathcal{W}_1 \cup \mathcal{W}_2$, $\#\mathcal{W}_1 = \#\mathcal{W}_2 = 3$, of its Weierstrass points.*

(4) *a smooth genus two curve Γ + an incidence diagram for the 16_6 configuration on its corresponding Kummer surface.*

The correspondence (1) $\leftrightarrow$ (2) *is established in two ways, namely J may be taken as the quotient of $\mathcal{T}^2$ using Λ_2 or as a cover of $\mathcal{T}^2$ using Λ_1 (or $\mathcal{W}_1$). Moreover, interchanging the components of the decomposition in* (2) *amounts to taking the dual $\hat{\mathcal{T}}^2$ of $\mathcal{T}^2$ in* (1). *J is the Jacobian of the curve Γ which appears in* (3) *and* (4) *and interchanging Λ_1 and Λ_2 in* (2) *amounts to interchanging $\mathcal{W}_1$ and $\mathcal{W}_2$ in* (3) *and taking the transpose of both square diagrams in the incidence diagram in* (4).

Summarizing we have the following commutative diagram, determined by $\mathcal{T}^2$ (only),

$$\begin{array}{ccc} J & \xrightarrow{\;\Lambda_2\;} & \hat{\mathcal{T}}^2 \\[4pt] {\scriptstyle\Lambda_1}\Big\downarrow & {\searrow}{\scriptstyle 2_J} & \Big\downarrow{\scriptstyle\Lambda_1} \\[4pt] T^2 & \xrightarrow[\;\Lambda_2\;]{} & J \end{array} \tag{4.20}$$

where 2_J denotes multiplication by 2 in J and a Λ_i labeling an arrow means that a projection is considered on the quotient torus that is obtained by doubling the sublattice Λ_i.

Proof

$(3) \rightarrow (2)$ Given a genus two curve Γ and a decomposition $\mathcal{W} = \mathcal{W}_1 \cup \mathcal{W}_2$ of its Weierstrass points, with $\#\mathcal{W}_i = 3$, let $\pi : \Gamma \rightarrow \mathbf{P}^1$ be any two-sheeted cover of $\mathbf{P}^1$. It is well known that π has branch points exactly at $\mathcal{W}$; the points in $\mathcal{W}$ as well as their projections under π will be denoted by $W_1, \ldots, W_6$, also $\pi(\mathcal{W}_i)$ will just be written as $\mathcal{W}_i$. If $\mathbf{P}^1$ is covered with connected open subsets U_1 and U_2 for which $\mathcal{W}_i \subset U_i$ and $U_1 \cap U_2 \cap \mathcal{W} = \emptyset$ then $H_1(\Gamma, \mathbf{Z})$ decomposes as $H_1 \oplus H_2$ where H_1 and H_2 are defined as

$$H_i = \{ \gamma \in H_1(\Gamma, \mathbf{Z}) \mid \pi_* \gamma \in H_1(U_i \setminus \mathcal{W}_i, \mathbf{Z}) \}.$$

Among the cycles in H_i there are those which come from simple closed curves in $U_i \setminus \mathcal{W}_i$ encircling two points in $\mathcal{W}_i$ and these generate H_i. Since any (different) of these intersect (once) the restriction $\sharp(\cdot)_{H_i}$ is non-degenerate, hence leads (upon using (4.19)) to a non-degenerate simple decomposition $\Lambda = \Lambda_1 \oplus \Lambda_2$ for the period lattice. Thus $\mathbf{C}^2/\Lambda$ and $\Lambda = \Lambda_1 \oplus \Lambda_2$ provide the corresponding data.

We now show that the constructed data only depend (up to isomorphism) on the isomorphism class of the data Γ, $\mathcal{W} = \mathcal{W}_1 \cup \mathcal{W}_2$. Let $\sigma : \Gamma \rightarrow \Gamma$ be an automorphism which permutes the Weierstrass points (such an automorphism only exists for special curves Γ). Then σ extends linearly to $\mathrm{Jac}(\Gamma) \cong \mathbf{C}^2/\Lambda$, hence also to the lattice Λ, giving a new decomposition $\Lambda = \sigma\Lambda_1 \oplus \sigma\Lambda_2$. The lattice $\sigma\Lambda_i$ contains the periods corresponding to the points $\sigma\mathcal{W}_i$ (with respect to the same basis of the space of holomorphic differential forms), hence $\Lambda = \sigma\Lambda_1 \oplus \sigma\Lambda_2$ corresponds to the decomposition $\mathcal{W} = \sigma\mathcal{W}_1 \cup \sigma\mathcal{W}_2$.

$(2) \rightarrow (3)$ By the classical Torelli Theorem, Γ can be reconstructed from its Jacobian, actually in dimension two, Γ is isomorphic to the theta divisor of $\mathrm{Jac}(\Gamma)$. The lattice $\Lambda \subset \mathbf{C}^2$ is the period lattice of Γ with respect to some basis $\vec{\omega} = \{\omega_1, \omega_2\}$ of the space of holomorphic differentials on Γ, which determines an isomorphism $\phi : \Lambda \rightarrow H_1(\Gamma, \mathbf{Z})$, which in turn leads to a decomposition $H_1(\Gamma, \mathbf{Z}) = H_1 \oplus H_2$ upon defining $H_i = \phi(\Lambda_i)$.

If we denote by $\mathcal{W}$ the set of Weierstrass points of Γ and by $\pi : \Gamma \rightarrow \mathbf{P}^1$ any two-sheeted cover as above, then H_i has a system of generators $\lambda_{i1}, \lambda_{i2}$, where $\pi_* \lambda_{ij}$ is a simple closed curve in $\mathbf{P}^1 \setminus \mathcal{W}$, encircling an even number of branch points $\mathcal{W}_i$, which reduces to two in this case (there are only six points $\mathcal{W}_i$ and encircling four points amounts to the same as encircling the other two points). Since the decomposition is non-degenerate, $\pi_* \lambda_{i1}$ and $\pi_* \lambda_{i2}$ encircle a common point, so we may take

$$\mathcal{W}_i = \pi^{-1}\{\text{points in } \mathcal{W} \text{ encircled by } \pi_* \lambda_{i1} \text{ or } \pi_* \lambda_{i2}\}.$$

Then $\#\mathcal{W}_1 = \#\mathcal{W}_2 = 3$ and it is easy to see that $\mathcal{W}_1 \cap \mathcal{W}_2 = \emptyset$.

We show again that the constructed data are independent of the choice of the basis $\{\omega_1, \omega_2\}$ and are well-defined up to isomorphism. To do this notice first that when the choice of basis $\vec{\omega} = {}^t(\omega_1, \omega_2)$ is not unique, say $\vec{\omega}'$ is another basis producing Λ, then $\vec{\omega} = A\vec{\omega}'$ for some $A \in GL(2, \mathbf{C})$, hence

$$\oint_\gamma \vec{\omega} = A \oint_\gamma \vec{\omega}'$$

for any $\gamma \in H_1(\Gamma, \mathbf{Z})$. We find that $\Lambda = A\Lambda$, i.e., Λ has a non-trivial symmetry group. Then $\mathrm{Jac}(\Gamma) = \mathbf{C}^2/\Lambda$ has a non-trivial automorphism group and the data $(\mathbf{C}^2/\Lambda, \Lambda = \Lambda_1 \oplus \Lambda_2)$ and $(\mathbf{C}^2/\Lambda, \Lambda = A\Lambda_1 \oplus A\Lambda_2)$ are isomorphic. Thus it suffices to show that the constructed data are well-defined up to isomorphism. This follows (as in the first part of the proof) at

208

once from the property that if Jac(Γ) has a non-trivial automorphism σ, then it is induced by an automorphism on Γ. To see this property (which is particular for the case in which the genus of Γ is 2) let Θ be a generic translate of the Riemann theta divisor passing through the origin O of Jac(Γ). Then $\sigma(\Theta)$ is another translate passing through O (since every curve in Jac(Γ) which is isomorphic to Γ is a translate of Θ) hence composing σ with this translate determines an automorphism of Γ. This shows the constructed data are well-defined.

(2) $\to$ (1)   Given $J = \mathbf{C}^2/\Lambda$ and $\Lambda = \Lambda_1 \oplus \Lambda_2$ we form the complex torus

$$\mathcal{T}^2 = \mathbf{C}^2/\Lambda' \quad \text{with} \quad \Lambda' = \frac{1}{2}\Lambda_1 \oplus \Lambda_2,$$

(i.e., the first lattice is doubled in both directions) and equip this torus with the polarization induced by the principal polarization on J. We claim that $\mathcal{T}^2$ is a $(1,4)$-polarized Abelian surface which belongs to $\tilde{\mathcal{A}}_{(1,4)}$. To show this, first notice that the cycles $\{\lambda_{11}, \lambda_{21}, \lambda_{12}, \lambda_{22}\}$ introduced above, form a symplectic basis of $H_1(\Gamma, \mathbf{Z})$, i.e., $^\#(\lambda_{1i} \cdot \lambda_{2i}) = 0$, $^\#(\lambda_{i1} \cdot \lambda_{i2}) = 1$, hence these cycles lead to a period matrix of the form (see [GH])

$$\begin{pmatrix} 1 & 0 & a & b \\ 0 & 1 & b & c \end{pmatrix}$$

satisfying the Riemann conditions. Since H_1 is spanned by λ_{11} and λ_{12} (which correspond to the first and third columns of this matrix) Λ' has in terms of slightly different coordinates the period matrix

$$\begin{pmatrix} 1 & 0 & a & 2b \\ 0 & 4 & 2b & 4c \end{pmatrix}$$

which leads immediately to the result that $\mathcal{T}^2$ is a $(1,4)$-polarized Abelian surface, $4:1$ isogeneous to J (notice that the right block of this matrix is positive definite). Since the original $J = \mathbf{C}^2/\Lambda$ is the canonical Jacobian of $\mathcal{T}^2$, we are in the generic case of Paragraph IV.5 which implies $\mathcal{T}^2 \in \tilde{\mathcal{A}}_{(1,4)}$.

Dually the surface is (up to isomorphism) also constructed by taking

$$\mathcal{T}^2 = \mathbf{C}^2/\Lambda'' \quad \text{with} \quad \Lambda'' = \Lambda_1 \oplus 2\Lambda_2,$$

but this decomposition induces a $4:1$ isogeny from J to (this) $\mathcal{T}^2$.

To show that the correspondence is well-defined, observe that

$$(\mathbf{C}^2/\Lambda, \Lambda = \Lambda_1 \oplus \Lambda_2) \cong (\mathbf{C}^2/\Lambda, \Lambda = \Lambda_1' \oplus \Lambda_2')$$

implies

$$\mathbf{C}^2 \Big/ \left(\frac{1}{2}\Lambda_1 \oplus \Lambda_2\right) \cong \mathbf{C}^2 \Big/ \left(\frac{1}{2}\Lambda_1' \oplus \Lambda_2'\right) \quad \text{and} \quad \mathbf{C}^2 \Big/ (\Lambda_1 \oplus 2\Lambda_2) \cong \mathbf{C}^2 \Big/ (\Lambda_1' \oplus 2\Lambda_2'),$$

the last two isomorphisms being isomorphism of polarized Abelian surfaces.

(1) $\to$ (2)   For given $\mathcal{T}^2 \in \mathcal{A}_{(1,4)}^0$, let J be its canonical Jacobian $J(\mathcal{T}^2)$. Then $\mathcal{T}^2 \to J$ is part of the isogeny $2_J : J \to J$ hence there is a unique complementary isogeny $J \to \mathcal{T}^2$ with kernel $\mathbf{Z}/2\mathbf{Z} \oplus \mathbf{Z}/2\mathbf{Z}$. Writing J as $J = \mathbf{C}^2/\Lambda$, the latter isogeny induces an injective lattice homomorphism $\phi : \Lambda \to \Lambda$ whose cokernel is isomorphic to $\mathbf{Z}/2\mathbf{Z} \oplus \mathbf{Z}/2\mathbf{Z}$. Then ϕ determines a unique decomposition $\Lambda_1 \oplus \Lambda_2$ of Λ for which $\phi_{|\Lambda_2}$ is an isomorphism and $\phi_{|\Lambda_1}$

is multiplication by 2. We have seen that such a decomposition is simple. It is also non-degenerate, since otherwise $\mathcal{T}^2$ would not have an induced $(1,4)$-polarization (see below).

Observe that in the exceptional case that $\mathcal{T}^2 \to J$ is another part of the isogeny 2_J, the two isogenies combine to an automorphism of J, leading to isomorphic data in (3).

$(3) \leftrightarrow (4)$ This is classical (see [Hud]); we prove it as follows. Given a decomposition of $\mathcal{W}$, say $\mathcal{W} = \{W_1, W_2, W_3\} \cup \{W_4, W_5, W_6\}$ the corresponding incidence diagram is taken as

$$
\begin{array}{cccc}
W_{11} & W_{12} & W_{23} & W_{13} \\
W_{45} & W_{36} & W_{16} & W_{26} \\
W_{46} & W_{35} & W_{15} & W_{25} \\
W_{56} & W_{34} & W_{14} & W_{24}
\end{array}
\qquad
\begin{array}{cccc}
\Gamma_{11} & \Gamma_{12} & \Gamma_{23} & \Gamma_{13} \\
\Gamma_{45} & \Gamma_{36} & \Gamma_{16} & \Gamma_{26} \\
\Gamma_{46} & \Gamma_{35} & \Gamma_{15} & \Gamma_{25} \\
\Gamma_{56} & \Gamma_{34} & \Gamma_{14} & \Gamma_{24}
\end{array}
$$

and obviously the decomposition of $\mathcal{W}$ is reconstructed from it at once. To show that every incidence diagram is of this form, notice at first that we have the freedom to permute the rows as well as the columns, so that we can put $W_{11} = \ldots = W_{66}$ in the upper left corner. The curves Γ_{ij} this point W_{11} belongs to are the entries in the first row and the first column (except Γ_{11}) of the square diagram on the right. If the origin belongs to $\Gamma_{ij} \cap \Gamma_{jk}$, $(j \neq k)$, then it also belongs to Γ_{ik}. Then Γ_{11} is easily identified as the image of the map $\Gamma \to \mathrm{Jac}(\Gamma)$ defined by

$$
P \mapsto \int_{W_i}^{P} \vec{\omega} + \int_{W_j}^{W_k} \vec{\omega} \pmod{\Lambda},
$$

and the other three curves are Γ_{lm}, Γ_{mn} and Γ_{ln} with $\{i, j, k, l, m, n\} = \{1, 2, 3, 4, 5, 6\}$. Hence the incidence diagram takes the above form from which the decomposition of $\mathcal{W}$ can be read off.

If the curve has non-trivial automorphisms, we define diagrams which correspond to such automorphisms as being isomorphic, so as to obtain the equivalence $(3) \leftrightarrow (4)$ at the level of isomorphism classes.

Finally we concentrate on the dual $\hat{\mathcal{T}}^2$ of $\mathcal{T}^2$ and its relation with the canonical Jacobian of $\mathcal{T}^2$. At first recall from [GH] that the period matrices of $\mathcal{T}^2$ and $\hat{\mathcal{T}}^2$ relate as

$$
\mathcal{T}^2 \sim \begin{pmatrix} 1 & 0 & a & 2b \\ 0 & 4 & 2b & 4c \end{pmatrix}
\qquad
\hat{\mathcal{T}}^2 \sim \begin{pmatrix} 4 & 0 & 4a & 2b \\ 0 & 1 & 2b & c \end{pmatrix} \sim \begin{pmatrix} 1 & 0 & c & 2b \\ 0 & 4 & 2b & 4a \end{pmatrix}
$$

showing that $\hat{\mathcal{T}}^2$ is constructed from J by taking $\Lambda_1 \oplus \frac{1}{2}\Lambda_2$ instead of taking $\frac{1}{2}\Lambda_1 \oplus \Lambda_2$ when constructing $\mathcal{T}^2$ from J. It follows that the isogeny 2_J can be factorized via $\hat{\mathcal{T}}^2$ as well and that taking the dual of $\mathcal{T}^2$ corresponds to interchanging the components of the decomposition of Λ. This finishes the proof of the proposition.   ∎

Remarks 4.4

1. If in (2) above one considers simple degenerate decompositions (instead of non-degenerate) then the decomposition in (3) is altered into $\mathcal{W} = \mathcal{W}_1 \cup \mathcal{W}_2 \cup \mathcal{W}_3$, $\#\mathcal{W}_i = 2$ and the order of the components in the decomposition of $\mathcal{W}$ is now irrelevant. The corresponding object in (1) is then a Jacobi surface (different from the one in (2)) from which the original Jacobi surface (or the curve) cannot be reconstructed.

2. Since $\binom{6}{3} = 20$, there are 20 different incidence diagrams and 20 possible decompositions of the isogeny $2_J : J \to J$, some of which are isomorphic if and only if J (hence Γ) has a non-trivial automorphism group (i.e., different from $\mathbf{Z}_2$). It follows from the above proposition that the 20 intermediate Abelian surfaces appear in 10 groups of dual pairs.

3. Let $\mathcal{C}^{(2)}$ denote the moduli space of all smooth curves of genus two. Then we have the following isomorphisms

$$\tilde{\mathcal{A}}_{(1,4)} \cong \{(\{W_1, W_2, W_3\}, \{W_4, W_5, W_6\}) \mid W_i \in \mathbf{P}^1, i \neq j \Rightarrow W_i \neq W_j\} \big/ \mathrm{mod}\ \mathbf{PGL}(2, \mathbf{C}),$$

$$\mathcal{C}^{(2)} \cong \{\{W_1, W_2, W_3, W_4, W_5, W_6\} \mid W_i \in \mathbf{P}^1, i \neq j \Rightarrow W_i \neq W_j\} \big/ \mathrm{mod}\ \mathbf{PGL}(2, \mathbf{C});$$

and both spaces are related by an obvious unramified cover $\tilde{\mathcal{A}}_{(1,4)} \to \mathcal{C}^{(2)}$. We have seen that $\tilde{\mathcal{A}}_{(1,4)}$ has a natural structure of an affine variety which is compactified in a natural way into its projective closure, which is the (singular) algebraic variety $\mathcal{M}^3$. On the other hand, $\mathcal{C}^{(2)}$ also has a natural compactification (the Mumford-Deligne compactification). It would be interesting to figure out how both compactifications are related.

4. Among the different ways to define (and characterize) the canonical Jacobian $J(\mathcal{T}^2)$ of $\mathcal{T}^2$, here is a final one: $J = J(\mathcal{T}^2)$ is the only Jacobian for which the diagram

commutes (2_T is multiplication by 2 on $\mathcal{T}^2$). The proof is easy using the ideas of the above proof. Observe that this diagram is (4.20) with $\mathcal{T}^2$ and J interchanged; we could drop a superfluous triangle since $\hat{J} = J$.

4.4. The relation with the canonical Jacobian made explicit

We have shown in Paragraph 4.3 that there is associated to an Abelian surface of type $(1, 4)$ the Jacobi surface of a genus two curve Γ and some additional data. Also we have seen (in Paragraph 4.1) that these Abelian surfaces appear as level of the integrable Hamiltonian system defined by one of the potentials $V_{\alpha\beta}$. This allows us to make this relation very explicit (in two different ways) and to calculate precisely the locus S in $\mathbf{P}^3$ for which the associated quartic fails to be a Kummer surface (and hence the associated $(1, 4)$-polarized Abelian surface fails to be birational to an octic). We know of no direct method (i.e., without using the theory of integrable systems) to do this.

Our calculations rely on the explicit construction of an embedding for $\mathcal{T}^2$ in projective space, which is found by using the Laurent solutions to the differential equations (4.2). Since we know that the potential $V_{\alpha\beta}$ is a.c.i. (for $\alpha \neq \beta$), the vector field X_H has a coherent tree of Laurent solutions (see Section V.3), in particular it has Laurent solutions depending on $\dim \mathbf{C}^4 - 1 = 3$ free parameters (*principal balances*). Moreover, since the divisor $\mathcal{D}_{fg}$ to be adjoined to a (general) fiber $\mathcal{F}_{fg}$ of the momentum map is irreducible, there is only one such family. Also q_1, q_2 and $q = q_1 p_2 - q_2 p_1$ have a simple pole along $\mathcal{D}_{fg}$ since their squares descend to $\mathrm{Jac}(\Gamma)$ with a double pole along (some translate of) its theta divisor. With this information the principal balance is given by

$$
\begin{aligned}
q_1 &= \frac{1}{t}\left[a + \frac{2}{3}((1 + a^2 - b^2)a + 2ab^2\beta)t^2 + bct^3 + \mathcal{O}(t^4)\right], \\
q_2 &= \frac{1}{t}\left[b + \frac{2}{3}((1 + b^2 - a^2)b + 2ba^2\alpha)t^2 - act^3 + \mathcal{O}(t^4)\right],
\end{aligned}
\tag{4.21}
$$

where $2a^2 + 2b^2 + 1 = 0$; the series for p_1 and p_2 are found by differentiation. Using the Laurent solutions it is easy to find an embedding of $\mathcal{T}_{fg}^2$ in projective space: since $2\mathcal{D}_{fg}$ induces a polarization of type $(2,8)$, it is very ample and this can be done using the sixteen functions with a double pole along $\mathcal{D}_{fg}$, to wit,

$$
\begin{aligned}
z_0 &= 1, & z_8 &= q_2^2, \\
z_1 &= q_1, & z_9 &= q_1 q, \\
z_2 &= q_2, & z_{10} &= q_2 q, \\
z_3 &= q = q_1 p_2 - q_2 p_1, & z_{11} &= (q_1^2 + q_2^2)q + \alpha q_1 p_2 - \beta q_2 p_1, \\
z_4 &= p_1, & z_{12} &= \{q_1, q\}, \\
z_5 &= p_2, & z_{13} &= \{q_2, q\}, \\
z_6 &= q_1^2, & z_{14} &= 2q_1 q_2(q_1^2 + q_2^2) + p_1 p_2, \\
z_7 &= q_1 q_2, & z_{15} &= q^2,
\end{aligned}
\tag{4.22}
$$

where $\{f_1, f_2\} = \dot{f}_1 f_2 - f_1 \dot{f}_2$, the Wronskian of f_2 and f_1. Since the embedding variables depend regularly on the base space (i.e., on α, β, f and g) it follows that this a.c.i. system is completable.

We compute the correspondence between the data by using the cover $J \to \mathcal{T}^2$; this can also be done using the cover $\mathcal{T}^2 \to J$ (see [Van3]). Recall from Paragraph 4.3 that given $\mathcal{T}^2 \in \tilde{\mathcal{A}}_{(1,4)}$ there is a unique Jacobian $J = J(\mathcal{T}^2)$ such that

yields a factorization of the map 2_J (multiplication by 2). This implies the existence of a singular divisor in $\mathcal{T}^2$ whose components are birational equivalent to $\Gamma = \Gamma(\mathcal{T}^2)$ as is shown in the following proposition.

Proposition 4.5 *The image $p_1(\mathcal{K})$ of Kummer's 16_6 configuration $\mathcal{K}$ consists of four curves, all passing through the half periods of $\mathcal{T}^2$; these points are the images of the sixteen points in the configuration and each of the four image curves has an ordinary three-fold point at one of these points, with tangents at this point, which are different from the tangents to the other curves. Each curve is birational equivalent to Γ and induces a $(1,4)$-polarization on $\mathcal{T}^2$. The image $p_2(p_1(\mathcal{K}))$ is one single curve, birational equivalent to Γ with an ordinary six-fold point.*

4. The Garnier potential

Proof

The map p_1 identifies all half-periods which appear in a row in the first square diagram of the incidence diagram which corresponds to $\mathcal{T}^2$. Therefore p_1 also identifies the curves which appear in a row in the second square diagram of this incidence diagram and we obtain four curves passing through the four image points, every curve having a three-fold point at the image of the three points in the same row (but not the same column) of the first square diagram. Since $\mathcal{K}$ induces a $(16, 16)$-polarization on J, $p_1(\mathcal{K})$ induces a $(4, 16)$-polarization on $\mathcal{T}^2$, hence each component induces a $(1, 4)$-polarization. The virtual genus of each component is thus five, and since each is obviously birational to Γ via p_1, the threefold point must be ordinary and there are no other singular points.

The intersection of two of these components is the self-intersection of one of them (since they are translates of each other), hence is by Theorem IV.3.7 equal to $2(5 - 1) = 8$; on the other hand, since each passes through the three-fold point of the other and since they have two simple points in common, this gives already $3 + 3 + 1 + 1 = 8$ so all tangents must be different and there are no other intersection points. The fact that $p_2(p_1(\mathcal{K}))$ has an ordinary six-fold point and is birational equivalent to Γ is shown in a similar way. ∎

The image $2_J(\Theta)$ is a divisor Δ with a six-fold point, first studied in [Van2] (where it was an essential ingredient in the construction of linearizing variables for integrable Hamiltonian systems) and $p_1(\mathcal{K})$ is nothing but $p_2^*\Delta$. We have also shown there that this divisor is the zero locus of the leading term in the equation of the Kummer surface of J (when normalized) as in the algorithm in Section V.4.

To apply this in the present case, we use the leading term (4.8) of the equation of the Kummer surface of $J\left(\mathcal{T}_{fg}^2\right)$ (which is expressed in terms of the original variables), and investigate its zero locus, i.e.,

$$(q_1^2 + q_2^2 + \alpha + \beta)^2 - 4(\alpha\beta + \beta q_1^2 + \alpha q_2^2) = 0.$$

This can be factorized completely as

$$\prod_{\epsilon_i = \pm 1} \left[q_2 - \epsilon_1 \sqrt{\alpha - \beta} - \epsilon_2 i q_1 \right] = 0.$$

reflecting the fact that $p_2^*\Delta$ is reducible. In order to find an equation for $\Gamma(\mathcal{T}_{fg}^2)$, let $q_2 = \epsilon_1 \sqrt{\alpha - \beta} + \epsilon_2 i q_1$ in the equations for $\mathcal{F}_{fg}$. Eliminating p_2 one finds an equation for the curve

$$\Delta_{\epsilon_1 \epsilon_2} : p_1^2 Q(q_1)(q_1 - \epsilon_1 \epsilon_2 i \sqrt{\alpha - \beta}) q_1 + P^2(q_1) = 0,$$

where

$$Q(x) = \epsilon_1 \epsilon_2 i (\alpha - \beta)^{3/2} x^3 + (\alpha - \beta)(2\alpha - \beta) x^2 + \epsilon_1 \epsilon_2 i \sqrt{\alpha - \beta}(h + \alpha(\beta - \alpha)) x - \frac{f}{2},$$

P is some polynomial of degree 3. This curve is clearly isomorphic to the curve

$$z^2 = x(x - i\epsilon_1\epsilon_2 \sqrt{\alpha - \beta}) Q(x). \tag{4.23}$$

213

In order to decide to which decomposition of the Weierstrass points this corresponds, let $P_1, \ldots, P_4$ be the following points in $\mathbf{P}^{15}$

$$P_1 = (0 : \cdots : 0 : -i\sqrt{\alpha - \beta} : -\sqrt{\alpha - \beta} : 1 : +i(\alpha - \beta)),$$
$$P_2 = (0 : \cdots : 0 : +i\sqrt{\alpha - \beta} : +\sqrt{\alpha - \beta} : 1 : +i(\alpha - \beta)),$$
$$P_3 = (0 : \cdots : 0 : +i\sqrt{\alpha - \beta} : -\sqrt{\alpha - \beta} : 1 : -i(\alpha - \beta)),$$
$$P_4 = (0 : \cdots : 0 : -i\sqrt{\alpha - \beta} : +\sqrt{\alpha - \beta} : 1 : -i(\alpha - \beta)),$$

and let q_δ denote the three roots of $Q(x)$. Then it is easily checked by picking local parameters around the points at infinity of $\Delta_{\epsilon_1 \epsilon_2}$ that the incidence relation of the P_i on the $\Delta_{\epsilon_1 \epsilon_2}$ is given by the following table:

	$q_1 \to 0$	$q_1 \to \infty$	$q_1 \to q_\delta$	$q_1 \to \epsilon_1 \epsilon_2 i \sqrt{\alpha - \beta}$
$\Delta_{+1,+1}$	P_1	P_4	$3P_3$	P_2
$\Delta_{-1,+1}$	P_2	P_3	$3P_4$	P_1
$\Delta_{+1,-1}$	P_3	P_2	$3P_1$	P_4
$\Delta_{-1,-1}$	P_4	P_1	$3P_2$	P_3

Table 9

The table is in agreement with the fact that each curve has a three-fold point and passes through the other singularities. Moreover it shows that the three points q_δ were identified under the map p_1 when going from J to T^2, hence these form the subset $\mathcal{W}_1$ in Proposition 4.3 and $\mathcal{W}_2 = \{0, \infty, \epsilon_1 \epsilon_2 i \sqrt{\alpha - \beta}\}$. If we substitute

$$x \mapsto \frac{x + \alpha}{\sqrt{\alpha - \beta}} i$$

in the equation (4.23) for the curves $\delta_{\epsilon_1 \epsilon_2}$ then we find the equation (4.11),

$$y^2 = (x + \alpha)(x + \beta)\left(x^3 + (\alpha + \beta)x^2 + (\alpha\beta - h)x + \frac{\beta f - \alpha g}{2(\alpha - \beta)}\right). \qquad (4.24)$$

Then the decomposition of $\mathcal{W}$ is given as follows: $\mathcal{W}_1$ contains the roots of $x^3 + (\alpha + \beta)x^2 + (\alpha\beta - h)x + (\beta f - \alpha g)/(2\alpha - 2\beta)$, and $\mathcal{W}_2 = \{\infty, -\alpha, -\beta\}$.

Suppose that $(T^2, \mathcal{L}) \in \tilde{\mathcal{A}}_{(1,4)}$ and let the surface be represented by a surface $\mathcal{F}_{(\alpha, \beta, f, g)}$, for some $\alpha \neq \beta$ (using (4.18)). Then the curve $\Gamma(T^2)$ corresponding to it under the basic bijection explained in Paragraph 4.3 must be smooth. Since we know from (4.24) that an equation for $\Gamma(T^2)$ is given by

$$y^2 = (x + \alpha)(x + \beta)P_3(x), \qquad P_3(x) = x^3 + (\alpha + \beta)x^2 + (\alpha\beta - h)x + \frac{\beta f - \alpha g}{2(\alpha - \beta)}, \qquad (4.25)$$

we conclude that $\mathrm{disc}(P_3(x)) \neq 0$ and $P_3(-\alpha) \neq 0$, $P_3(-\beta) \neq 0$, the last condition meaning just that $f \neq 0$ and $g \neq 0$. Conversely, both conditions together are sufficient to guaranty that the curve is smooth and the corresponding Abelian surface is in $\tilde{\mathcal{A}}_{(1,4)}$. In order to state this result in terms of the coordinates μ_i for $\mathbf{P}^3$, use (4.18) to rewrite (4.25) in the simple form $y^2 = x(x-1)P_3^\mu(x)$ where

$$P_3^\mu(x) = 4\mu_2 x^3 - (\mu_0 + 2\mu_1 + 6\mu_2 + 2\mu_3)x^2 + (\mu_0 - 2\mu_1 + 2\mu_2 + 6\mu_3)x - 4\mu_3,$$

(x and y are slightly rescaled); in this representation $\mathcal{W}_2 = \{0, 1, \infty\}$ and $\mathcal{W}_1$ contains the roots of $P_3^\mu(x)$. The condition for $(\mu_0 : \mu_1 : \mu_2 : \mu_3)$ to correspond to a surface in $\tilde{\mathcal{A}}_{(1,4)}$ is now that $\mu_1 \mu_2 \mu_3 \neq 0$ and $\mathrm{disc}(P_3^\mu(x)) \neq 0$. It shows that the locus S' is given by the four divisors $\mu_1 \mu_2 \mu_3 = 0$ and $\mathrm{disc}(P_3^\mu(x)) = 0$ and the exceptional locus S is found immediately from it by substituting λ_i^2 for μ_i in these equations[27]. Combining this with Proposition 4.1 we have shown the following proposition.

Proposition 4.6 *The surface $\mathcal{F}_{(\alpha, \beta, f, g)}$ is (isomorphic to) an affine part $T^2 \setminus \mathcal{D}$ of an Abelian surface $(T^2, [\mathcal{D}]) \in \tilde{\mathcal{A}}_{(1,4)}$ if and only if $\alpha \neq \beta$, $f \neq 0, g \neq 0$ and $\mathrm{disc}(P_3(x)) \neq 0$. Equivalently $(\mu_0 : \mu_1 : \mu_2 : \mu_3) \in \mathbf{P}^3$ are moduli coming from the birational map[28] $\phi_{\mathcal{L}} : T^2 \to \mathbf{P}^3$ with $(T^2, \mathcal{L}) \in \tilde{\mathcal{A}}_{(1,4)}$ if and only if $\mu_1 \mu_2 \mu_3 \neq 0$ and $\mathrm{disc}(P_3^\mu(x)) \neq 0$. The curve $\Gamma(T^2)$ corresponding to the canonical Jacobian of T^2 is then written as*

$$y^2 = x(x-1)\left(4\mu_2 x^3 - (\mu_0 + 2\mu_1 + 6\mu_2 + 2\mu_3)x^2 + (\mu_0 - 2\mu_1 + 2\mu_2 + 6\mu_3)x - 4\mu_3\right),$$

when the coordinates x for $\mathbf{P}^1$ is taken such that $\mathcal{W}_2 = \{0, 1, \infty\}$. Conversely the equation of the octic (IV.5.2) is written down at once when giving the equation of the genus two curve and a decomposition $W = \mathcal{W}_1 \cup \mathcal{W}_2$ of its set of Weierstrass points: the coefficients of the octic are $\lambda_i = \sqrt{\mu_i}$ where μ_i are essentially the symmetric functions of $\mathcal{W}_2$ when the coordinate x for $\mathbf{P}^1$ is taken such that $\mathcal{W}_2 = \{0, 1, \infty\}$.

Taking also the non-generic case into account, there is an Abelian surface $\mathcal{F}_{(\alpha, \beta, f, g)}$ corresponding to each point in the image $\psi\left(\mathcal{A}_{(1,4)}\right) = (\mathcal{M}^3 \setminus \mathcal{D}) \cup (C \setminus \{P, Q\})$. ∎

The following corollary follows at once from this proposition.

Corollary 4.7 *For any Abelian surface $(T^2, [\mathcal{D}]) \in \tilde{\mathcal{A}}_{(1,4)}$ the affine variety $T^2 \setminus \mathcal{D}$ is (isomorphic to) a complete intersection of two quartics in $\mathbf{C}^4$.*

Remarks 4.8

1. Recalling the description of $\tilde{\mathcal{A}}_{(1,4)}$ from Remark 4.3.2 one has the following description of the moduli space $\tilde{\mathcal{A}}_{(1,4)}$:

$$\tilde{\mathcal{A}}_{(1,4)} \cong \left\{(\{W_1, W_2, W_3\}, \{W_4, W_5, W_6\}) \mid W_i \in \mathbf{P}^1, \, i \neq j \Rightarrow W_i \neq W_j\right\} \Big/ \mathrm{mod}\ \mathbf{PGL}(2, \mathbf{C})$$

$$\cong \left\{\{W_4, W_5, W_6\} \mid W_i \in \mathbf{C} \setminus \{0, 1\}, \, i \neq j \Rightarrow W_i \neq W_j\right\} \Big/ S_3,$$

---

[27] These equations for S can in principle be found purely algebraic, but the calculations are very tedious and some cases are easily overlooked. In fact it is claimed (without proof) in [BLS] that the only condition is $\mu_1 \mu_2 \mu_3 \neq 0$, the more subtle condition $\mathrm{disc}(P_3^\mu(x)) \neq 0$ being overlooked.

[28] Recall that $\mu_i = \lambda_i^2$, where λ_i are taken from (IV.5.2).

where the action of S_3 consists of permuting $0, 1$ and ∞ in the equation $y^2 = x(x-1)(x - W_4)(x - W_5)(x - W_6)$, i.e., it is generated by replacing x by $1/x$ and $1 - x$ in this equation. Obviously the ring of invariants of the symmetric functions of W_4, W_5 and W_6 is just the cone $\mathcal{M}^3$, which explains why $\tilde{A}_{(1,4)}$ has such a nice structure. Using Tables 7 and 8, this leads to a geometric interpretation of the "intermediate" moduli space $\mathbf{P}^3 \setminus S'$, namely

$$\mathbf{P}^3 \setminus S' \cong \{\{W_4, W_5, W_6\} \mid W_i \in \mathbf{C} \setminus \{0,1\}, \ i \neq j \Rightarrow W_i \neq W_j\}.$$

To explain this, notice that taking the basis vectors mod 2 in the first column of Table 8 determines an ordering for the 4 half-periods on the canonical Jacobian which correspond to the lattice Λ_2, which in turn induce an ordering in the points in $\mathcal{W}_2$; on the other hand, all elements in the first column of Table 7 are the same mod 2.

2. In the classical literature one defines a *Rosenhain tetrahedron* for a Kummer surface as a tetrahedron in $\mathbf{P}^3$ with singular planes of the surface as faces and singular points of it as vertices. In [Hud] it is shown that the equation for the Kummer surface with respect to a Rosenhain tetrahedron is written as the quartic (4.5). It then follows from Proposition 4.6 how to read off from the equation of a Kummer surface with respect to a Rosenhain tetrahedron, an equation for the curve corresponding to this Kummer surface and vice versa. It seems that this result is not known in the classical or recent literature.

4.5. The central Garnier potentials

In this final paragraph we concentrate on the potentials $V_{\alpha\alpha}$ which were excluded up to now. It is interesting to compare the classical linearization of the central potential $V_{\alpha\alpha}$ which uses polar coordinates with the $\alpha = \beta$ limit of the linearization of the perturbed potential $V_{\alpha\beta}$ ($\alpha \neq \beta$): they will be seen to coincide. We will also construct a Lax pair for this limiting case and discuss the geometry of its level manifolds.

At first, consider for general values of h, k the level surface $\mathcal{F}_{hk}$ defined by

$$\mathcal{F}_{hk} : \begin{cases} h = \dfrac{1}{2}\left(p_1^2 + p_2^2\right) + \left(q_1^2 + q_2^2\right)^2 + \alpha\left(q_1^2 + q_2^2\right), \\ k = q_1 p_2 - q_2 p_1, \end{cases}$$

which in terms of polar coordinates (ρ, θ) becomes

$$h = \frac{1}{2}\left(\dot{\rho}^2 + \rho^2\dot{\theta}^2\right) + \rho^4 + \alpha\rho^2,$$
$$k = \rho^2\dot{\theta},$$

leading to

$$-\frac{1}{2}\rho^2\dot{\rho}^2 = \rho^6 + \alpha\rho^4 - h\rho^2 + \frac{k^2}{2}.$$

This suggests setting $\sigma = \rho^2$, yielding

$$-\frac{\dot{\sigma}^2}{8} = \sigma^3 + \alpha\sigma^2 - h\sigma + \frac{k^2}{2}. \tag{4.26}$$

216

4. The Garnier potential

Secondly the transformation (4.9) reduces for $\alpha = \beta$ to

$$\begin{aligned} x_1 + x_2 &= -\left(q_1^2 + q_2^2 + 2\alpha\right), \\ x_1 x_2 &= \alpha^2 + \alpha q_1^2 + \alpha q_2^2, \end{aligned} \tag{4.27}$$

and (4.10) becomes

$$\dot{x}_i^2 = \frac{8(x_i + \alpha)^2\left(x_i^3 + 2\alpha x_i^2 + (\alpha^2 - h)x_i - (h\alpha + f/2)\right)}{(x_1 - x_2)^2}. \tag{4.28}$$

The equivalence of (4.26) and (4.28) becomes clear after the simple translation $x_i = x_i + \alpha$ on the curve; indeed (4.27) becomes

$$\begin{aligned} s_1 + s_2 &= -\left(q_1^2 + q_2^2\right), \\ s_1 s_2 &= 0, \end{aligned}$$

so that only one of the s_i differs from zero, say $0 \neq s_1 = -(q_1^2 + q_2^2) = -s$, (the last equality is a definition), which matches the linearizing variable σ introduced above. In terms of s (4.28) is reduced to one equation which reads

$$-\frac{\dot{s}^2}{8} = s^3 + \alpha s^2 - hs + \frac{f}{2},$$

which is exactly (4.26) since $f = (q_1 p_2 - q_2 p_1)^2 = k^2$.

It is also interesting that the Lax pair gives in the limit $\alpha = \beta$ a Lax pair for the potential $V_{\alpha\alpha}$. The polynomials $u(\lambda), v(\lambda)$ and $w(\lambda)$ are now all divisible by $(\lambda + \alpha)$,

$$\begin{aligned} u(\lambda) &= (\lambda + \alpha)\left(\lambda + q_1^2 + q_2^2 + \alpha\right), \\ v(\lambda) &= \frac{1}{\sqrt{2}}(\lambda + \alpha)\left(q_1 p_1 + q_2 p_2\right), \\ w(\lambda) &= (\lambda + \alpha)\left(\lambda^2 + (\alpha - q_1^2 - q_2^2)\lambda - \frac{1}{2}\left(p_1^2 + p_2^2\right) - \alpha\left(q_1^2 + q_2^2\right)\right), \end{aligned}$$

which leads to a simpler Lax pair by canceling the factor $(\lambda + \alpha)$.

Finally we describe the level surfaces for the central potentials $V_{\alpha\alpha}$. These turn out to be $\mathbf{C}^*$-bundles over the elliptic curves (4.26), as described in the following proposition.

Proposition 4.9 *For any $k, h \in \mathbf{C}$, let $\mathcal{F}_{hk}$ denote the affine surface defined by*

$$\mathcal{F}_{hk}: \begin{cases} h = \dfrac{1}{2}\left(p_1^2 + p_2^2\right) + \left(q_1^2 + q_2^2\right)^2 + \alpha\left(q_1^2 + q_2^2\right), \\ k = q_1 p_2 - q_2 p_1. \end{cases} \tag{4.29}$$

If $k \neq 0$ then $\mathcal{F}_{hk}$ is a $\mathbf{C}^*$-bundle over the elliptic curve

$$E_{hk}: -\frac{\tau^2}{2} = \sigma^3 + \alpha\sigma^2 - h\sigma + \frac{k^2}{2}. \tag{4.30}$$

Moreover the $\mathbf{C}^*$-action on $\mathcal{F}_{hk}$ is a Hamiltonian action, the Hamiltonian function corresponding to it being the momentum $q_1 p_2 - q_2 p_1$.

Proof

The linearizing variables, calculated above suggest to consider the map $\xi : \mathbf{C}^4 \to \mathbf{C}^2$ given by

$$(q_1, q_2, p_1, p_2) \mapsto (\sigma, \tau) = \left(q_1^2 + q_2^2, q_1 p_1 + q_2 p_2\right).$$

Our first claim is that the image $\xi(\mathcal{F}_{hk})$ is given by the plane elliptic curve (4.30). Indeed, one easily obtains for $q_1^2 + q_2^2 \neq 0$,

$$p_1 = \frac{q_2 k - q_1 \tau}{q_1^2 + q_2^2},$$

$$p_2 = -\frac{q_1 k + q_2 \tau}{q_1^2 + q_2^2},$$

which leads by direct substitution in the first equation of (4.29) immediately to

$$-\frac{\tau^2}{2} = \sigma^3 + \alpha \sigma^2 - h\sigma + \frac{k^2}{2}.$$

For $q_1^2 + q_2^2 = 0$, i.e., $q_2 = \pm i q_1$ one gets

$$h = \frac{1}{2}(p_1^2 + p_2^2),$$

$$k = q_1(p_2 \mp i p_1),$$

$$\tau = q_1(p_1 \pm i p_2),$$

from which we deduce $\tau = \pm i k$, giving the point $(\sigma, \tau) = (0, \pm i k)$ on E_{hk}, proving the first claim.

Secondly, we determine the fiber $\xi^{-1}(\sigma, \tau)$ over each point on E_{hk}. To do this, observe that the multiplicative group of non-zero complex numbers,

$$\mathbf{C}^* \cong \mathbf{SO}(2, \mathbf{C}) = \left\{ \begin{pmatrix} a & b \\ -b & a \end{pmatrix} \mid a^2 + b^2 = 1 \right\}$$

acts on $\mathcal{F}_{hk}$ by

$$\begin{pmatrix} a & b \\ -b & a \end{pmatrix} \cdot \begin{pmatrix} q_1 & p_1 \\ q_2 & p_2 \end{pmatrix} = \begin{pmatrix} aq_1 + bq_2 & ap_1 + bp_2 \\ aq_2 - bq_1 & ap_2 - bp_1 \end{pmatrix}$$

and the surjective map ξ is $\mathbf{C}^*$-invariant. It is proved by direct calculation that the action is free, hence each fiber of ξ consists of one or more $\mathbf{C}^*$'s. If $(\sigma, \tau) \in E_{hk}$ then p_1 and p_2 are determined from q_1 and q_2 (at least if $q_1^2 + q_2^2 \neq 0$), which themselves are determined (up to the action of $\mathbf{C}^*$) by $q_1^2 + q_2^2 = \rho$, so exactly one $\mathbf{C}^*$ lies over each point (q_1, q_2, p_1, p_2) for which $q_1^2 + q_2^2 \neq 0$; in the special case that $q_1^2 + q_2^2 = 0$, the same is true, since p_1 and p_2 are determined (up to the action of $\mathbf{C}^*$) by $p_1^2 + p_2^2 = 2h$, and q_1, q_2 are uniquely determined from p_1 and p_2. It follows that $\mathcal{F}_{hk}$ is a $\mathbf{C}^*$-bundle over the elliptic curve E_{hk}.

Finally, observe that the Hamiltonian vector field corresponding to the momentum $q_1 p_2 - q_2 p_1$ is given by

$$\dot{q}_1 = -q_2, \qquad \dot{p}_1 = -p_2,$$

$$\dot{q}_2 = q_1, \qquad \dot{p}_2 = p_1,$$

from which it is seen that the complex flow of this vector field is given by the $\mathbf{C}^*$-action, proving the last claim in the proposition. ∎

Let us define (and calculate) the moduli (in $\mathbf{P}^{(1,2,2,3,4)}$) corresponding to a level surface $\mathcal{F}_{hk}$ of a central potential for $k \neq 0$ as the limit[29]

$$\lim_{\alpha \to \beta} \tilde{\psi}(\mathcal{T}^2_{(\alpha,\beta,f,g)}), \qquad f = k^2.$$

Then an easy computation shows that this limit exists, is independent of $f \neq 0$, h and $\alpha = \beta$ and moreover is exactly equal to the special point P at the boundary of $\psi\left(\mathcal{A}_{(1,4)}\right)$ defined in Proposition 4.2. Namely for $f \to g$ and $\alpha \to \beta$ one finds

$$(\mu_0 : \mu_1 : \mu_2 : \mu_3) = (-4 : 1 : 0 : 1)$$

so that

$$(f_0, f_1, f_2, f_3, f_4) = (-4 : 0 : 3 : -2 : 0)$$

hence by weight homogeneity the associated moduli correspond to P. Notice that the point is independent of $\alpha = \beta$ as well as of $f = g$, so the map ψ does not distinguish between any of the level surfaces of any central potential $V_{\alpha\alpha}$.

---

[29] Recall that $f - g = 2(\beta - \alpha)h$.

5. An integrable geodesic flow on SO(4)

5.1. The geodesic flow on SO(4) for metric II

It was shown by Adler and van Moerbeke (unpublished proof) that there exist three classes of left-invariant metrics on $SO(4)$ for which the geodesic flow reduces to an algebraic completely integrable system (a.c.i. system) on its Lie algebra $\mathfrak{so}(4)$. In the sequel, we will consider the second case, known as the case of metric II. In suitable coordinates, the first vector field $\mathcal{X}_1$ of this a.c.i. system is given by the differential equations

$$\dot{z}_1 = 2z_5z_6, \quad \dot{z}_2 = 2z_3z_4, \quad \dot{z}_3 = z_5(z_1 + z_4),$$
$$\dot{z}_4 = 2z_2z_3, \quad \dot{z}_5 = z_3(z_1 + z_4), \quad \dot{z}_6 = 2z_1z_5. \tag{5.1}$$

The second vector field $\mathcal{X}_2$, commuting with $\mathcal{X}_1$, is given by the differential equations

$$z_1' = z_2z_6, \quad z_2' = z_4(2z_3 - z_6), \quad z_3' = z_4z_5,$$
$$z_4' = z_2(2z_3 - z_6), \quad z_5' = z_3z_4, \quad z_6' = z_1z_2; \tag{5.2}$$

the vector fields $\mathcal{X}_1$ and $\mathcal{X}_2$ admit four independent quadratic invariants, given by the following functions:

$$\begin{aligned}
F_1 &= z_3^2 - z_5^2, \\
F_2 &= z_1^2 - z_6^2, \\
F_3 &= z_2^2 - z_4^2, \\
F_4 &= (z_1 + z_4)^2 + 4(z_3^2 - z_2z_5 - z_3z_6).
\end{aligned} \tag{5.3}$$

It is easy to verify that there exist precisely three linearly independent linear Poisson structures on $\mathbf{C}^6$ with respect to which $\mathcal{X}_1$ and $\mathcal{X}_2$ are Hamiltonian; moreover, these Poisson structures are compatible, implying that the integrable system admits a tri-Hamiltonian structure. Explicitly, for any $(\alpha, \beta, \gamma) \in \mathbf{C}^3$, the matrix

$$\begin{pmatrix}
0 & \alpha z_6 & -\beta z_5 & 0 & -\beta z_3 - 2\gamma z_6 & \beta(z_2 - 2z_5) \\
-\alpha z_6 & 0 & 2\gamma z_4 & \alpha(z_6 - 2z_3) & 0 & -\alpha z_1 - \beta z_4 \\
\beta z_5 & -2\gamma z_4 & 0 & -\alpha z_5 - 2\gamma z_2 & -\gamma(z_1 + z_4) & 0 \\
0 & \alpha(2z_3 - z_6) & \alpha z_5 + 2\gamma z_2 & 0 & \alpha z_3 & -\beta z_2 \\
\beta z_3 + 2\gamma z_6 & 0 & \gamma(z_1 + z_4) & -\alpha z_3 & 0 & 2\gamma z_1 \\
\beta(2z_5 - z_2) & \alpha z_1 + \beta z_4 & 0 & \beta z_2 & -2\gamma z_1 & 0
\end{pmatrix}$$

is the Poisson matrix of a Poisson structure $P_{\alpha\beta\gamma}$ on $\mathbf{C}^6$. If $(\alpha, \beta, \gamma) \neq (0, 0, 0)$ then $P_{\alpha\beta\gamma}$ generates the Hamiltonian vector fields $\mathcal{X}_1$ and $\mathcal{X}_2$ as described in the following table; a system of generators of the algebra of Casimirs of these structures $P_{\alpha\beta\gamma}$ also follow from the table.

	F_1	F_2	F_3	F_4
P_{100}	0	0	$2\mathcal{X}_2$	$-2\mathcal{X}_1$
P_{010}	0	$2(\mathcal{X}_1 - \mathcal{X}_2)$	0	$2\mathcal{X}_1$
P_{001}	$2\mathcal{X}_1$	0	0	$8\mathcal{X}_2$

Table 10

5. Geodesic flow on SO(4)

It was shown by Adler and van Moerbeke in [AM7] that, for any $f = (f_1, f_2, f_3, f_4)$ which belongs to some[30] Zariski open subset $\mathcal{H}$ of $\mathbf{C}^4$, the affine surface

$$\mathcal{A}_f = \{z \in \mathbf{C}^6 \mid F_i(z) = g_i, \ i = 1, \ldots, 4\}$$

is isomorphic to an affine part of the Jacobian of a compact Riemann surface $\bar{\Gamma}_f$ of genus two (which depends on $f \in \mathcal{H}$), $\mathcal{A}_f \cong \text{Jac}(\bar{\Gamma}_f) \setminus \mathcal{D}_f$ and that the vector fields $\mathcal{X}_1$ and $\mathcal{X}_2$ are linear, thereby proving that the above system is algebraic completely integrable. The affine part $\mathcal{A}_f$, the divisor $\mathcal{D}_f$ and the Riemann surface $\bar{\Gamma}_f$ can be described as follows. First notice that the group $\mathfrak{T}$ of involutions, generated by

$$\begin{aligned}
\sigma_1(z_1, \ldots, z_6) &= (-z_1, -z_2, z_3, -z_4, -z_5, z_6), \\
\sigma_2(z_1, \ldots, z_6) &= (-z_1, z_2, -z_3, -z_4, z_5, -z_6),
\end{aligned} \tag{5.4}$$

commutes with the vector fields $\mathcal{X}_1$ and $\mathcal{X}_2$ and leaves the affine surfaces $\mathcal{A}_f$ invariant; in fact they generate, for any $f \in \mathcal{H}$, a group $\mathfrak{T}_f$ of translations over half periods in the tori $\text{Jac}(\bar{\Gamma}_f)$. As a consequence, the divisors $\mathcal{D}_f$ are also stable under these translations. For a more precise description of the divisors $\mathcal{D}_f$ one applies Painlevé analysis to the vector field $\mathcal{X}_1$ (or any combination of $\mathcal{X}_1$ and $\mathcal{X}_2$). It has has precisely four principal balances, labeled by $\epsilon_1 = \pm 1$, $\epsilon_2 = \pm 1$, whose first few terms are explicitly given as follows ($a, b, \ldots, e$ are the free parameters).

$$\begin{aligned}
z_1 &= \frac{(a-1)\epsilon_1}{t}\left(1 - bt + (b^2 - d - e)t^2 + O(t^3)\right), \\
z_2 &= \frac{\epsilon_1\epsilon_2}{t}\left(a - abt + ((a-1)(ae - c - ab^2) + a^2d)t^2 + O(t^3)\right), \\
z_3 &= \frac{\epsilon_2}{2t}\left(1 + bt - ((a-1)e + ad - c - ab^2)t^2 + O(t^3)\right), \\
z_4 &= \frac{\epsilon_1}{t}\left(-a + abt + ct^2 + O(t^3)\right), \\
z_5 &= \frac{\epsilon_1\epsilon_2}{2t}\left(1 + bt + dt^2 + O(t^3)\right), \\
z_6 &= \frac{(a-1)\epsilon_2}{t}\left(-1 + bt - et^2 + O(t^3)\right).
\end{aligned} \tag{5.5}$$

When any of these families of Laurent solutions is substituted in the equations $F_i(z) = f_i$, $i = 1, \ldots, 4$, the resulting expressions are independent of t. This leads to four algebraic equations in the five free parameters, giving explicit equations for an affine part Γ_f of $\bar{\Gamma}_f$. Each of these equations is easily rewritten as

$$y^2 = x(1-x)\left[4x^3 f_1 - (4f_1 + f_4)x^2 + (f_4 - f_3 - f_2)x + f_3\right]. \tag{5.6}$$

In what follows, we will refer to the curve in $\mathbf{C}^2$, given by (5.6), as the curve Γ_f. In order to recover the Riemann surface $\bar{\Gamma}_f$ from it one has to adjoin one point which we denote by ∞_f. Since there are four families of Laurent solutions (5.5), the divisor $\mathcal{D}_f$ consists of four copies $\bar{\Gamma}_f(\epsilon_1, \epsilon_2)$, $\epsilon_1^2 = \epsilon_2^2 = 1$, of the curve $\bar{\Gamma}_f$, i.e.,

$$\mathcal{D}_f = \bar{\Gamma}_f(1,1) + \bar{\Gamma}_f(1,-1) + \bar{\Gamma}_f(-1,1) + \bar{\Gamma}_f(-1,-1).$$

---

[30] Explicit equations for $\mathcal{H}$ will be given in the next paragraph.

The Laurent solutions can also be used to compute an explicit embedding of the tori $\mathrm{Jac}(\bar{\Gamma}_f)$ in $\mathbf{P}^{15}$: the sections of the line bundle on $\mathrm{Jac}(\bar{\Gamma}_f)$, defined by $\mathcal{D}_f$, correspond to the meromorphic functions on $\mathrm{Jac}(\bar{\Gamma}_f)$ with a simple pole (at worst) at the divisor $\mathcal{D}_f$ and, in turn, these are found by constructing those polynomials on $\mathbf{C}^6$ which have a simple pole in t (at worst) when any of the four families of Laurent solutions are substituted in them (see Chapter V). Apart from the constant function $z_0 = 1$ and the functions z_i, $i = 1, \ldots, 6$, one easily finds the following independent functions with this property:

$$
\begin{aligned}
z_7 &= z_5(2z_3 - z_6) - z_2 z_3, \\
z_8 &= z_1(2z_3 - z_6) - z_4 z_6, && z_{12} = z_1 z_2 z_3 - z_4 z_5 z_6, \\
z_9 &= z_4(2z_5 - z_2) - z_1 z_2, && z_{13} = z_2 z_3 z_6 - z_1 z_4 z_5, \\
z_{10} &= (2z_5 - z_2)^2 - z_6^2, && z_{14} = z_2 z_5 z_6 - z_1 z_3 z_4, \\
z_{11} &= (2z_3 - z_6)^2 - z_2^2, && z_{15} = z_1 z_2 z_5 - z_3 z_4 z_6.
\end{aligned}
\tag{5.7}
$$

The embedding of $\mathrm{Jac}(\bar{\Gamma}_f)$ in $\mathbf{P}^{15}$ is given on the affine part $\mathcal{A}_f$ by the map

$$
\phi : \mathcal{A}_f \to \mathbf{P}^{15} : P = (z_1, \ldots, z_6) \mapsto (1 : z_1(P) : \cdots : z_{15}(P)).
$$

These functions will be used later to construct two maps which are similar to ϕ and which map two different quotients of $\mathcal{A}_f$ birationally into $\mathbf{P}^3$.

5.2. Linearizing variables

In this paragraph we show that from the point of view of moduli, the family of affine surfaces $\mathcal{A}_f$, $f \in \mathcal{H}$, can be replaced by a family of polarized Abelian surfaces of type $(1, 4)$. In order to do this we will first construct an explicit map from the affine surface $\mathcal{A}_f$ ($f \in \mathcal{H}$) to an affine part of $\mathrm{Jac}(\bar{\Gamma}_f)$. Following [BV4] we do this by following the algorithm which was outlined in Section V.4.

We define $\mathcal{H}$ to be the set of those $f = (f_1, f_2, f_3, f_4) \in \mathbf{C}^4$ for which the curve (5.6) is a non-singular curve of genus two, i.e., that its right hand side is of degree 5 and has no multiple roots; notice that this entails in particular that $f_1 f_2 f_3 \neq 0$ for all $f \in \mathcal{H}$. It will follow from our construction that, for every $f \in \mathcal{H}$, $\mathcal{A}_f$ is indeed an affine part of the Jacobian, thereby justifying the notation $\mathcal{H}$. In order to apply the procedure described in Section V.4, we fix an arbitrary element $f \in \mathcal{H}$ and we choose one component, say $C = \bar{\Gamma}_f(1, -1)$, of the divisor $\mathcal{D}_f$ on $\mathrm{Jac}(\bar{\Gamma}_f)$. The meromorphic functions on $\mathrm{Jac}(\bar{\Gamma}_f)$ which have at worst a double pole along the divisor C can be obtained by constructing those polynomials on $\mathbf{C}^6$ which have at worst a double pole in t when the Laurent solutions (5.5) corresponding to $\epsilon_1 = 1$, $\epsilon_2 = -1$ are substituted into them (and no poles when the other solutions are substituted). It is easily computed that the space of such polynomials is spanned by

$$
\chi_0 = 1, \quad \chi_1 = (z_2 + z_4)(z_3 + z_5), \quad \chi_2 = (z_3 + z_5)(z_1 + z_6), \quad \chi_3 = (z_1 + z_6)(z_2 + z_4), \tag{5.8}
$$

where we think of these polynomials as being restricted to $\mathcal{A}_f$. The mapping ϕ, given on $\mathrm{Jac}(\bar{\Gamma}_f) \setminus C$ by

$$
\phi : \mathrm{Jac}(\bar{\Gamma}_f) \setminus C \to \mathbf{P}^3 : P = (z_1, z_2, \ldots, z_6) \mapsto (\chi_0(P) : \chi_1(P) : \chi_2(P) : \chi_3(P))
$$

maps the surface $\mathrm{Jac}(\bar{\Gamma}_f)$ to its Kummer surface, which is a singular quartic in $\mathbf{P}^3$. An equation for this quartic surface can be computed by eliminating the variables $z_1, \ldots, z_6$ from the equations (5.3) and (5.8): solving the equations (5.8) and the first three equations in (5.3) for the variables $z_1, z_2, \ldots, z_6$ and substituting these values in the remaining equation, the equation for the Kummer surface of $\mathrm{Jac}(\bar{\Gamma}_f)$ can be written in the form

$$\chi_3^2((\chi_1 + \chi_2 - 2f_1)^2 + 8f_1\chi_1) + f_3(\chi_1, \chi_2)\chi_3 + f_4(\chi_1, \chi_2) = 0, \qquad (5.9)$$

where f_3 (respectively f_4) is a polynomial of degree three (respectively four) in χ_1 and χ_2.

It follows from (5.9) that a system of linearizing variables (x_1, x_2) is given by the equations

$$-2f_1(x_1 + x_2) = \chi_1 + \chi_2 - 2f_1, \qquad -2f_1 x_1 x_2 = \chi_1. \qquad (5.10)$$

This is checked in the present case as follows. First make use of (5.8), to rewrite the equations (5.10) as

$$(z_3 + z_5)(z_2 + z_4) = -2f_1 x_1 x_2, \qquad (z_3 + z_5)(z_1 + z_6) = 2f_1(x_1 - 1)(x_2 - 1). \qquad (5.11)$$

Since $f \in \mathcal{H}$ the variables x_1 and x_2 are both different from 1 and from 0 so that below we can divide by x_i and by $x_i - 1$ as necessary. Deriving the equations (5.11) with respect to the vector field $\mathcal{X}_1$ given by (5.1) we find that

$$\begin{aligned} \dot{x}_1 x_1^{-1} + \dot{x}_2 x_2^{-1} &= z_1 + z_4 + 2z_3, \\ \dot{x}_1(x_1 - 1)^{-1} + \dot{x}_2(x_2 - 1)^{-1} &= z_1 + z_4 + 2z_5. \end{aligned} \qquad (5.12)$$

Then we can solve the first three equations of (5.3), together with (5.11) and the difference of the two equations in (5.12) for $z_1, \ldots, z_6$. Substituting these values in the second equation of (5.12) we find that

$$\left(\frac{\dot{x}_1}{x_1(x_1 - 1)}\right)^2 - \left(\frac{\dot{x}_2}{x_2(x_2 - 1)}\right)^2 = \frac{1}{x_1 - x_2}\left[4f_1 + \frac{f_2}{(x_1 - 1)(x_2 - 1)} + \frac{f_3}{x_1 x_2}\right]. \qquad (5.13)$$

Notice that this equation is linear in $\dot{x}_1^2$ and $\dot{x}_2^2$. Finally we substitute the values for $z_1, \ldots, z_6$ in the fourth equation of (5.3) to find another equation which is linear in $\dot{x}_1^2$ and $\dot{x}_2^2$, leading to

$$\dot{x}_i^2 = \frac{g(x_i)}{(x_1 - x_2)^2}, \qquad (i = 1, 2),$$

where

$$g(x) = x(1 - x)[4f_1 x^3 - (4f_1 + f_4)x^2 + (f_4 - f_2 - f_3)x + f_3].$$

(We note that the curve $y^2 = g(x)$ is precisely the curve Γ_f given by (5.6).) It follows that, in terms of the coordinates x_1, x_2 given by (5.10), the differential equations (5.1) reduce to the Jacobi form

$$\frac{\dot{x}_1}{\sqrt{g(x_1)}} + \frac{\dot{x}_2}{\sqrt{g(x_2)}} = 0, \qquad \frac{x_1 \dot{x}_1}{\sqrt{g(x_1)}} + \frac{x_2 \dot{x}_2}{\sqrt{g(x_2)}} = 1,$$

so that x_1 and x_2 are indeed linearizing variables.

The construction of these linearizing variables leads to an explicit map into the Jacobian $\text{Jac}(\bar{\Gamma}_f)$ by defining

$$u(x) = x^2 + \left(\frac{z_1 + z_2 + z_4 + z_6}{2(z_3 - z_5)} - 1\right)x - \frac{z_2 + z_4}{2(z_3 - z_5)}. \tag{5.14}$$

and by defining a polynomial v of degree at most 1 as follows

$$v(0) = u(0)(z_1 + z_4 + 2z_3), \quad v(1) = u(1)(z_1 + z_4 + 2z_5). \tag{5.15}$$

Indeed, $g(x) - v^2(x)$ is divisible by $u(x)$, as can be checked by a direct computation, so that the above formulas indeed define a point of $\text{Jac}(\bar{\Gamma}_f) \setminus \Theta_f$, where Θ_f is (a translate of) the theta divisor of $\text{Jac}(\bar{\Gamma}_f)$. Notice that as such this does not define a map to the odd Mumford system because g is not monic. Since $f \in \mathcal{H}$, $f_1 \neq 0$ and hence $z_3 - z_5 \neq 0$, showing that the above map is regular; moreover it is birational because (5.15) gives

$$z_3 - z_5 = \frac{1}{2}\left(\frac{v(0)}{u(0)} - \frac{v(1)}{u(1)}\right), \tag{5.16}$$

while, using (5.14), $z_2 + z_4$ and $z_1 + z_6$ can be rewritten as follows:

$$\begin{aligned} z_2 + z_4 &= \left(\frac{v(1)}{u(1)} - \frac{v(0)}{u(0)}\right)u(0), \\ z_1 + z_6 &= \left(\frac{v(0)}{u(0)} - \frac{v(1)}{u(1)}\right)u(1). \end{aligned} \tag{5.17}$$

Using the invariants F_1, F_2 and F_3 one easily finds formulas for $z_3 + z_5$, $z_2 - z_4$ and $z_1 - z_6$ showing that the map is birational. On the one hand this proves that when $f \in \mathcal{H}$, i.e., when Γ_f is a non-singular curve of genus two, then $\mathcal{A}_f$ is isomorphic to an affine part of $\text{Jac}(\bar{\Gamma}_f)$. On the other hand it leads to explicit solutions for (5.1) with respect to initial conditions which correspond to a point $f \in \mathcal{H}$, in terms of theta functions,

$$u(0) = c_0\left(\frac{\vartheta[\delta_0](At + B)}{\theta[\delta](At + B)}\right)^2, \quad u(1) = c_1\left(\frac{\vartheta[\delta_1](At + B)}{\theta[\delta](At + B)}\right)^2,$$

as follows from (VI.4.9). The expressions for $v(0)$ and $v(1)$ in terms of theta functions follow from it because they are the derivatives of $u(0)$ and $u(1)$ with respect to t.

We see that the inverse map, given by (5.16) and (5.17), is holomorphic away from the divisors $u(0) = 0$, $u(1) = 0$ and $u(1)v(0) - u(0)v(1) = 0$. When $u(0) = 0$ then 0 is one of the roots of u so that the corresponding divisors are of the form $W_0 + P$, where W_0 stands for the Weierstrass point over 0, $x(W_0) = 0$ and $P \in \bar{\Gamma}_f$. Similarly, $u(1) = 0$ corresponds to the divisors $W_1 + P$, where W_1 stands for the Weierstrass point over 1. In order to avoid a rather involved explicit computation for the third divisor we appeal to the fact that the divisor at infinity $\mathcal{D}_f$ is invariant for the group $\mathfrak{T}_f$. Knowing that $\mathcal{D}_f$ consists of the theta divisor (consisting of divisors $\infty_f + P$) besides the two divisors that we have just determined we can identify the elements of $\mathfrak{T}_f$ as translations over $[W_1 - W_0]$, $[\infty_f - W_1]$ and $[W_0 - \infty_f]$. Thus, the divisor $u(1)v(0) - u(0)v(1) = 0$ corresponds to the effective divisors in $[W_0 + W_1 + P - \infty_f]$. It is now easy to see that the four points $2\infty_f$, $\infty_f + W_0$, $\infty_f + W_1$ and $W_0 + W_1$ (which constitute a single $\mathfrak{T}_f$ orbit) each belong to exactly three of the four curves and that these four curves have no other intersection points. Thus, as a byproduct, we have recovered[31] the following intersection pattern of the components of the divisor $\mathcal{D}_f$.

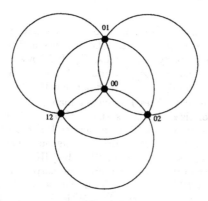

Figure 8

We will now use the above results to study the moduli space $\mathcal{M}$ defined by

$$\mathcal{M} = \{\mathcal{A}_f \mid f \in \mathcal{H}\}/\text{isomorphism},$$

where isomorphism means isomorphism of affine algebraic surfaces. We will relate this moduli space to the cone $\mathcal{M}^3$, introduced in Section 4. In the following two propositions we show how $\mathcal{M}$ and $\mathcal{M}_{(1,4)}$ are related.

Proposition 5.1 *For any $f \in \mathcal{H}$ the quotient $\mathcal{A}_f/\mathfrak{T}_f$ is an affine part of an Abelian surface $\mathcal{T}_f$. The line bundle $\mathcal{L}_f = [\mathcal{D}_f/\mathfrak{T}_f]$ induces a polarization of type $(1,4)$ on $\mathcal{T}_f$ and the induced map $\phi_{\mathcal{L}_f} : \mathcal{T}_f \to \mathbf{P}^3$ is birational to its image.*

Proof
    We have shown in Proposition 4.3 that the quotient of a Jacobi surface by a group of translations of the form $\{0, [W_2 - W_1], [W_1 - W_0], [W_0 - W_2]\}$, where W_0, W_1 and W_2 are Weierstrass points on the underlying curve, is an Abelian surface of type $(1,4)$, more precisely it belongs to $\tilde{\mathcal{A}}_{(1,4)}$. The divisor $\mathcal{D}_f$ descends to the irreducible divisor $\mathcal{D}_f/\mathfrak{T}_f$ which has a triple point which corresponds to the singular points of $\mathcal{D}_f$. Since $\mathcal{D}_f$ induces a polarization of type $(4,4)$ on $\text{Jac}(\bar{\Gamma}_f)$, $\mathcal{D}_f/\mathfrak{T}_f$ induces a polarization of type $(1,4)$ on $\mathcal{T}_f$. In order to see that the induced map $\phi_{\mathcal{L}_f}$ is birational onto its image one considers $\mathcal{T}_f/K_f$ where K_f is the group of two-torsion elements inside the kernel of the natural isogeny from $\mathcal{T}_f$ to its dual Abelian surface $\hat{\mathcal{T}}_f$. Since $\mathcal{T}_f = \text{Jac}(\bar{\Gamma}_f)/\mathfrak{T}_f$ the map $\text{Jac}(\bar{\Gamma}_f) \to \mathcal{T}_f/K_f$ is an isogeny whose kernel consists of the sixteen half periods of $\text{Jac}(\bar{\Gamma}_f)$. This means that this isogeny is multiplication by 2 in $\text{Jac}(\bar{\Gamma}_f)$ and hence that $\mathcal{T}_f/K_f$ is a Jacobi surface. This implies that the map $\phi_{\mathcal{L}_f} : \mathcal{T}_f \to \mathbf{P}^3$ is birational to its image. ∎

---

[31] This intersection pattern was first determined in [AM9] by using the Laurent solutions to the vector field $\mathcal{X}_1$.

Proposition 5.2 *The above correspondence between affine surfaces $\mathcal{A}_f$ and Abelian surfaces $\mathcal{T}$ induces a bijection $\chi : \mathcal{M} \to \tilde{\mathcal{A}}_{(1,4)}$.*

Proof

For $f \in \mathcal{H}$ we know that $\mathfrak{T}_f$ is a group of four translations of $\mathrm{Jac}(\bar{\Gamma}_f)$ over half periods leaving $\mathcal{D}_f$ invariant. Since the group of translations over half periods acts transitively on the set of theta curves this property characterizes $\mathfrak{T}_f$. It follows that isomorphic surfaces $\mathcal{A}_f$ and $\mathcal{A}_k$ lead to isomorphic quotients $\mathcal{A}_f/\mathfrak{T}_f$ and $\mathcal{A}_k/\mathfrak{T}_k$ and hence to isomorphic polarized Abelian surfaces $(\mathcal{T}_f, \mathcal{L}_f)$ and $(\mathcal{T}_k, \mathcal{L}_k)$. This shows that the given correspondence between affine surfaces $\mathcal{A}_f$ and Abelian surfaces $\mathcal{T}$ induces a map $\chi : \mathcal{M} \to \tilde{\mathcal{A}}_{(1,4)}$.

Starting from any polarized Abelian surface $(\mathcal{T}, \phi_\mathcal{L})$ of type $(1, 4)$ for which the induced map is birational there exists a Riemann surface $\bar{\Gamma}$ and a partition $\mathcal{W} = \mathcal{W}_1 \cup \mathcal{W}_2 = \{W_0, W_1, W_2\} \cup \{W_3, W_4, W_5\}$ of its Weierstrass points such that $\mathcal{T} = \mathrm{Jac}(\bar{\Gamma})/\mathfrak{T}$, where $\mathfrak{T}$ is the group of translations, given by $\mathfrak{T} = \{0, [W_0 - W_1], [W_1 - W_2], [W_2 - W_0]\}$. Moreover the triple $(\bar{\Gamma}, \mathcal{W}_1, \mathcal{W}_2)$ is uniquely determined up to isomorphism (see Proposition 4.3). Let us pick one particular triple $(\bar{\Gamma}, \mathcal{W}_1, \mathcal{W}_2)$ and let us choose coordinates for $\mathbf{P}^1$ such that the image of $\mathcal{W}_1$ under the natural double cover $\bar{\Gamma} \to \mathbf{P}^1$ is given by $0, 1$ and ∞ (in some order). Then we find an equation of the form

$$y^2 = x(1 - x)(Ax^3 + Bx^2 + Cx + D)$$

in which the right hand side has no double roots. Obviously we can find then at least one $f \in \mathcal{H}$ such that this above curve corresponds to the curve Γ_f, given by (5.6). By construction (the isomorphism class of) the affine surface $\mathcal{A}_f$ is contained in the fiber $\chi^{-1}(\mathcal{T}, \mathcal{L})$, showing the surjectivity of χ. Finally, a triple $(\bar{\Gamma}', \mathcal{W}_1', \mathcal{W}_2')$ which is isomorphic to $(\bar{\Gamma}, \mathcal{W}_1, \mathcal{W}_2)$ leads to an isomorphic surface $\mathcal{A}_k$ because $\mathcal{A}_f$ is intrinsically described in terms of the triple $(\bar{\Gamma}, \mathcal{W}_1, \mathcal{W}_2)$ as being the affine part of the Jacobian of $\bar{\Gamma}$, obtained by removing the translates of the theta divisor, corresponding to the half periods $\{0, [W_0 - W_1], [W_1 - W_2], [W_2 - W_0]\}$, where $\mathcal{W}_1 = \{W_0, W_1, W_2\}$. ∎

5.3. The map $\mathcal{M} \to \mathcal{M}^3$

It follows from Section 5.2 that for any $f \in \mathcal{H}$ the line bundle $\mathcal{L}_f$ which corresponds to $\mathcal{D}_f/\mathfrak{T}_f$ defines a birational map $\phi_{\mathcal{L}_f}$ from $\mathcal{T}_f$ to an octic surface in $\mathbf{P}^3$. We will compute an equation of this octic because the coefficients of this equation, which depend on f, will allow us to construct explicitly the map $\mathcal{M} \to \mathcal{M}^3$. Since $\mathcal{T}_f = \mathrm{Jac}(\bar{\Gamma}_f)/\mathfrak{T}_f$ the vector space of functions which provide this map consists of the $\mathfrak{T}_f$-invariant functions on $\mathrm{Jac}(\bar{\Gamma}_f)$ with a simple pole along $\mathcal{D}_f$ (at worst), i.e., the $\mathfrak{T}$-invariant functions in the span of $\{z_0, \ldots, z_{15}\}$. Using (5.4) and (5.7) one finds the following four independent invariant functions:

$$\begin{aligned}
\theta_0 &= z_0 = 1, \\
\theta_1 &= z_{10} = (z_2 - 2z_5)^2 - z_6^2, \\
\theta_2 &= z_{11} = (2z_3 - z_6)^2 - z_2^2, \\
\theta_3 &= z_{12} = z_1 z_2 z_3 - z_4 z_5 z_6.
\end{aligned} \tag{5.18}$$

In order to compute an equation for the octic it suffices — in principle — to eliminate the variables $z_1, \ldots, z_6$ from the equations (5.3) and (5.18). In practice, doing the calculation in a straightforward way leads to disastrous results, even when using a computer algebra package such as MuPad or Maple. Therefore we will describe in some detail how this computation can be done. As a first step we notice that the octic which we want to compute is isomorphic to the variety defined by the following equations:

$$
\begin{aligned}
f_1 &= X_3 - X_5, & 0 &= X_1 X_4 - Z_1^2, \\
f_2 &= X_1 - X_6, & 0 &= X_2 X_5 - Z_2^2, \\
f_3 &= X_2 - X_4, & 0 &= X_3 X_6 - Z_3^2, \\
f_4 &= X_1 + X_4 + 2Z_1 + 4X_3 - 4Z_2 - 4Z_3, \\
\theta_1 &= 4X_5 - 4Z_2 + X_2 - X_6, \\
\theta_2 &= 4X_3 - 4Z_3 + X_6 - X_2, \\
\theta_3^2 &= X_1 X_2 X_3 + X_4 X_5 X_6 - 2Z_1 Z_2 Z_3.
\end{aligned}
\tag{5.19}
$$

To see this, we consider a regular map φ from the variety given by (5.3) and (5.18) to the variety given by (5.19). The map φ is given by $X_i = z_i^2$ and $Z_j = z_j z_{j+3}$, where $i = 1, \ldots, 6$ and $j = 1, 2, 3$. On the one hand φ is constant on the orbits of $\mathfrak{T}$ because all X_i and Z_j are $\mathfrak{T}$-invariant; on the other hand it is easy to check that every fiber of φ contains precisely four points, hence the degree of φ is four. This shows that (5.19) represents the image of $\mathcal{A}_f / \mathfrak{T}_f$ in projective space, obtained by using the sections of the line bundle associated to $\mathcal{D}_f / \mathfrak{T}_f$.

Six of the equations in (5.19) are linear and we can use these equations to eliminate X_2, X_3, X_5, X_6, Z_2 and Z_3 from the four non-linear equations. Apart from $X_1 X_4 = Z_1^2$, this leaves us with the following three equations (we have used $X_1 X_4 = Z_1^2$ to simplify them)

$$
\begin{aligned}
& 2\left[f_3 X_1^2 - f_2 X_4^2 - (f_2 - f_3 - \theta_1 - \theta_2)Z_1^2\right] - 2(4f_1 + f_2 - f_3 - f_4 + \theta_1)X_4 Z_1 \\
& \quad - 2(f_2 - f_3 - f_4 + \theta_2)X_1 Z_1 + 2f_3(4f_1 - f_4 + \theta_1 + \theta_2)X_1 + 2f_2(f_4 - \theta_1 - \theta_2)X_4 \\
& \quad - (f_2 + f_3 - f_4 + \theta_2)(4f_1 - f_2 - f_3 - f_4 + \theta_1)Z_1 - 8\theta_3^2 = 0, \\
& 4f_3 X_1 - 4(f_2 - \theta_1)(f_3 + X_4) + 4X_1 X_4 - (f_2 - f_3 - f_4 + \theta_2 + 2Z_1)^2 = 0, \\
& 4(f_3 + \theta_2)(X_1 - f_2) - 4X_4 f_2 + 4X_1 X_4 - (4f_1 + f_2 - f_3 - f_4 + \theta_1 + 2Z_1)^2 = 0.
\end{aligned}
\tag{5.20}
$$

The first trick that we use to make the rest of the computation feasible stems from the following observation. If we multiply the second equation by X_1 and the third equation by X_4 to remove from the first equation in (5.20) those terms which contain X_1^2 and X_4^2, then the resulting equation is a linear equation in X_1, X_4 and Z_1 (the relation $X_1 X_4 = Z_1^2$ is again used to simplify this expression) so that (5.20) is equivalent to a linear system of equations in X_1, X_4 and Z_1, which is solved at once. An equation for the octic is then given by substituting the expressions for X_1, X_4 and Z_1 in the only remaining equation $X_1 X_4 = Z_1^2$.

The resulting equation is monstrous (it has 2441 terms), in contrast with the equation (IV.5.2) for the octic, corresponding to an Abelian surface of type $(1, 4)$. By a recaling of some of the coordinates by roots of -1 we write the octic in the following more symmetric form:

$$
\begin{aligned}
& \mu^2 y_0^2 y_1^2 y_2^2 y_3^2 + \mu_1^2 (y_0^4 y_1^4 + y_2^4 y_3^4) + \mu_2^2 (y_1^4 y_3^4 + y_0^4 y_2^4) + \mu_3^2 (y_0^4 y_3^4 + y_1^4 y_2^4) + \\
& \quad - 2\mu_1 \mu_2 (y_0^2 y_1^2 + y_2^2 y_3^2)(y_0^2 y_2^2 - y_3^2 y_1^2) - 2\mu_2 \mu_3 (y_0^2 y_2^2 + y_3^2 y_1^2)(y_0^2 y_3^2 - y_1^2 y_2^2) \\
& \quad - 2\mu_3 \mu_1 (y_0^2 y_3^2 + y_1^2 y_2^2)(y_0^2 y_1^2 - y_2^2 y_3^2) = 0.
\end{aligned}
\tag{5.21}
$$

The difference between these two equations lies of course in the choice of coordinates. In order to compute the coordinate transformation which reduces our equation to the symmetric form (5.21) we use the following geometric fact. Since the octic that we obtained has the form $A\theta_3^4 + B\theta_3^2 + C^2 = 0$ the octic has a singular point of order four at $(0 : 0 : 0 : 1)$ and such a singular point necessarily comes from four of the sixteen half periods on $\mathrm{Jac}(\bar{\Gamma}_f)$. Clearly, (5.21) also has a singular point of order four at $(0 : 0 : 0 : 1)$. On the other hand, we see that the tangent cone to (5.21) at $(0 : 0 : 0 : 1)$, is the union of four hyperplanes because the zero locus of the coefficient of y_3^4 in (5.21) has the form

$$(Y_0 + Y_1 + Y_2)(Y_0 - Y_1 + Y_2)(Y_0 + Y_1 - Y_2)(Y_0 - Y_1 - Y_2)$$

where $Y_0 = \sqrt{\mu_3}y_0$, $Y_1 = \sqrt{\mu_2}y_1$ and $Y_2 = i\sqrt{\mu_1}y_2$ (the particular choices made for each square root are irrelevant). The coefficient A of θ_3^4 in our equation for the octic must also be the product of four linear factors, but these are harder to determine because this can only be done by passing to an extension field of the field $\mathbf{C}[f_1, f_2, f_3, f_4]$. However, if one uses the sections of a symmetric line bundle to map a Jacobian in projective space, then symmetric equations for the image are usually obtained by explicitly introducing the Weierstrass points on the curve, rather than working with the coefficients of a polynomial that defines the underlying curve (see [PV2]). In view of the equation (5.6) for Γ_f we are therefore led to defining[32]

$$4x^3 f_1 - (4f_1 + f_4)x^2 + (f_4 - f_3 - f_2)x + f_3 = \lambda(x - \lambda_1)(x - \lambda_2)(x - \lambda_3).$$

Indeed, in terms of the λ_i one finds the following factorization for A,

$$A = \theta_0 \prod_{i=1}^{3} [\lambda\lambda_i(\lambda_i - 1)\theta_0 - \lambda_i\theta_1 - (\lambda_i - 1)\theta_2].$$

In order to find the required coordinate transformation, we can now use the following ansatz:

$$\begin{aligned}
Y_0 + Y_1 + Y_2 &= \theta_0, \\
-Y_0 - Y_1 + Y_2 &= \kappa_1(\lambda\lambda_1(\lambda_1 - 1)\theta_0 - \lambda_1\theta_1 - (\lambda_1 - 1)\theta_2), \\
-Y_0 + Y_1 - Y_2 &= \kappa_2(\lambda\lambda_2(\lambda_2 - 1)\theta_0 - \lambda_2\theta_1 - (\lambda_2 - 1)\theta_2), \\
Y_0 - Y_1 - Y_2 &= \kappa_3(\lambda\lambda_3(\lambda_3 - 1)\theta_0 - \lambda_3\theta_1 - (\lambda_3 - 1)\theta_2).
\end{aligned} \tag{5.22}$$

The coefficients κ_i are uniquely determined by the compatibility equations, which stem from the vanishing of the sum of the left hand sides of these four equations. If we denote $\Lambda(x) = \lambda(x - \lambda_1)(x - \lambda_2)(x - \lambda_3)$ then the solution to the compatibility equations is given by $\kappa_i = -1/\Lambda'(\lambda_i)$, $(i = 1, \ldots, 3)$. Substituting these values for κ_i in (5.22) we can rewrite our equation for the octic in terms of the coordinates $Y_0, \ldots, Y_3$. Putting $Y_i = \rho_i y_i$ we can determine the ρ_i such that we obtain precisely (5.21). It gives the following values for μ, μ_1, μ_2, μ_3:

$$\begin{aligned}
\mu_i^2 &= \lambda_i(1 - \lambda_i)(\lambda_{i+1} - \lambda_{i+2})^3, \\
\mu^2 &= 12(\sigma_2^2 - \sigma_1^2\sigma_3) + 2(\sigma_2 - \sigma_1)(\sigma_1\sigma_2 + 9\sigma_3),
\end{aligned} \tag{5.23}$$

---

[32] The final result will be symmetric in $\lambda_1, \lambda_2, \lambda_3$, hence does not depend on the order of these parameters.

where σ_i is the i-th symmetric function of $\lambda_1, \lambda_2, \lambda_3$ and $\lambda_4 = \lambda_1$, $\lambda_5 = \lambda_2$. This determines the parameters μ_i explicitly in terms of the Weierstrass points of the curve $\bar{\Gamma}_f$. The sign of the parameters μ and μ_i is not important. Indeed, the coefficients $(\mu, \mu_1, \mu_2, \mu_3)$ are only intermediate moduli for Abelian surfaces of type $(1, 4)$, the moduli themselves being given by the following expressions which realize the moduli space as the cone $\mathcal{C} : f_4^2 = f_1(4f_2^3 - 27f_3^2)$ in weighted projective space $\mathbf{P}^{(1,2,2,3,4)}$:

$$f_0 = \mu^2,$$
$$f_1 = (\mu_1^2 + \mu_2^2 + \mu_3^2)^2,$$
$$f_2 = \mu_1^4 + \mu_2^4 + \mu_3^4 - \mu_1^2\mu_2^2 - \mu_2^2\mu_3^2 - \mu_3^2\mu_1^2,$$
$$f_3 = (\mu_2^2 - \mu_1^2)(\mu_3^2 - \mu_2^2)(\mu_1^2 - \mu_3^2),$$
$$f_4 = (\mu_1^2 + \mu_2^2 + \mu_3^2)(\mu_1^2 + \mu_2^2 - 2\mu_3^2)(\mu_2^2 + \mu_3^2 - 2\mu_1^2)(\mu_3^2 + \mu_1^2 - 2\mu_2^2).$$

The standard action of the symmetric group S_3 on $\mathbf{C}[\lambda_1, \lambda_2, \lambda_3]$ induces on $\mathbf{C}[\mu_1^2, \mu_2^2, \mu_3^2]$ an action which is determined by $(1,2)\cdot(\mu_1^2, \mu_2^2, \mu_3^2) = (-\mu_2^2, -\mu_1^2, -\mu_3^2)$ and $(1,2,3)\cdot(\mu_1^2, \mu_2^2, \mu_3^2) = (\mu_2^2, \mu_3^2, \mu_1^2)$. Therefore, every symmetric function in $\mathbf{C}[\mu_1^2, \mu_2^2, \mu_3^2]$ is either invariant or anti-invariant with respect to this induced action. It follows that the above polynomials $f_0, \ldots, f_4$ are symmetric in $\lambda_1, \lambda_2, \lambda_3$. They are easily expressed in terms of $f = (f_1, f_2, f_3, f_4)$, giving an explicit formula for the map $\mathcal{M} \to \mathcal{M}^3 \subset \mathbf{P}^{(1,2,2,3,4)}$.

6. The Hénon-Heiles hierarchy

6.1. The cubic Hénon-Heiles potential

On $\mathbf{C}^4$ we take q_1, q_2, p_1 and p_2 as coordinates and the standard Poisson bracket given by $\{q_1, q_2\} = \{p_1, p_2\} = 0$, $\{q_i, p_j\} = \delta_{ij}$. We denote by $\mathcal{A}_3$ the algebra $\mathbf{C}[K_3, L_3]$ where

$$
K_3 = \frac{1}{2}(p_1^2 + p_2^2) + 8q_2^3 + 4q_1^2 q_2,
$$
$$
L_3 = -q_2 p_1^2 + q_1 p_1 p_2 + q_1^2(q_1^2 + 4q_2^2).
$$

They are in involution and determine the following two (commuting) vector fields:

$$
\begin{aligned}
\dot{q}_1 &= p_1, & q_1' &= -2q_2 p_1 + q_1 p_2, \\
\dot{q}_2 &= p_2, & q_2' &= q_1 p_1, \\
\dot{p}_1 &= -8q_1 q_2, & p_1' &= -p_1 p_2 - 4q_1(q_1^2 + 2q_2^2), \\
\dot{p}_2 &= -4q_1^2 - 24q_2^2, & p_2' &= p_1^2 - 8q_1^2 q_2.
\end{aligned}
$$

Clearly, for any closed point in $\operatorname{Spec}\mathcal{A}_3$ its corresponding level set has dimension two and is irreducible hence $\mathcal{A}_3$ is complete (by Proposition II.3.7). It follows that $\mathcal{A}_3$ is integrable.

We now define a regular map $\phi_3 : \mathbf{C}^4 \to \mathbf{C}^6$, where $\mathbf{C}^6$ is the phase space of the genus 2 odd Mumford system, which we view again as the affine space of Lax operators (2.1). The map ϕ_3 is given by

$$
(q_1, q_2, p_1, p_2) \mapsto
\begin{cases}
u(\lambda) = \lambda^2 - 2q_2 \lambda - q_1^2, \\
v(\lambda) = \dfrac{i}{\sqrt{2}}(p_2 \lambda + q_1 p_1), \\
w(\lambda) = \lambda^3 + 2q_2 \lambda^2 + (q_1^2 + 4q_2^2)\lambda - \dfrac{p_1^2}{2}.
\end{cases}
$$

It is verified by direct substitution that

$$
\begin{aligned}
\phi_3^*(-u_1^2 + u_0 + w_1) &= 0, \\
\phi_3^*(-u_1 u_0 + u_1 w_1 + w_0 + v_1^2) &= -K_3, \\
\phi_3^*(u_1 w_0 + u_0 w_1 + 2v_1 v_0) &= -L_3, \\
\phi_3^*(u_0 w_0 + v_0^2) &= 0.
\end{aligned}
$$

In terms of the system of generators $H_0, \ldots, H_3$ of the integrable algebra $\mathcal{A}$ for the genus 2 odd Mumford system, this means $\phi_3^* H_0 = \phi_3^* H_3 = 0$, $\phi_3^* H_2 = -K_3$, $\phi_3^* H_1 = -L_3$. It follows that $\phi_3^* \mathcal{A} = \mathcal{A}_3$ and the Poisson structure on $\mathbf{C}^6$ which makes ϕ_3 into a morphism (of integrable Hamiltonian systems) is the one given by (2.4), denoted here by $\{\cdot, \cdot\}_2$.

This morphism is neither injective nor surjective. It is however finite and its image is given by $H_0 = H_3 = 0$. Thus we know from Corollary II.2.16 that ϕ is the composition of a surjective and an injective morphism. Indeed, since the Poisson structure $\{\cdot, \cdot\}_2$ has H_0 and H_3 as Casimirs it restricts to a Poisson structure on the level over any point of $\operatorname{Cas}(\{\cdot, \cdot\}_2)$, in particular to the level $\mathcal{F}$ defined by $H_0 = H_3 = 0$. Since $\mathcal{F}$ is irreducible it is an affine Poisson variety and since it is four-dimensional $(\mathcal{F}, \{\cdot,\cdot\}_2, \mathcal{A})$ is an integrable Hamiltonian system (here $\{\cdot,\cdot\}_2$ and $\mathcal{A}$ denote their restrictions to $\mathcal{F}$). Thus we can see ϕ_3 as a surjective morphism

$$\phi_3 : (\mathbf{C}^4, \{\cdot,\cdot\}, \mathcal{A}_3) \to (\mathcal{F}, \{\cdot,\cdot\}_2, \mathcal{A})$$

to a trivial subsystem of the genus 2 odd Mumford system.

We decompose further this surjective morphism ϕ_3. Consider the action τ of $\mathbf{Z}_2$ on $\mathbf{C}^4$ given by

$$\tau(q_1, q_2, p_1, p_2) = (-q_1, q_2, -p_1, p_2).$$

It is an automorphism of the Hénon-Heiles system since it is a Poisson map and $\tau^* \mathcal{A}_3 = \mathcal{A}_3$; in fact even more is true: $\tau^* f = f$ for all $f \in \mathcal{A}_3$. By Proposition II.3.25 we have a quotient integrable Hamiltonian system which we denote as $(\mathbf{C}^4/\tau, \{\cdot,\cdot\}_0, \mathcal{A}_3)$ (since all elements of $\mathcal{A}_3$ go to the quotient). Notice that in this case the level sets are quotients of the original level sets. Moreover the action leaves ϕ_3 invariant and ϕ_3 can be factorized via the quotient $\mathbf{C}^4/\tau$, i.e., we have a morphism (between affine varieties) $\bar\phi_3 : \mathbf{C}^4/\tau \to \mathcal{F}$ which makes the following diagram commutative.

Since ϕ_3 and π are Poisson the same is true for $\bar\phi_3$ and $\bar\phi_3^* \mathcal{A} = \mathcal{A}_3$ by surjectivity of π. Thus

$$\bar\phi_3 : (\mathbf{C}^4/\tau, \{\cdot,\cdot\}_0, \mathcal{A}_3) \to (\mathcal{F}, \{\cdot,\cdot\}_2, \mathcal{A})$$

is a surjective morphism of integrable Hamiltonian systems.

It follows from Example II.2.31 that $Q_1 = q_1^2$, $Q_2 = q_2$, $P_1 = p_1^2$, $P_2 = p_2$, $R = q_1 p_1$ generate the algebra of invariant functions on $\mathbf{C}^4$ for the action of τ; they generate the algebra of regular functions on the quotient, which therefore is a cone with equation $R^2 = Q_1 P_1$. The map $\bar\phi_3$ is then given by

$$(Q_1, Q_2, P_1, P_2, R) \mapsto \begin{cases} u(\lambda) = \lambda^2 - 2Q_2\lambda - Q_1, \\[2mm] v(\lambda) = \dfrac{i}{\sqrt{2}}(P_2\lambda + R), \\[2mm] w(\lambda) = \lambda^3 + 2Q_2\lambda^2 + (Q_1 + 4Q_2^2)\lambda - \dfrac{P_1}{2}, \end{cases}$$

and it is important to note here that this map is biregular. The first conclusion is that the Hénon-Heiles potential admits an automorphism of order two whose quotient is isomorphic to a trivial subsystem of the genus 2 even Mumford system. It is easy to see that the quotient map is even unramified.

Since $\bar\phi_3$ is an isomorphism we have seen that $(\mathbf{C}^4/\tau, \{\cdot,\cdot\}, \mathcal{A}_3)$ is an a.c.i. system (whose general fiber is an affine part of a Jacobian). To conclude that the Hénon-Heiles system is a.c.i. we use the explicit expressions (VI.4.9) of the variables u_i in terms of theta functions. It follows from that formula that u_0 is the square of a quotient of two theta functions, hence q_1, which is its square root, is a quotient of two theta functions. It follows that the generators of the ideal which define the level sets are expressible in terms of theta functions, hence they define an affine part of an Abelian variety.

6.2. The quartic Hénon-Heiles potential

We now turn to the Hénon-Heiles potential V_4. It is defined on $\mathbf{C}^4$ with the same Poisson structure as the cubic Hénon-Heiles potential, the integrable algebra now being given by $\mathcal{A}_4 = \mathbf{C}[K_4, L_4]$ where

$$K_4 = \frac{1}{2}(p_1^2 + p_2^2) + q_1^4 + 12q_1^2 q_2^2 + 16q_2^4,$$

$$L_4 = -q_2 p_1^2 + q_1 p_1 p_2 + q_1^2(8q_2^3 + 4q_1^2 q_2).$$

$\mathcal{A}_4$ is also contained in the algebra of invariant functions (for τ) and leads also to a quotient system $(\mathbf{C}^4/\tau, \{\cdot, \cdot\}_0, \mathcal{A}_4)$. The analog of the map ϕ_3 is in this case the map $\phi_4 : \mathbf{C}^4 \to \mathbf{C}^7$ defined by

$$(q_1, q_2, p_1, p_2) \mapsto \begin{cases} u(\lambda) = \lambda^2 - 2q_2 \lambda - q_1^2, \\[2mm] v(\lambda) = \dfrac{i}{\sqrt{2}}(p_2 \lambda + q_1 p_1), \\[2mm] w(\lambda) = \lambda^4 + 2q_2 \lambda^3 + (q_1^2 + 4q_2^2)\lambda^2 + 4q_2(q_1^2 + 2q_2^2) - \dfrac{p_1^2}{2}. \end{cases}$$

Here $\mathbf{C}^7$ is interpreted as the phase space of the genus 2 even Mumford system. The map ϕ_4 is a Poisson morphism when the Poisson structure (2.8) is chosen for it. As in the previous paragraph it follows that the quotient system is isomorphic to a trivial subsystem of the genus 2 even Mumford system, the level now being defined by $H_4 = H_3 = H_0 = 0$. In particular the quotient system is an a.c.i. system (with an affine part of a two-dimensional Jacobian as the general fiber of its momentum map).

Our argument which showed that the Hénon-Heiles potential is a.c.i. is not valid in this case. To see this let us fix one Jacobian from the trivial subsystem of the genus 2 even Mumford system (or from the isomorphic quotient system) and recall from Section VI.4 that the divisor to be adjoined to the general fiber of the momentum map consists of two translates of the theta divisor and that u_0 has on each of these a simple pole. If the quotient map, restricted to a general level set, extends to an unramified map π over the whole Jacobian then $\pi^* u_0$ has a simple pole on the inverse images of the two translates of the theta divisor; since π is unramified these inverse images are reduced. But then $\pi^* u_0$ cannot be the square of a meromorphic function on the inverse image of the level set (which is a level set of the quartic potential). It follows that π must be ramified and since it is unramified on the affine part it must be ramified at the divisor at infinity. Thus the completed general fiber of the momentum map of the quartic potential is a ramified cover of a two-dimensional Jacobians.

A first example of an integrable Hamiltonian system with this property was given by Bechlivanidis and van Moerbeke in [BM]. The algebraic invariants of the surfaces which appear as ramified covers of Jacobians in the context of integrable Hamiltonian systems were computed by L. Piovan (see [Pio2]).

6.3. The Hénon-Heiles hierarchy

For the higher potentials the situation is even worse. We can still construct a quotient system, but even these quotient systems are not a.c.i. They are however still trivial subsystems of the (two-dimensional) hyperelliptic systems which we considered in Paragraph III.2.4; these hyperelliptic systems correspond to higher genus curves and are not a.c.i. (recall that only if $g = d$ they are a.c.i.; here $d = 2$ and $g \geq 3$).

We are still considering the standard Poisson structure on $\mathbf{C}^4$ and consider now the integrable algebra $\mathcal{A}_n = [K_n, L_n]$ which generalizes the algebras $\mathcal{A}_3$ and $\mathcal{A}_4$. The polynomials K_n and L_n are given by

$$K_n = \frac{1}{2}(p_1^2 + p_2^2) + V_n,$$

$$L_n = -q_2 p_1^2 + q_1 p_1 p_2 + q_1^2 V_{n-1},$$

where V_n is defined by

$$V_n = \sum_{k=0}^{[n/2]} 2^{n-2k} \binom{n-k}{k} q_1^{2k} q_2^{n-2k}.$$

and satisfies the recursion relation

$$V_{i+2} = 2q_2 V_{i+1} + q_1^2.$$

We still have τ defining an automorphism of order two and we can consider the quotient system. The analog of the morphisms ϕ_3 and ϕ_4 is the map ϕ_n defined by

$$(q_1, q_2, p_1, p_2) \mapsto \begin{cases} u(\lambda) = \lambda^2 - 2q_2\lambda - q_1^2, \\[2mm] v(\lambda) = \dfrac{i}{\sqrt{2}}(p_2\lambda + q_1 p_1), \\[2mm] w(\lambda) = \displaystyle\sum_{i=0}^{n-1} V_i \lambda^{n-i} - \dfrac{p_1^2}{2}. \end{cases}$$

Using the recursion relation it is easy to compute that

$$u(\lambda)w(\lambda) + v^2(\lambda) = \lambda^{n+2} - \lambda^2 K_n - \lambda L_n.$$

This can be seen as in the previous cases as a morphism to a trivial subsystem of one of the hyperelliptic systems discussed in Paragraph III.2.4 (namely the one associated to the family of polynomials $F(x,y) = y^2 - x^{n+2} + ax^2$ with parameter a). These are also multi-Hamiltonian and the Poisson structure which has to be taken (in order to have a Poisson morphism) is again the one for which all but the coefficients of λ and λ^2 are Casimirs; this determines the brackets of the coefficients of $w(\lambda)$ in terms of the brackets of the coefficients of $u(\lambda)$ and $v(\lambda)$, which are all zero except $\{u_1, v_1\} = 1$, $\{u_0, v_0\} = -u_0$ (which corresponds to $\varphi = x$ in the notation of Chapter III).

We are merely interested here in the level sets of the momentum map of the potential V_n ($n > 4$). Clearly these surfaces are unramified $2:1$ covers of the level set $\mathcal{F}_{F,2}$ (defined in (III.3.1)) where $F = y^2 - x^{n+2} - \lambda^2 K_n - \lambda L_n$. We determine these level sets for F of the more general form $F = y^2 - f(x)$ and of any dimension lower than the genus of the curve Γ_F (which is assumed non-singular) in the following proposition.

Proposition 6.1 *In the hyperelliptic case* $F(x,y) = y^2 - f(x)$, *the level set* $\mathcal{F}_{F,d}$ *is for* $d \le g$ *biholomorphic to a (smooth) affine part of a distinguished d-dimensional subvariety* W_d *of* $\mathrm{Jac}(\bar{\Gamma}_F)$, *namely*

$$\mathcal{F}_{F,d} \cong W_d \setminus W_{d-1} \qquad\qquad \deg f(x) \ odd,$$
$$\mathcal{F}_{F,d} \cong W_d \setminus (W_{d-1} \cup (\vec{e} + W_{d-1})) \qquad \deg f(x) \ even,$$

where $\vec{e} \in \mathrm{Jac}(\bar{\Gamma}_F)$ *is given by* $\vec{e} = \mathrm{Ab}(\infty_1 - \infty_2) = \int_{\infty_2}^{\infty_1} \vec{\omega} \bmod \Lambda_{\bar{\Gamma}_F}$. *Also*

$$W_g = \mathrm{Jac}(\bar{\Gamma}_F),$$
$$W_{g-1} = \ theta \ divisor \ \Theta \subset \mathrm{Jac}(\bar{\Gamma}_F),$$
$$\vdots$$
$$W_1 = \ curve \ \bar{\Gamma}_F \ embedded \ in \ \mathrm{Jac}(\bar{\Gamma}_F),$$
$$W_0 = \ origin \ of \ \mathrm{Jac}(\bar{\Gamma}_F).$$

Proof

We prove the proposition only for the case in which $\deg f(x)$ is odd. In this case Γ_F is compactified by adding one point which we call ∞. We choose this point as the base point for the Abel-Jacobi map (on the symmetric product) and define W_k for $k = 1, \ldots, g$ as $W_k = \mathrm{Ab}_k(\mathrm{Sym}^k \bar{\Gamma}_F)$. By Jacobi's Theorem $W_g = \mathrm{Jac}(\bar{\Gamma}_F)$ and by Riemann's Theorem, W_{g-1} is (a translate of) the Riemann theta divisor. Clearly for each $k \le g$, W_{k-1} is a divisor in W_k and, $W_k \setminus W_{k-1}$ is smooth. We claim that

$$\mathrm{Ab}_d(\mathrm{Sym}^d \Gamma_F \setminus \mathcal{D}_{F,d}) = W_d \setminus W_{d-1},$$

more precisely Ab_d realizes a holomorphic bijection between these smooth varieties. Namely,

$$\langle P_1, \ldots, P_d \rangle \in \mathrm{Sym}^d \Gamma_F \setminus \mathcal{D}_{F,d}$$
$$\Longleftrightarrow \forall i \ P_i \ne \infty \ \text{and} \ \exists i \ne j : x(P_i) = x(P_j) \Longrightarrow \left(\begin{array}{l} P_i = P_j \ \text{and} \ P_i \ \text{is not} \\ \text{a ramification point of} \ x \end{array} \right)$$
$$\Longleftrightarrow \mathrm{Ab}_d(P_1, \ldots, P_d) \notin W_d \setminus W_{d-1},$$

where we used Abel's Theorem in the last step. It follows that $\mathrm{Sym}^d \Gamma_F \setminus \mathcal{D}_{F,d}$ and $W_d \setminus W_{d-1}$ are biholomorphic, hence by Proposition III.3.3, $\mathrm{Ab}_d \circ \phi_{F,d}$ is a biholomorphism and the manifolds $\mathcal{F}_{F,d}$ and $W_d \setminus W_{d-1}$ are biholomorphic. ∎

Applied to the case of the Hénon-Heiles hierarchy the proposition says that the general fiber of the momentum map of the n-th potential of this hierarchy is an unramified $2 : 1$ cover of an affine part of the W_2 stratum of a hyperelliptic curve of genus $\lfloor \frac{n+1}{2} \rfloor$; this affine part is completed into the W_2 stratum by adding one copy of the curve if n is odd, otherwise two copies need to be added. As before, the morphism ϕ_n to the Mumford systems lead to a Lax representation (with spectral parameter) for the Hénon-Heiles hierarchy. Notice that although we have a Lax equation with spectral parameter the system is not a.c.i.: its general level surface corresponds to a non-linear subvariety of a higher-dimensional hyperelliptic Jacobian.

7. The Toda lattice

In this paragraph we look at the $\mathfrak{sl}(3)$ periodic Toda lattice and some of its variants. For a generalization to other Lie algebras, Lax equations and a physical interpretation see [OP2] and [AM8]. For the non-periodic case, see [FH].

7.1. Different forms of the Toda lattice

Consider the following Poisson matrix on $\mathbf{C}^6$:

$$\begin{pmatrix} 0 & T \\ -{}^tT & 0 \end{pmatrix} \quad \text{where} \quad T = \begin{pmatrix} 0 & -t_1 & t_1 \\ t_2 & 0 & -t_2 \\ -t_3 & t_3 & 0 \end{pmatrix};$$

we denote the corresponding Poisson structure on $\mathbf{C}^6$ by $\{\cdot,\cdot\}$. One easily find that $t_1 t_2 t_3$ and $t_4 + t_5 + t_6$ are Casimirs. We show that the algebra of Casimirs is generated by these two elements, in particular it is a polynomial algebra.

Lemma 7.1 $\mathrm{Cas}(\mathbf{C}^6, \{\cdot,\cdot\}) = \mathbf{C}[t_1 t_2 t_3, t_4 + t_5 + t_6]$.

Proof

Let us denote $a = t_1 t_2 t_3$ and $b = t_4 + t_5 + t_6$. Then $F \in \mathrm{Cas}(M, \{\cdot,\cdot\})$ can be written in terms of $t_1, \ldots, t_5$ and the Casimir b by replacing t_6 with $b - t_4 - t_5$. We call the resulting polynomial F_1. Since F is a Casimir the same holds true for F_1 and we find from $\{t_1, F_1\} = \{t_2, F_1\} = 0$ that

$$\frac{\partial F_1}{\partial t_5} = \frac{\partial F_1}{\partial t_4} = 0,$$

hence F_1 depends on t_1, t_2 and t_3 only. If F_1 is symmetric in these variables it is a polynomial in $v_1 = t_1 + t_2 + t_3$ and $v_2 = t_1 t_2 + t_2 t_3 + t_3 t_1$ and $v_3 = t_1 t_2 t_3$, so $F_1(t_1, t_2, t_3) = F_2(v_1, v_2, v_3)$. Since F_2 and v_3 are Casimirs we find from $\{t_4, F_2\} = 0$ that

$$\frac{\partial F_2}{\partial v_1} = -t_1 \frac{\partial F_2}{\partial v_2}.$$

Since both derivatives are polynomials in v_i only it follows from this that these derivatives are actually zero, hence F_1 is a polynomial in $t_1 t_2 t_3$ only. If F_1 is not symmetric then one symmetrizes it,

$$F_2(v_1, v_2, v_3) = F_1(t_1, t_2, t_3) + F_1(t_2, t_3, t_1) + F_1(t_3, t_1, t_2),$$

and as above one finds that F_2 depends on v_3 only. This implies that F_1 is a polynomial in $t_1 t_2 t_3$ only and we are done. ∎

Recall that we computed the invariant polynomial of this Poisson structure in Example II.2.56. We define

$$T_1 = t_1 t_2 t_3,$$
$$T_2 = t_4 + t_5 + t_6,$$
$$T_3 = \frac{1}{2}(t_4^2 + t_5^2 + t_6^2) + t_1 + t_2 + t_3,$$
$$T_4 = t_4 t_5 t_6 - t_1 t_4 - t_2 t_5 - t_3 t_6,$$

and $\mathcal{A} = \mathbf{C}[T_1, T_2, T_3, T_4]$. Since T_1 and T_2 are Casimirs and $\{T_3, T_4\} = 0$ the algebra $\mathcal{A}$ is involutive. We do not show completeness here because it will follow automatically from Paragraph 7.2.

Let us look at the level sets of the Casimirs over closed points. They are given by

$$t_1 t_2 t_3 = a,$$
$$t_4 + t_5 + t_6 = b,$$

where $a, b \in \mathbf{C}$ are arbitrary; we denote this level set by $\mathcal{F}_{ab}$. If $a \neq 0$ then $\mathcal{F}_{ab}$ is irreducible, four-dimensional and the Poisson structure has rank four (even at every point). If $a = 0$ then $\mathcal{F}_{ab}$ has three irreducible components, each of which is a four-dimensional plane and the Poisson structure has rank four. It follows as in Proposition II.2.42 that if we restrict the Poisson structure to any of these levels then its algebra of Casimirs is still maximal. Similarly the integrable algebra leads to an integrable Hamiltonian system on (the irreducible components) of these level sets, as in Proposition II.3.19. The original Toda lattice corrsponds to $a = 1$ and $b = 0$, but often it is just as easy to work on the larger space $\mathbf{C}^6$, or on the hyperplane $\mathcal{H} \subset \mathbf{C}^6$ defined by $t_4 + t_5 + t_6 = 0$. One may however also want to consider all possible levels of the Casimirs at once, except the reducible ones, i.e., the ones for which $a = 0$. This is easily done by using Proposition II.3.29. Then the phase space is taken as the affine variety defined by $t_0 t_1 t_2 t_3 = 1$ in $\mathbf{C}^7$. Since $t_1 t_2 t_3$ is a Casimir the same holds true for t_0. The algebra of Casimirs is now given by

$$\mathrm{Cas}(\mathbf{C}^7) = \mathbf{C}[T_0, T_1, T_2] / \mathrm{idl}(T_0 T_1 - 1)$$

and the corresponding integrable algebra is the tensor product of it with $\mathbf{C}[T_3, T_4]$. We will come back to this form of the Toda lattice in Paragraph 7.2.

As is even apparent from the physical origin of the problem, the Toda lattice has an automorphism of order three, which is given by the map $\tau : \mathbf{C}^6 \to \mathbf{C}^6$,

$$\tau : (t_1, t_2, t_3, t_4, t_5, t_6) = (t_2, t_3, t_1, t_5, t_6, t_4).$$

As a preparation of the computation of the quotient by this automorphism (in Paragraph 7.3) we do a simple linear transformation giving coordinates which are diagonal for the action of τ. Let ϵ denote a fixed cubic root of unity and define

$$x_1 = t_1 + t_2 + t_3, \qquad y_1 = t_4 + t_5 + t_6,$$
$$x_2 = t_1 + \epsilon t_2 + \epsilon^2 t_3, \qquad y_2 = t_4 + \epsilon t_5 + \epsilon^2 t_6,$$
$$x_3 = t_1 + \epsilon^2 t_2 + \epsilon t_3, \qquad y_3 = t_4 + \epsilon^2 t_5 + \epsilon t_6.$$

Then τ is in diagonal form and is given by

$$\tau(x_1, x_2, x_3, y_1, y_2, y_3) = (x_1, \epsilon^2 x_2, \epsilon x_3, y_1, \epsilon^2 y_2, \epsilon y_3).$$

The integrable algebra $\mathcal{A}$ is now given by $\mathcal{A} = \mathbf{C}[X_1, X_2, X_3, X_4]$ where

$$X_1 = x_1^3 + x_2^3 + x_3^3 - 3 x_1 x_2 x_3,$$
$$X_2 = y_1,$$
$$X_3 = 3 x_1 + y_2 y_3,$$
$$X_4 = y_2^3 + y_3^3 + 9(x_2 y_3 + x_3 y_2).$$

The Poisson structure is (up to a factor $\sqrt{-3}$) given in these coordinates by

$$\begin{pmatrix} 0 & X \\ -{}^t\!X & 0 \end{pmatrix} \quad \text{where} \quad X = \begin{pmatrix} 0 & -x_2 & x_3 \\ 0 & -x_3 & x_1 \\ 0 & -x_1 & x_2 \end{pmatrix}$$

and the algebra of Casimirs reduces to $\mathbf{C}[X_1, X_2]$.

7.2. A morphism to the genus 2 even Mumford system

We now describe a morphism from the $\mathfrak{sl}(3)$ periodic Toda lattice to the Bechlivanidis-van Moerbeke system which we have shown to be isomorphic to the genus 2 even Mumford system (on $\mathbf{C}^7$). For a generalization, giving a morphism from the $\mathfrak{sl}(g+1)$ periodic Toda lattice to the genus g even Mumford system, see [FV]. For explicitness we recall that we consider the Toda lattice as being defined on the hyperplane $\mathcal{H}$ defined by $t_4 + t_5 + t_6 = 0$ in $\mathbf{C}^6$. If we do not suppose this then we do not have a morphism to the Bechlivanidis-van Moerbeke system, but still to the genus 2 even Mumford system, which we now have to take as being defined on $\mathbf{C}^8$, namely we cannot assume anymore that the coefficient of x^5 in the equation $y^2 = f(x)$ vanishes. Since the formulas are simpler in this case and since there is no phenomenologic difference we will not consider this more general case.

Consider the map $\phi : \mathcal{H} \subset \mathbf{C}^6 \to \mathbf{C}^7$ defined by

$$s_1 = t_5 t_6 - t_1,$$
$$s_2 = -t_4/2, \qquad s_{4,5} = (t_2 \pm t_3)/8,$$
$$s_3 = -t_2 t_3/16, \qquad s_{6,7} = (t_3 t_6 \pm t_2 t_5)/8.$$

Then

$$\phi^*(s_1 - 4s_2^2 - 8s_4) = -T_3,$$
$$\phi^*(s_1 s_2 + 4s_6) = -T_4/2,$$
$$\phi^*(s_4^2 - s_5^2 + s_3) = 0, \tag{7.1}$$
$$\phi^*(s_2 s_3 + s_4 s_6 + s_5 s_7) = 0,$$
$$\phi^*(s_7^2 - s_6^2 - s_1 s_3) = -T_1/16.$$

Thus ϕ is a regular map and $\phi^* \mathcal{A}' = \mathcal{A}$ where $\mathcal{A}'$ denotes the Bechlivanidis-van Moerbeke system. Moreover it is easy to check that the Poisson structure which has to be taken on the latter system in order for ϕ to be a Poisson map is the one given by (2.14) and we have obtained a morphism of integrable Hamiltonian systems.

As in the Hénon-Heiles example this morphism is again neither injective nor surjective. As a first attempt to cure the non-surjectivity we restrict the Bechlivanidis-van Moerbeke system to the level set $\mathcal{F}$ (of the Casimirs) given by

$$s_4^2 - s_5^2 + s_3 = 0,$$
$$s_2 s_3 + s_4 s_6 + s_5 s_7 = 0,$$

which is irreducible and of codimension two, hence it is an affine Poisson variety; clearly it contains the image of ϕ. Let us compute the invariant polynomial $\rho(\mathcal{F})$ of the Poisson bracket on $\mathcal{F}$. The matrix 2.14 has rank 4 for $x_3 \neq 0$, hence by irreducibility of $\mathcal{F}$ and since

$\dim \mathcal{F} = 5$, the leading term of $\rho(\mathcal{F})$ is $R^2 S^5$. The rank of (2.14) can only go down at those points where $s_3 = 0$ and (at the same time) the rank of the matrix

$$\begin{pmatrix} -4(s_5 s_6 + s_4 s_7) & 2s_7 & -2s_6 & 2s_1 s_5 + 4s_2 s_7 & 4s_2 s_6 - 2s_1 s_4 \\ 0 & -s_5 & -s_4 & s_7 & s_6 \end{pmatrix} \tag{7.2}$$

is less than 2, i.e., all 2×2 determinants must vanish. Since on $\mathcal{F}$, $s_4^2 - s_5^2 + s_3 = 0$ it splits up in the cases $s_5 = \pm s_4$ and we are left with

$$\begin{pmatrix} -4s_4(s_7 \pm s_6) & 2s_7 & -2s_6 & 4s_2 s_7 \pm 2s_1 s_4 & 4s_2 s_6 - 2s_1 s_4 \\ 0 & \mp s_4 & -s_4 & s_7 & s_6 \end{pmatrix}.$$

If $s_7 \pm s_6 = 0$ then the rank of this matrix is smaller than 2 if and only if

$$\det \begin{pmatrix} -2s_6 & \mp 2(2s_2 s_6 - s_1 s_4) \\ -s_4 & \mp s_6 \end{pmatrix} = \pm 2(s_6^2 - s_4(2s_2 s_6 - s_1 s_4)) = 0.$$

This means that we have found two irreducible components which are defined (in $\mathbf{C}^7$) by $s_3 = 0$, $s_5 = \pm s_4$, $s_7 = \mp s_6$ and $s_6^2 - s_4(2s_2 s_6 - s_1 s_4)$; each of these components has dimension 3. If $s_7 \pm s_6 \neq 0$ then the rank of (7.2) can only be smaller than 2 if $s_4 = s_5 = s_6 = s_7 = 0$, a locus which is contained in both irreducible components, hence this one does not contribute to $\rho(\mathcal{F})$. Finally the rank is zero if and only if $s_3 = s_4 = s_5 = s_6 = s_7 = 0$, a two-dimensional plane inside $\mathcal{F}$. The upshot is that $\rho(\mathcal{F})$ is given by

$$\rho(\mathcal{F}) = R^2 S^5 + 2RS^3 + S^2.$$

This should be compared with the invariant polynomial for the Poisson structure of the Toda lattice on $\mathcal{H}$, which we denote by $\rho(\mathcal{H})$. The invariant polynomial for the Poisson structure of the Toda lattice on $\mathbf{C}^6$ was found in Example II.2.56 and computed as $R^2 S^6 + 3RS^4 + S^3$; it follows at once (devide by S) that its restriction to the hyperplane $\mathcal{H}$ (given by $t_4 + t_5 + t_6 = 0$) equals

$$\rho(\mathcal{F}) = R^2 S^5 + 3RS^3 + S^2.$$

The small difference between $\rho(\mathcal{F})$ and $\rho(\mathcal{H})$ suffices to conclude that ϕ is not a biregular map (although it is regular and dominant). The polynomials actually indicate that something goes wrong at the rank 2 level. Let us denote the subsets of $\mathcal{F}$ (resp. $\mathcal{H}$) were the rank is at most 2 by $\mathcal{F}_1$ (resp. $\mathcal{H}_1$). Thus ϕ restricts[33] to a regular map $\phi_1 : \mathcal{H}_1 \to \mathcal{F}_1$ and $\mathcal{H}_1$ has three irreducible components while $\mathcal{F}_1$ has two. Thus either (at least) two irreducible components are identified or (at least) one of them is mapped completely inside $\mathcal{F}_0$, the locus inside $\mathcal{F}$ of points where the rank is zero. On the component $t_2 = t_3 = 0$ of $\mathcal{H}_1$ the map ϕ is given by

$$\begin{aligned} s_1 &= t_5 t_6 - t_1, \\ s_2 &= -t_4/2, \\ s_3 &= s_4 = s_5 = s_6 = s_7 = 0, \end{aligned}$$

hence the latter case occurs[34] and we conclude that ϕ is not a finite map.

---

[33] Recall that the rank at an image point is never larger than the rank at the point, hence restricts indeed.

[34] The other two components of $\mathcal{H}_1$ are mapped neatly to the two components of $\mathcal{F}_1$.

Our way to deal with this is to cut away the bad piece: from the Toda side we remove the (reducible) divisor of $T_1 = t_1 t_2 t_3$ and from the Bechlivanidis-van Moerbeke side we remove the zero locus of $S_5 = s_7^2 - s_6^2 - s_1 s_3$: it follows from (7.1) that these correspond under ϕ. Since both T_1 and S_5 are Casimirs we obtain from Proposition II.3.29 two integrable Hamiltonian systems $(M_1, \{\cdot, \cdot\}_1, \mathcal{A}_1)$ and $(M_2, \{\cdot, \cdot\}_2, \mathcal{A}_2)$ where M_1 and M_2 are given by

$$M_1 = \{(t_0, t_1, \ldots, t_6) \mid t_0 t_1 t_2 t_3 = 1, \ t_4 + t_5 + t_6 = 0\},$$
$$M_2 = \{(s_0, s_1, \ldots, s_7) \mid s_0(s_7^2 - s_6^2 - s_1 s_3) = 1, \ s_4^2 - s_5^2 + s_3 = s_2 s_3 + s_4 s_6 + s_5 s_7 = 0\},$$

and $\{\cdot, \cdot\}_i$ and $\mathcal{A}_i$ are obtained accordingly. The morphism $M_1 \to M_2$ which corresponds to ϕ (and which is also a morphism of integrable Hamiltonian systems) will be denoted by the same letter ϕ. Now both invariant polynomials are the same (being given by $R^2 S^5$), however $\phi : M_1 \to M_2$ is not surjective, since it is obviously missing the points of M_2 where s_3 vanishes. Since s_3 does not belong to $\mathcal{A}_2$ and $(M_2, \{\cdot, \cdot\}_2, \mathcal{A}_2)$ satisfies the conditions of Proposition II.3.7, removing the zero locus of s_3 leads to another integrable Hamiltonian system $(M_3, \{\cdot, \cdot\}_3, \mathcal{A}_3)$ where M_3 is given by

$$M_3 = \{(s_0, s_0', s_1, \ldots, s_7) \mid s_0(s_7^2 - s_6^2 - s_1 s_3) = s_0' s_3 = 1, \ s_4^2 - s_5^2 + s_3 = s_2 s_3 + s_4 s_6 + s_5 s_7 = 0\},$$

and $\{\cdot, \cdot\}_3$ and $\mathcal{A}_3$ derive at once from $\{\cdot, \cdot\}_2$ and $\mathcal{A}_2$.

Finally we have a biregular map! It is explicitly given by

$$
\begin{aligned}
s_0 &= -16 t_0, & t_0 &= -s_0/16, \\
s_0' &= -16 t_0 t_1, & t_1 &= s_0'(s_7^2 - s_6^2) - s_1, \\
s_1 &= t_5 t_6 - t_1, & t_2 &= 4(s_4 + s_5), \\
s_2 &= -t_4/2, & t_3 &= 4(s_4 - s_5), \\
s_3 &= -t_2 t_3/16, & t_4 &= -2 s_2, \\
s_{4,5} &= (t_2 \pm t_3)/8, & t_5 &= -s_0(s_4 - s_5)(s_6 - s_7), \\
s_{5,6} &= (t_3 t_6 \pm t_2 t_5)/8 & t_6 &= -s_0(s_4 + s_5)(s_6 + s_7).
\end{aligned}
$$

Now it follows at once that the Toda lattice (on $\mathcal{H}$) is a.c.i.: since the Bechlivianidis-van Moerbeke system is a.c.i. the same is true for the system we constructed on M_3 since we just removed a divisor (with two components), however the level sets have changed since we removed the divisor of a function which does not belong to the integrable algebra. Thus our system on M_1 is a.c.i. and since it contains the general level set of the original Toda lattice (on $\mathcal{H}$) as its general level set, the Toda lattice is a.c.i. and its general level set is an affine part of a Jacobian; using the order three automorphism one easily recovers the well-known fact that this affine part is obtained from the Jacobian by removing three translates of a genus two curve, each pair of which is tangent at their intersection point. Each of these curves induces a principal polarisation on its Jacobian, hence the Toda lattice is an a.c.i. system of polarization type $(3, 3)$.

7.3. Toda and Abelian surfaces of type (1,3)

Having shown that the Toda lattice is a.c.i. of polarization type $(3,3)$ we are now ready to construct a new a.c.i. system of polarization type $(1,3)$. Using Proposition V.2.5 it is obtained as follows: the order three automorphism τ fixes all fibers of the momentum map (since $\tau^* f = f$ for all $f \in \mathcal{A}$). Therefore the general level set of the quotient a.c.i. system is isogeneous to a Jacobian and the isogeny is a $3:1$ (unramified) map, showing that this level set carries a polarization of type $(1,3)$. It is also easy to determine the divisor which is to be adjoined to the general fiber (of the momentum map) in the quotient system: it is the quotient of the Toda divisor by τ which acts as a translation permuting the three components of this divisor (and the three intersection points). Hence the quotient is an irreducible divisor, birationally equivalent to the genus two curve and having one singular point which is a tacnode. It follows from IV.3.4 that its virtual genus equals 4 which is by Theorem IV.3.7 consistent with the fact that it induces a polarization of type $(1,3)$ on the surface.

If one wants to have an explicit realization of the quotient system, then the main object to be computed is $\mathbf{C}^6/\tau$ (or $\mathcal{H}/\tau$): the Poisson structure and the involutive algebra are obtained at once. Thus we need to construct a system of generators of for $\mathcal{O}(\mathbf{C}^6/\tau)$ (or $\mathcal{O}(\mathcal{H}/\tau)$) as well as a generating set of relations between these generators. To do this, we use the coordinates x_i, y_i, constructed in Paragraph 7.1, which are diagonal with respect to the action of τ. Obviously the following elements are invariant with respect to τ.

$$X_1 = x_1, \; Y_1 = y_1,$$
$$X_2 = x_2 x_3, \; Y_2 = y_2 y_3, \; X_3 = x_3 y_2, \; Y_3 = x_2 y_3,$$
$$X_4 = x_2^3, \; Y_4 = y_2^3, \; X_5 = x_3^3, \; Y_5 = y_3^3,$$
$$X_6 = x_2^2 y_2, \; Y_6 = x_2 y_2^2, \; X_7 = x_3^2 y_3, \; Y_7 = x_3 y_3^2.$$

We claim that these fourteen elements generate $\mathcal{O}(\mathbf{C}^6/\tau)$. To show this, let F be an invariant polynomial. Writing it as a polynomial in x_1 and y_1 it suffices to show that an invariant polynomial in x_2, x_3, y_2 and y_3 can be written as a polynomial in the above candidate generators. Because of the elements X_4, Y_4, X_5 and Y_5 it suffices to check this for a polynomial of degree less than three in x_2, x_3, y_2 and y_3. Since the action is diagonal it suffices to check it for monomials; there are 27 of these, namely the monomials

$$x_2^{i_1} x_3^{i_2} y_2^{i_3} y_3^{i_4} \quad \text{with} \quad i_1 + 2i_2 + i_3 + 2i_4 = 0 \pmod 3.$$

It is easy to check that they all depend on the fourteen ones above.

Finding all relations requires more work: there are quite a lot of them. We give half of the explicit list, the other half is found by interchanging X and Y in our list; our list is ordered by the degree of the monomials (in the x_i, y_i) they come from. In order four there is just one,

$$X_2 Y_2 = X_3 Y_3,$$

in order five there are (two times) six,

$$
\begin{array}{ll}
X_3 X_4 = X_2 X_6, & X_2 Y_4 = X_3 Y_6, \\
X_2 X_7 = X_5 Y_3, & X_6 Y_2 = Y_3 Y_6, \\
X_2 Y_5 = Y_3 Y_7, & X_7 Y_2 = X_3 Y_7,
\end{array}
$$

240

finally, in order six there are (ten plus) twelve,

$$X_2^3 = X_4X_5, \qquad X_2^3Y_3 = X_4X_7,$$
$$X_3^3 = X_5Y_4, \qquad X_2Y_3^2 = X_4Y_7,$$
$$X_4Y_4 = X_6Y_6, \qquad X_2^2X_3 = X_5X_6,$$
$$X_5Y_5 = X_7Y_7, \qquad X_2X_3Y_3 = X_6X_7,$$
$$X_6^2 = Y_4Y_6, \qquad X_2Y_2Y_3 = X_6Y_7,$$
$$X_7^2 = X_5Y_7, \qquad X_2X_3^2 = Y_5Y_6.$$

More important than this list is how to prove that it is complete; we do this as before by reducing it to a finite list. Every relation must come from an identity in the variables x_i, y_i, since x_1 and y_1 are invariant themselves we may forget about these as before. Second, since the variables x_i and y_i are diagonal for the action these identities come from identities between monomials, i.e., from identities which are written in terms of the x_i, y_i as

$$x_2^{i_1} x_3^{j_1} y_2^{k_1} y_3^{l_1} \cdot x_2^{i_2} x_3^{j_2} y_2^{k_2} y_3^{l_2} = x_2^{i_3} x_3^{j_3} y_2^{k_3} y_3^{l_3} \cdot x_2^{i_4} x_3^{j_4} y_2^{k_4} y_3^{l_4}. \tag{7.3}$$

with $i_s + 2j_s + k_s + 2l_s = 0 \pmod 3$ by invariance ($s = 1, \ldots, 4$). All powers in this equation may be supposed smaller than three; to check this for the powers of x_2, suppose that x_2 appears on the left (hence also on the right) with a power at least three, then this must come from one of the following terms (or a multiple of it):

$$X_2^3, \ X_2^2Y_3, \ X_2^2Y_6, \ X_2Y_3^2, \ X_2Y_3Y_6, \ X_2Y_6^2, \ Y_3^3, \ Y_3^2Y_6, \ Y_3Y_6^2, \ Y_6^3.$$

Now check that by using the relations one can always factorize X_4 (on both sides) thereby reducing the order, for example

$$X_2^2Y_6 = X_2X_3X_6 = X_3^2X_4.$$

Since now all exponents in (7.3) may be supposed smaller than three we have a finite list to check. In this way it is verified that all relations are a consequence of the relations which we have given and we have an explicit description of the a.c.i. system.

References

[CGH] ARBARELLO, E., CORNALBA, M., GRIFFITHS, P.A., HARRIS, J., *Geometry of algebraic curves I*, Grundlehren der Mathematischen Wissenschaften, 267, Springer-Verlag (1985)

[AD] ATIYAH, M.F., MCDONALD, I.G., *Introduction to commutative algebra*, Addison-Wesley Publishing Company (1969)

[Adl1] ADLER, M., *Some finite dimensional integrable systems and their scattering behavior*, Comm. Math. Phys., **55** (1977) 195–230

[Adl2] ADLER, M., *On a trace functional for formal pseudo-differential operators and the symplectic structure of the Korteweg-de Vries type equations*, Invent. Math., **50** (1979) 219–248

[AHH] ADAMS, M.R., HARNAD, J., HURTUBISE, J., *Isospectral Hamiltonian flows in finite and infinite dimensions, II. Integration of flows*, Comm. Math. Phys., **134** (1990) 555–585

[AHM] ADLER, M., HAINE, L., VAN MOERBEKE, P., *Limit matrices for the Toda flow and periodic flags for loop groups*, Math. Ann., **296** (1993) 1–33

[AHP] ADAMS, M.R., HARNAD, J., PREVIATO, E., *Isospectral Hamiltonian flows in finite and infinite dimensions, I. Generalized Moser systems and moment maps into loop algebras*, Comm. Math. Phys., **117** (1988) 451–500

[AL] AUDIN, M., LAFONTAINE, J., *Holomorphic curves in symplectic geometry*, Progr. Math., 117, Birkhäuser (1994)

[AM1] ABRAHAM, R., MARSDEN, J.E., *Foundations of mechanics*, Benjamin/Cummings Publishing Co. (1978)

[AM2] ADLER, M., VAN MOERBEKE, P., *Completely integrable systems, Euclidean Lie algebras, and curves*, Adv. Math., **38** (1980) 267–317

[AM3] ADLER, M., VAN MOERBEKE, P., *Linearization of Hamiltonian systems, Jacobi varieties and representation theory*, Adv. Math., **38** (1980) 318–379

[AM4] ADLER, M., VAN MOERBEKE, P., *The algebraic integrability of geodesic flow on $SO(4)$*, Invent. Math., **67** (1982) 297–331

[AM5] ADLER, M., VAN MOERBEKE, P., *The intersection of four quadrics in $\mathbf{P}^6$, Abelian surfaces and their moduli*, Math. Ann., **279** (1987) 25–85

[AM6] ADLER, M., VAN MOERBEKE, P., *The Kowalewski and Hénon-Heiles motions as Manakov geodesic flows on $SO(4)$ — a two-dimensional family of Lax pairs*, Comm. Math. Phys., **113** (1988) 659–700

[AM7] ADLER, M., VAN MOERBEKE, P., *The complex geometry of the Kowalewski-Painlevé analysis*, Invent. Math., **97** (1989) 3–51

References

[AM8] ADLER, M., VAN MOERBEKE, P., *The Toda lattice, Dynkin diagrams, singularities and Abelian varieties*, Invent. Math., **103** (1991) 223–278

[AM9] ADLER, M., VAN MOERBEKE, P., *Algebraic completely integrable systems: a systematic approach*, Academic Press (in preparation)

[AM10] ALEKSEEV, A.Y., MALKIN, A.Z., *Symplectic structure of the moduli space of flat connections on a Riemann surface*, Comm. Math. Phys., **169** (1995) 99–119

[AMR] ANDERSEN, J.E. MATTES, J. RESHETIKHIN, N., *The Poisson structure on the moduli space of flat connections and chord diagrams*, Topology, **35** (1996) 1069–1083

[AN] ARNOLD, V.I., NOVIKOV, S.P., *Dynamical systems IV*, Encyclopaedia of Mathematical Sciences 4, Springer-Verlag (1990)

[Arn] ARNOLD, V.I., *Mathematical methods of classical mechanics*, Graduate Texts in Mathematics 60, Springer-Verlag (1978)

[AS] AUDIN, M., SILHOL, R., *Variétés abéliennes réelles et toupie de Kowalevski*, Composito Math., **87** (1993) 153–229

[Aud1] AUDIN, M., *Courbes algébriques et systèmes intégrables: géodésiques des quadriques*, Exposition. Math., **12** (1994) 193–226

[Aud2] AUDIN, M., *Vecteurs propres de matrices de Jacobi*, Ann. Inst. Fourier, **44** (1994) 1505–1515

[Aud3] AUDIN, M., *Spinning tops. A course on integrable systems.*, Cambridge Studies in Advanced Mathematics, 51, Cambridge University Press (1996)

[Bar] BARTH, W., *Affine parts of Abelian surfaces as complete intersections of four quadrics*, Math. Ann., **278** (1987) 117–131

[Bau] BAUER, T., *Projective images of Kummer surfaces*, Math. Ann., **299** (1994) 155–170

[BCK1] BABELON, O., CARTIER, P., KOSMANN-SCHWARZBACH, Y., *Lectures on integrable systems. In memory of Jean-Louis Verdier. Proceedings of the CIMPA School on Integrable Systems held in Sophia-Antipolis, June 10-28, 1991.*, World Scientific Publishing Co. (1994)

[BCK2] BABELON, O., CARTIER, P., KOSMANN-SCHWARZBACH, Y., *Integrable systems. The Verdier Memorial Conference Proceedings of the International Conference held in Luminy, July 1-6 1991*, Progress in Mathematics, 115, Birkhäuser (1993)

[Bea1] BEAUVILLE, A., *Complex algebraic surfaces*, London Mathematical Society Lecture Note Series, 68, Cambridge University Press (1983)

[Bea2] BEAUVILLE, A., *Jacobiennes des courbes spectrales et systèmes hamiltoniens complètement intégrables*, Acta Math., **164** (1990) 211–235

[BL] BIRKENHAKE, CH., LANGE, H., *Cubic theta relations*, J. Reine Angew. Math., **407** (1990) 166–177

[BLS] BIRKENHAKE, CH., LANGE, H., VAN STRATEN, D., *Abelian surfaces of type* (1,4), Math. Ann., **285** (1989) 625–646

[BM] BECHLIVANIDIS, C., VAN MOERBEKE, P., *The Goryachev-Chaplygin top and the Toda lattice*, Comm. Math. Phys., **110** (1987) 317–324

[Bog] BOGOYAVLENSKY, O.I., *On perturbations of the periodic Toda lattice*, Comm. Math. Phys. **51** (1976) 201–209

References

[Bot] BOTTACIN, F., *Poisson structures on Hilbert schemes of points of a surface and integrable systems*, Manuscripta Math., **97** (1998) 517–527

[Bro] BROUZET, R., *Géométrie des systèmes bihamiltoniens en dimension 4*, Thèse, Université Montpellier II (1991)

[BRS] BOBENKO, A.I., REYMAN, A.G., SEMENOV-TIAN-SHANSKY, M.A., *The Kowalewski top 99 years later: A Lax pair, generalizations and explicit solitons*, Comm. Math. Phys., **122** (1989) 321–354

[Bue] BUEKEN, P., *Multi-Hamiltonian formulation for a class of degenerate completely integrable systems*, J. Math. Phys., **37** (1996) 2851–2862

[BV1] BABELON, O., VIALLET, C-M., *Hamiltonian structures and Lax equations*, Phys. Lett. B, **237** (1990) 411–416

[BV2] BERTIN, J., VANHAECKE, P., *The even master system and generalized Kummer surfaces*, Math. Proc. Camb. Phil. Soc., **116** (1994) 131–142

[BV3] BIRKENHAKE, CH., VANHAECKE, P., *The order of vanishing of the Riemann theta function in the KP direction: a geometric proof*, preprint

[BV4] BUEKEN, P., VANHAECKE, P., *The moduli problem for integrable systems: the example of a geodesic flow on $SO(4)$*, J. London Math. Soc., **62** (2000) 357–369

[Cal] CALOGERO, F., *Solution of the one-dimensional n-body problem with quadratic and/or inversely quadratic pair potentials*, J. Math. Phys., **12** (1971) 419–436

[Cas] CASTELNUOVO, G., *Sulle congruenze del 3° ordine del spazio a 4 dimensioni*, Memoria Atti Istituto Venoto (1888)

[CC] CHOODNOVSKY, D.V., CHOODNOVSKY, G.V., *Completely integrable class of mechanical systems connected with Korteweg-de Vries and multicomponent Schrödinger equations*, Lett. Nuovo Cimento, **22** (1978) 47–51

[CGR] CABOZ, R., GAVRILOV, L., RAVOSON, V., *Bi-Hamiltonian structure of an integrable Hénon-Heiles system*, J. Phys. A, **24** (1991) L523–L525

[CMP] CASATI, P., MAGRI, F., PEDRONI, M., *The bi-Hamiltonian approach to integrable systems*, Modern group analysis: Advanced analytical and computational methods in mathematical physics, Kluwer Academic Publishers, Dordrecht, (1993) 101–110

[CW] CANNAS DA SILVA, A., WEINSTEIN, A., *Geometric models for noncommutative algebras*, Berkeley Mathematics Lecture Notes, 10, American Mathematical Society (1999)

[Dam] DAMIANOU, P.A., *Master symmetries and R-matrices for the Toda lattice*, Lett. Math. Phys., **20** (1990) 101–112

[DGR] DORIZZI, B., GRAMMATICOS, B., RAMANI, A., *Painlevé conjecture revisited*, Phys. Rev. Lett., **49** (1982) 1539–1541

[Dho] DHOOGHE, P.F., *Completely integrable systems of the KdV type related to isospectral periodic regular difference operators*, Acta Appl. Math., **5** (1986) 181–194

[Die] VAN DIEJEN, J., *Families of commuting difference operators*, PhD-thesis, Universiteit van Amsterdam (1994)

[Dub] DUBROVIN, B.A., *Theta functions and non-linear equations*, Russian Math. Surveys, **36** (1981) 11–80

References

[Dui] DUISTERMAAT, J.J., *On global action-angle coordinates*, Comm. Pure Appl. Math., **32** (1980) 687–706

[ES] ERCOLANI, N., SIGGIA, E.D., *Painlevé propriety and geometry*, Phys. D, **34** (1989) 303–346

[Eul] EULER, L., *Du mouvement de rotation des corps solides autour d'un axe variable*, Histoire d' Acad. Roy. des Sci. Berlin, **14** (1758–1765) 154–193

[Fai] FAIRBANKS, L.D., *Lax equation representation of certain completely integrable systems*, Com positio Math., **68** (1988) 31–40

[Fas] FASTRÉ, J., *A Grassmannian version of the Darboux transformation*, Bull. Sci. math., **12** (1999) 181–232

[FG] FRANCAVIGLIA, M., GRECO, S.,, *Integrable systems and quantum groups, Lectures given a the First 1993 C.I.M.E. Session held in Montecatini Terme, June 14–22, 1993*, Springer-Verla (1985)

[FH] FLASCHKA, H., HAINE, L., *Variétés de drapeaux et réseaux de Toda*, Math. Z., **208** (1991 545–556

[Fla1] FLASCHKA, H., *The Toda lattice. I. Existence of integrals*, Phys. Rev. B, **9** (1974) 1924–192

[Fla2] FLASCHKA, H., *The Toda lattice. II. Inverse scattering solution*, Progr. Theoret. Phys., **5** (1974) 703–716

[FS] FERNANDES, R., SANTOS, J.P., *Integrability of the periodic KM system*, Reports on Mathe matical Physics, **40** (1997) 475–484

[FT] FADDEEV, L.D., TAKHTAJAN, L.A., *Hamiltonian methods in the theory of solitons*, Springe Series in Soviet Mathematics, Springer-Verlag (1987)

[FV] FERNANDES, R.L., VANHAECKE, P., *Hyperelliptic Prym varieties and integrable system* preprint

[Gar] GARNIER, R., *Sur une classe de systèmes différentiels abéliens déduits de la théorie d équations linéaires*, Rend. Circ. Mat. Palermo, **43** (1919) 155–191

[Gav1] GAVRILOV, L., *Bifurcations of invariant manifolds in the generalized Hénon-Heiles systen* Phys. D, **34** (1989) 223–239

[Gav2] GAVRILOV, L., *Generalized Jacobians of spectral curves and completely integrable system* Math. Z., **230** (1999) 487–508

[GH] GRIFFITHS, P.A., HARRIS, J., *Principles of algebraic geometry*, Pure and Applied Mathema ics, Wiley-Interscience (1978)

[GMP] GRABOW, J., MARMO, G., PERELOMOV, A.M., *Poisson structures: towards a classificatio* Modern Phys. Lett. A, **18** (1994) 1719–1733

[Gol] GOLDMAN, W.M., *Invariant functions on Lie groups and Hamiltonian flows of surface grou representations*, Invent. Math., **85** (1986) 263–302

[Gri] GRIFFITHS, P.A., *Linearizing flows and cohomological interpretation of Lax equations*, Ame J. Math., **107** (1985) 1445–1483

[GS] GUILLEMIN, V. STERNBERG, S., *Symplectic techniques in physics*, Cambridge University Pre (1990)

References

[Gun] GUNNING, R.C., *Lectures on Riemann surfaces, Jacobi varieties*, Princeton Mathematical Notes, Princeton University Press (1972)

[GZ] GAVRILOV, L., ZHIVKOV, A., *The complex geometry of the Lagrange top*, l'Enseignement Mathématique, **44** (1998) 133–170

[Hai1] HAINE, L., *Geodesic flow on SO(4) and Abelian surfaces*, Math. Ann., **263** (1983) 435–472

[Hai2] HAINE, L., *The algebraic complete integrability of geodesic flow on SO(n)*, Comm. Math. Phys., **94** (1984) 271–287

[Har] HARTSHORNE, R., *Algebraic geometry*, Graduate Texts in Mathematics, 52, Springer-Verlag (1977)

[HH1] HÉNON, M., HEILES, C., *The applicability of the third integral of motion: Some numerical examples*, Astronom. J., **69** (1964) 73–78

[HH2] HAINE, L. HOROZOV, E, *A Lax pair for Kowalevski's top*, Phys. D, **29** (1987) 173–180

[HI] HAINE, L., ILIEV, P., *Commutative rings of difference operators and an adelic flag manifold.*, Internat. Math. Res. Notices, **6** (2000) 281–323

[Hie] HIETARINTA, J., *Direct methods for the search of the second invariant*, Phys. Rep., **147** (1987) 87–154

[Hit] HITCHIN, N.J., *Stable bundles and integrable systems*, Duke Math. J., **54** (1987) 91–114

[HM] HOROZOV, E., VAN MOERBEKE, P, *The full geometry of Kowalevski's top and (1,2)-Abelian surfaces*, Comm. Pure Appl. Math., **42** (1989) 357–407

[Hud] HUDSON, R., *Kummers quartic surface*, Cambridge Mathematical Library, Cambridge University Press (1990)

[Hue] HUEBSCHMANN, J., *Symplectic and Poisson structures of certain moduli spaces I*, Duke Math. J., **80** (1995) 737–756

[Ium] HUMPHREYS, J.E., *Linear algebraic groups*, Graduate Texts in Mathematics, 21, Springer-Verlag (1990)

[Hur] HURTUBISE, J., *Integrable systems and algebraic surfaces*, Duke Math. J., **83** (1996) 19–50

[Jac1] JACOBI, C., *Sur le mouvement d'un point et sur un cas particulier du problème des trois corps*, Compt. Rend., **3** (1836) 59–61

[Jac2] JACOBI, C., *Note von der geodätischen Linie auf einem Ellipsoid und den verschiedenen Anwendungen einer merkwürdigen analytischen Substitution*, J. Reine Angew. Math., **19** (1839) 309–313

[JW] JEFFREY, L., WEITSMAN, J., *Bohr-Sommerfeld orbits in the moduli spaces of flat connections and the Verlinde dimension formula*, Comm. Math. Phys., **150** (1992) 593–630

[Kem] KEMPF, G.R., *Complex Abelian varieties and theta functions*, Universitext, Springer-Verlag (1991)

[KGT] KOSMANN-SCHWARZBACH, Y. GRAMMATICOS, B. TAMIZHMANI, K.M., *Integrability of nonlinear systems. Proceedings of the CIMPA International School on Nonlinear Systems, held at Pondicherry University, Pondicherry, January 8-26, 1996*, Lecture Notes in Physics, 495, Springer-Verlag (1997)

References

[KM] McKEAN, H.P., VAN MOERBEKE, P., *The spectrum of Hill's equation*, Invent. Math., **3** (1975) 217–274

[Knö] KNÖRRER, H., *Integrable Hamiltonsche Systeme und Algebraische Geometrie*, Jber. d. D Math.-Verein, **88** (1986) 82–103

[Kos] KOSTANT, B., *The solution to a generalized Toda lattice and representation theory*, Adv. Math **39** (1979) 195–338

[Kow1] KOWALEVSKI, S., *Sur une propriété de système d'équations différentielles qui définit la rotatio d'un corps solide autour d'un point fixe*, Acta Math., **12** (1889) 81–93

[Kow2] KOWALEVSKI, S., *Sur le problème de la rotation d'un corps solide autour d'un point fixe*, Act Math., **12** (1889) 177–232

[Kri1] KRICHEVER, I., *Methods of algebraic geometry in the theory of non-linear equations*, Rus Math. Surveys, **32** (1977) 185-213

[Kri2] KRICHEVER, I.M., *Elliptic solutions of the Kadomtsev-Pethiashvili equation and many-bod problems*, Functional Anal. Appl., **14** (1980) 282–290

[KV] KIMURA, M., VANHAECKE, P., *Commuting matrix differential operators and loop algebra* preprint

[Lag] LAGRANGE, J.-L., *Recherches sur le mouvement d'un corps qui est attiré vers deux centr fixes*, Auc. Mem. de Turin, **4** (1766–1769) 118–215

[Lax] LAX, P., *Integrals of nonlinear evolution equations and solitary waves*, Comm. Pure App Math., **21** (1968) 467–490

[LB] LANGE, H., BIRKENHAKE, CH., *Complex Abelian varieties*, Grundlehren der Mathematische Wissenschaften, 302, Springer-Verlag (1992)

[Lio1] LIOUVILLE, J., *Sur quelques cas particuliers où les équations du mouvement d'un point matéri peuvent s'intégrer*, J. Math. Pures Appl., **11** (1846) 345–378

[Lio2] LIOUVILLE, J., *L'intégration des équations différentielles du mouvement d'un nombre que conque de points matérielles*, J. Math. Pures Appl., **14** (1849) 257–299

[LM1] LEPRÉVOST, F., MARKUSHEVICH, D., *A tower of genus two curves related to the Kowalevs top*, J. Reine Angew. Math., **514** (1999) 103–111

[LM2] LI, Y., MULASE, M., *Prym varieties and integrable systems*, Comm. Anal. Geom., **5** (199′ 279–332

[LM3] LIBERMANN, P., MARLE, C.-M., *Symplectic geometry and analytic mechanics*, Mathemati and its Applications, 35, D. Reidel Publishing Co. (1987)

[LP] LI, L.C., PARMENTIER, S., *Nonlinear Poisson structures and r-matrices*, Comm. Math. Phy **125** (1989) 545–563

[Man] MANAKOV, S.V., *Note on the integration of Euler's equations of the dynamics of an dimensional rigid body*, Functional Anal. Appl., **10** (1976) 328–329

[MM] MUMFORD, D., VAN MOERBEKE, P., *The spectrum of difference operators and algebra curves*, Acta Math., **143** (1979) 93–154

[Moe] VAN MOERBEKE, P., *The spectrum of Jacobi matrices*, Invent. math., **37** (1976) 45–81

References

[Mos1] MOSER, J., *Three integrable Hamiltonian systems connected with isospectral deformations*, Adv. Math., **16** (1975) 197–220

[Mos2] MOSER, J., *Finitely many mass points on the line under the influence of an exponential potential — an integrable system*, Lecture Notes in Physics, Springer-Verlag, **38** (1975) 97–101

[Mul1] MULASE, M., *Cohomological structure in soliton equations and Jacobian varieties*, J. Differential Geom., **19** (1984) 403–430

[Mul2] MULASE, M., *Category of vector bundles on algebraic curves and infinite dimensional Grassmannians*, Internat. J. of Math., **1** (1990) 293–342

[Mul3] MULASE, M, *Algebraic theory of the KP equations*, Conf. Proc. Lecture Notes Math. Phys., Internat. Press, Cambridge, (1994) 151–217

[Mum1] MUMFORD, D., *Geometric invariant theory*, Ergebnisse der Mathematik und ihrer Grenzgebiete, Neue Folge, 34, Springer-Verlag (1965)

[Mum2] MUMFORD, D., *Abelian varieties*, Oxford University Press (1970)

[Mum3] MUMFORD, D., *Curves and their Jacobians*, Ann Arbor, The University of Michigan Press (1975)

[Mum4] MUMFORD, D., *An algebro-geometric construction of commuting operators and of solutions to the Toda lattice equation, Korteweg-de Vries equation and related non-linear equations*, Int. Symp. on Algebraic Geometry, (1977) 115–153

[Mum5] MUMFORD, D., *Tata lectures on theta II*, Progress in Mathematics, 43, Birkhäuser (1984)

[Ngu] NGUYEN, T.Z., *Symplectic topology of integrable Hamiltonian systems. I. Arnold-Liouville with singularities*, Compositio Math., **101** (1996) 179–215

[Nov] NOVIKOV, S.P., *Solitons and geometry*, Lezioni Fermiane, Cambridge University Press (1992)

[NV] NOVIKOV, S.P., VESELOV, A.P., *Poisson brackets and complex tori*, Proc. Steklov Inst. Math., **3** (1985) 53–65

[Oka] OKAMOTO, K., *Isomonodromic deformation and Painlevé equations, and the Garnier system*, J. Fac. Sci. Univ. Tokyo Sect. IA Math., **33** (1986) 575–618

[OP1] OLSHANETSKY, M.A., PERELOMOV, A.M., *Completely integrable Hamiltonian systems connected with semisimple Lie algebras*, Invent. Math., **37** (1976) 93–108

[OP2] OLSHANETSKY, M.A., PERELOMOV, A.M., *Explicit solutions of classical generalized Toda models*, Invent. Math., **54** (1979) 261–269

[Pai] PAINLEVÉ, P., *Sur les fonctions qui admettent un théorème d'addition*, Acta Math., **25** (1902) 1–54

[Per1] PERELOMOV, A.M., *Lax representations for the systems of S. Kowalewskaya type*, Comm. Math. Phys., **81** (1981) 239–244

[Per2] PERELOMOV, A.M., *Integrable systems of classical mechanics and Lie algebras*, Birkhäuser (1990)

[Pio1] PIOVAN, L.A., *Algebraically complete integrable systems and Kummer varieties*, Math. Ann., **290** (1991) 349–403

[Pio2] PIOVAN, L.A., *Cyclic coverings of Abelian surfaces and some related integrable systems*, Math. Ann., **294** (1992) 755–764

References

[Pio3] PIOVAN, L.A., *Canonical system on elliptic curves*, Proc. Amer. Math. Soc., **119** (1993) 1323–1329

[Poi] POISSON, S., *Mémoire sur la variation des constantes arbitraires dans les questions de mécanique*, J. Ecole Polytec., **8** (1809) 266–344

[Pre1] PREVIATO, E., *Hyperelliptic quasiperiodic and soliton solutions of the nonlinear Schrödinger equation*, Duke Math. J., **52** (1985) 329–377

[Pre2] PREVIATO, E., *Generalized Weierstrass ℘-functions and KP flows in affine space*, Comment. Math. Helv., **62** (1987) 292–310

[PS] PRESSLEY, A., SEGAL, G., *Loop groups and their representations*, Oxford Mathematical Monographs, Clarendon Press (1986)

[PV1] PEDRONI, M., VANHAECKE, P., *A Lie algebraic generalization of the Mumford system, its symmetries and its multi-Hamiltonian structure*, Regul. Chaotic Dyn., **3** (1998) 132–160

[PV2] PIOVAN, L.A., VANHAECKE, P., *Integrable systems and projective images of Kummer surfaces*, Ann. Scuola Norm. Sup. Pisa, **29** (2000) 351–392

[Rat] RATIU, T., *The motion of the free n-dimensional rigid body*, Indiana Univ. Math. J., **29** (1980) 609–629

[RS1] REIMANN, A.G., SEMENOV-TIAN-SHANSKY, M.A., *Reduction of Hamiltonian systems, affine Lie algebras and Lax equations. I*, Invent. Math., **54** (1979) 81–100

[RS2] REIMANN, A.G., SEMENOV-TIAN-SHANSKY, M.A., *Reduction of Hamiltonian systems, affine Lie algebras and Lax equations. II*, Invent. Math., **63** (1981) 423–432

[RSF] REIMANN, A.G., SEMENOV-TIAN-SHANSKI, M.A., FRENKEL, I.E., *Graded Lie algebras and completely integrable dynamical systems*, Soviet Math. Dokl., **20** (1979) 811–814

[Sch] SCHLESINGER, L., *Über eine Klasse von Differentialsystemen beliebiger Ordnung mit fester kritischen Punkten*, Journal für Mathematik, **141** (1912) 96–145

[Seg] SEGRE, M., *Sulle variata cubiche dello spazio a 4 dimensioni*, Memorie Acad. Torino (1888)

[Sem] SEMENOV-TIAN-SHANSKY, M.A., *What is a classical r-matrix?*, Functional Anal. Appl., **17** (1983) 259–272

[Sha] SHAFAREVICH, I.R., *Basic Algebraic Geometry*, Die Grundlehren der mathematischen Wissenschaften, 213, Springer-Verlag (1974)

[Shi] SHIOTA, T., *Characterization of Jacobian varieties in terms of soliton equations*, Invent. Math., **83** (1986) 333–382

[Sil] SILHOL, R., *Real algebraic surfaces*, Lecture Notes in Mathematics, 1392, Springer-Verlag (1989)

[Sin] SINGER, S., *Some maps from the full Toda lattice are Poisson*, Phys. Lett. A, **174** (1993) 66–70

[Skl1] SKLYANIN, E.K., *Some algebraic structures connected with the Yang-Baxter equation*, Functional Anal. Appl., **16** (1982) 263–270

[Skl2] SKLYANIN, E.K., *Separation of variables—new trends*, Progr. Theoret. Phys. Suppl., **118** (1995) 35–60

[Spr] SPRINGER, T.A., *Invariant theory*, Lecture Notes in Mathematics, 585, Springer-Verlag (1977)

References

[SS] SATO, M., SATO, Y., *Soliton equations as dynamical systems on infinite dimensional Grassmann manifold*, Lecture Notes Numer. Appl. Anal., **5** (1982) 259–271

[SW] SEGAL, G., WILSON, G., *Loop groups and equations of KdV type*, Publ. I.H.E.S., **61** (1985) 5–65

[Sze] SZEMBERG T., *On principally polarized Adler-van Moerbeke surfaces*, Math. Nach, **185** (1997) 239–260

[Tod] TODA, M., *Vibration of a chain with nonlinear interaction*, J. Phys. Soc. Japan, **20** (1967) 431–436

[Tre] TREIBICH, A., *Des solitons elliptiques aux revêtements tangentiels*, Thèse d'Etat, Université de Rennes I (1991)

[TV] TREIBICH, A, VERDIER, J.L., *Au delà des potentiels et revêtements tangentiels hyperelliptiques exceptionnels*, C.R. Acad. Sci. Paris Sér. I Math., **325** (1997) 1101–1106

[Vai] VAISMAN, I., *Lectures on the Geometry of Poisson Manifolds*, Progress in Mathematics, 118, Birkhaüser (1994)

[Van1] VANHAECKE, P., *Explicit techniques for studying two-dimensional integrable systems*, PhD-thesis, Katholieke Universiteit Leuven (1991)

[Van2] VANHAECKE, P., *Linearising two-dimensional integrable systems and the construction of action-angle variables*, Math. Z., **211** (1992) 265–313

[Van3] VANHAECKE, P., *A special case of the Garnier system, (1,4)-polarised Abelian surfaces and their moduli*, Compositio Math., **92** (1994) 157–203

[Van4] VANHAECKE, P., *Stratifications of hyperelliptic Jacobians and the Sato Grassmannian*, Acta Appl. Math., **40** (1995) 143–172

[Van5] VANHAECKE, P., *Integrable Hamiltonian systems associated to families of curves and their bi-Hamiltonian structure*, Integrable systems and foliations/Feuilletages et systèmes intégrables (Montpellier, 1995), Progr. Math., **145**, Birkhaüser, (1997) 188–212

[Van6] VANHAECKE, P., *Integrable systems and symmetric products of curves*, Math. Z., **227** (1998) 93–127

[Ver] VERDIER, J.L., *Algèbres de Lie, systèmes hamiltoniens, courbes algébriques*, Astérisque, **566** (1980/81) 1–10

[Wei1] WEIL, A., *Courbes algébriques et variétés abéliennes*, Hermann (1948)

[Wei2] WEINSTEIN, A., *The local structure of Poisson manifolds*, J. Differential Geom., **18** (1983) 523–557

[Wei3] WEINSTEIN, A., *Errata and addenda*, J. Differential Geom., **22** (1985) 255

[Whi] WHITTAKER, E.T., *A treatise on the analytical dynamics of particles & rigid bodies*, Cambridge Mathematical Library, Cambridge University Press (1988)

Index

Lecture Notes in Mathematics

For information about Vols. 1–1580
please contact your bookseller or Springer-Verlag

Vol. 1726: V. Marić, Regular Variation and Differential Equations. X, 127 pages. 2000.

Vol. 1727: P. Kravanja, M. Van Barel, Computing the Zeros of Analytic Functions. VII, 111 pages. 2000.

Vol. 1728: K. Gatermann, Computer Algebra Methods for Equivariant Dynamical Systems. XV, 153 pages. 2000.

Vol. 1729: J. Azéma, M. Émery, M. Ledoux, M. Yor, Séminaire de Probabilités XXXIV. VI, 431 pages. 2000.

Vol. 1730: S. Graf, H. Luschgy, Foundations of Quantization for Probability Distributions. X, 230 pages. 2000.

Vol. 1731: T. Hsu, Quilts: Central Extensions, Braid Actions, and Finite Groups,. XII, 185 pages. 2000.

Vol. 1732: K. Keller, Invariant Factors, Julia Equivalences and the (Abstract) Mandelbrot Set. X, 206 pages. 2000.

Vol. 1733: K. Ritter, Average-Case Analysis of Numerical Problems. IX, 254 pages. 2000.

Vol. 1734: M. Espedal, A. Fasano, A. Mikelić, Filtration in Porous Media and Industrial Applications. Cetraro 1998. Editor: A. Fasano. 2000.

Vol. 1735: D. Yafaev, Scattering Theory: Some Old and New Problems. XVI, 169 pages. 2000.

Vol. 1736: B. O. Turesson, Nonlinear Potential Theory and Weighted Sobolev Spaces. XIV, 173 pages. 2000.

Vol. 1737: S. Wakabayashi, Classical Microlocal Analysis in the Space of Hyperfunctions. VIII, 367 pages. 2000.

Vol. 1738: M. Émery, A. Nemirovski, D. Voiculescu, Lectures on Probability Theory and Statistics. XI, 356 pages. 2000.

Vol. 1739: R. Burkard, P. Deuflhard, A. Jameson, J.-L. Lions, G. Strang, Computational Mathematics Driven by Industrial Problems. Martina Franca, 1999. Editors: V. Capasso, H. Engl, J. Periaux. VII, 418 pages. 2000.

Vol. 1740: B. Kawohl, O. Pironneau, L. Tartar, J.-P. Zolesio, Optimal Shape Design. Tróia, Portugal 1999. Editors: A. Cellina, A. Ornelas. IX, 388 pages. 2000.

Vol. 1741: E. Lombardi, Oscillatory Integrals and Phenomena Beyond all Algebraic Orders. XV, 413 pages. 2000.

Vol. 1742: A. Unterberger, Quantization and Non-holomorphic Modular Forms. VIII, 253 pages. 2000.

Vol. 1743: L. Habermann, Riemannian Metrics of Constant Mass and Moduli Spaces of Conformal Structures. XII, 116 pages. 2000.

Vol. 1744: M. Kunze, Non-Smooth Dynamical Systems. X, 228 pages. 2000.

Vol. 1745: V. D. Milman, G. Schechtman, Geometric Aspects of Functional Analysis. VIII, 289 pages. 2000.

Vol. 1746: A. Degtyarev, I. Itenberg, V. Kharlamov, Real Enriques Surfaces. XVI, 259 pages. 2000.

Vol. 1747: L. W. Christensen, Gorenstein Dimensions. VIII, 204 pages. 2000.

Vol. 1748: M. Růžička, Electrorheological Fluids: Modeling and Mathematical Theory. XV, 176 pages. 2001.

Vol. 1749: M. Fuchs, G. Seregin, Variational Methods for Problems from Plasticity Theory and for Generalized Newtonian Fluids. VI, 269 pages. 2001.

Vol. 1750: B. Conrad, Grothendieck Duality and Base Change. X, 296 pages. 2001.

Vol. 1751: N. J. Cutland, Loeb Measures in Practice: Recent Advances. XI, 111 pages. 2001.

Vol. 1752: Y. V. Nesterenko, P. Philippon, Introduction to Algebraic Independence Theory. XIII, 256 pages. 2001.

Vol. 1753: A. I. Bobenko, U. Eitner, Painlevé Equations in the Differential Geometry of Surfaces. VI, 120 pages. 2001.

Vol. 1754: W. Bertram, The Geometry of Jordan and Lie Structures. XVI, 269 pages. 2001.

Vol. 1755: J. Azéma, M. Émery, M. Ledoux, M. Yor, Séminaire de Probabilités XXXV. VI, 427 pages. 2001.

Vol. 1756: P. E. Zhidkov, Korteweg de Vries and Nonlinear Schrödinger Equations: Qualitative Theory. VII, 147 pages. 2001.

Vol. 1757: R. R. Phelps, Lectures on Choquet's Theorem. VII, 124 pages. 2001.

Vol. 1758: N. Monod, Continuous Bounded Cohomology of Locally Compact Groups. X, 214 pages. 2001.

Vol. 1759: Y. Abe, K. Kopfermann, Toroidal Groups. VIII, 133 pages. 2001.

Vol. 1760: D. Filipović, Consistency Problems for Heath-Jarrow-Morton Interest Rate Models. VIII, 134 pages. 2001.

Vol. 1761: C. Adelmann, The Decomposition of Primes in Torsion Point Fields. VI, 142 pages. 2001.

Vol. 1762: S. Cerrai, Second Order PDE's in Finite and Infinite Dimension. IX, 330 pages. 2001.

Vol. 1763: J.-L. Loday, A. Frabetti, F. Chapoton, F. Goichot, Dialgebras and Related Operads. IV, 132 pages. 2001.

Vol. 1764: A. Cannas da Silva, Lectures on Symplectic Geometry. XII, 217 pages. 2001.

Recent Reprints and New Editions

Vol. 1200: V. D. Milman, G. Schechtman, Asymptotic Theory of Finite Dimensional Normed Spaces - Corrected Second Printing 2001. X, 156 pages. 1986.

Vol. 1618: G. Pisier, Similarity Problems and Completely Bounded Maps - Second, Expanded Edition VII, 198 pages. 2001.

Vol. 1629: J. D. Moore, Lectures on Seiberg-Witten Invariants - Second Edition. VIII, 121 pages. 2001.

Vol. 1638: P. Vanhaecke, Integrable Systems in the realm of Algebraic Geometry - Second Edition. X, 256 pages. 2001.

Vol. 1702: J. Ma, J. Yong, Forward-Backward Stochastic Differential Equations and Their Applications - Corrected Second Printing 2000. XIII, 270 pages. 1999.

4. Lecture Notes are printed by photo-offset from the master-copy delivered in camera-ready form by the authors. Springer-Verlag provides technical instructions for the preparation of manuscripts. Macro packages in T_EX, L^AT_EX2e, $L^AT_EX2.09$ are available from Springer's web-pages at

http://www.springer.de/math/authors/b-tex.html.

Careful preparation of the manuscripts will help keep production time short and ensure satisfactory appearance of the finished book.

The actual production of a Lecture Notes volume takes approximately 12 weeks.

5. Authors receive a total of 50 free copies of their volume, but no royalties. They are entitled to a discount of 33.3 % on the price of Springer books purchase for their personal use, if ordering directly from Springer-Verlag.

Commitment to publish is made by letter of intent rather than by signing a formal contract. Springer-Verlag secures the copyright for each volume. Authors are free to reuse material contained in their LNM volumes in later publications: A brief written (or e-mail) request for formal permission is sufficient.

Addresses:

Professor J.-M. Morel
CMLA, Ecole Normale Supérieure de Cachan
61 Avenue du Président Wilson
94235 Cachan Cedex France
E-mail: Jean-Michel.Morel@cmla.ens-cachan.fr

Professor B. Teissier
Université Paris 7
UFR de Mathématiques
Equipe Géométrie et Dynamique
Case 7012
2 place Jussieu
75251 Paris Cedex 05
E-mail: Teissier@ens.fr

Professor F. Takens, Mathematisch Instituut,
Rijksuniversiteit Groningen, Postbus 800,
9700 AV Groningen, The Netherlands
E-mail: F.Takens@math.rug.nl

Springer-Verlag, Mathematics Editorial, Tiergartenstr. 17
D-69121 Heidelberg, Germany
Tel.: *49 (6221) 487-701
Fax: *49 (6221) 487-355
E-mail: lnm@Springer.de